鸿蒙应用开发入门与实践

刘陈◎编著

清华大学出版社

北京

内 容 简 介

鸿蒙操作系统（HarmonyOS）是华为公司开发的一款分布式操作系统，旨在实现各种设备之间的智能互联和协同工作。本书详细讲解了开发 HarmonyOS 应用程序的知识，共分为 13 章，依次讲解了 HarmonyOS 开发基础，搭建开发环境，HarmonyOS 应用模型，Java UI 开发，Ark UI 开发，图形、图像开发，多媒体开发，相机开发，网络程序开发，数据管理，电话和短信服务，设备管理，新闻客户端开发（Node.js 服务端 + HarmonyOS 客户端）。全书内容循序渐进，深入讲解了每个知识点的具体细节，并穿插了大量的实例来演示每个知识点的用法，引领读者全面掌握 HarmonyOS 应用开发技术。

本书封面贴有清华大学出版社防伪标签，无标签者不得销售。

版权所有，侵权必究。举报：010-62782989，beiqinquan@tup.tsinghua.edu.cn。

图书在版编目 (CIP) 数据

鸿蒙应用开发入门与实践 / 刘陈编著. -- 北京：
清华大学出版社，2024.9. -- ISBN 978-7-302-66907-4
Ⅰ . TN929.53
中国国家版本馆 CIP 数据核字第 20243MJ708 号

责任编辑：魏　莹
封面设计：李　坤
版式设计：方加青
责任校对：马荣敏
责任印制：沈　露

出版发行：清华大学出版社
网　　址：https://www.tup.com.cn，https://www.wqxuetang.com
地　　址：北京清华大学学研大厦 A 座　　邮　　编：100084
社 总 机：010-83470000　　邮　　购：010-62786544
投稿与读者服务：010-62776969，c-service@tup.tsinghua.edu.cn
质 量 反 馈：010-62772015，zhiliang@tup.tsinghua.edu.cn
印 装 者：三河市龙大印装有限公司
经　　销：全国新华书店
开　　本：185mm×260mm　　印　　张：24.25　　字　　数：590 千字
版　　次：2024 年 9 月第 1 版　　印　　次：2024 年 9 月第 1 次印刷
定　　价：99.00 元

产品编号：099023-01

前 言
PREFACE

背景介绍

HarmonyOS 的横空出世，推动了智能设备领域的发展，迈向了全场景、全连接的新时代。作为一款分布式操作系统，HarmonyOS 通过其独特的分布式架构、多设备支持和开发者友好的特性，实现了设备之间的无缝协作。这意味着用户可以在不同类型的设备上实现一致的使用体验，无论是智能手机、智能家居、智能汽车还是其他智能终端，都能更加流畅地互联互通。HarmonyOS 的意义还在于打破了设备间的壁垒，将不同品牌、不同类型的设备集成为一个统一的生态系统。这为用户提供了更广泛、更灵活的选择，也为开发者提供了更广泛的应用场景，促进了产业的协同创新。同时，HarmonyOS 注重隐私和安全性，通过分布式架构将数据存储在本地设备上，有效保障了用户数据的安全性，符合当今数字化社会对隐私保护的重要需求。

随着 HarmonyOS 的不断普及，开发者对于深入了解和掌握其应用程序开发技术的需求日益增长。目前，市场上对 HarmonyOS 应用程序开发的实用指南相对较少，而开发者迫切需要一本全面而深入的教材来帮助他们快速融入这个新兴的开发领域。本书填补了这一空白，以其系统性、实战性和易读性为特点，满足了广大开发者在 HarmonyOS 应用开发方面的学习和实践需求。无论是初学者还是有一定经验的开发者，都能在本书中找到适合自己的学习路径和实践案例，从而更好地应对 HarmonyOS 应用开发的挑战。

本书针对 HarmonyOS 应用开发领域，系统深入地研究了 HarmonyOS 的开发基础、搭建开发环境、应用模型、UI 开发、多媒体开发、网络程序开发、数据管理等方面的知识。作为一本全面而实用的指南，本书致力于为开发者提供清晰、深入的指导，使他们能够更容易地掌握 HarmonyOS 开发技能，从而更加高效地构建和部署应用程序。

本书的特色

1. 零基础学习

本书以零基础读者为目标受众，通过清晰易懂的语言和逐步深入的教学方式，从介绍 HarmonyOS 的发展历程、优势和架构开始，逐步引导读者理解。每一章节都以通俗易懂的方式呈现基础概念，确保读者能够轻松理解和消化复杂的技术知识，为深入学习打下坚实的基础。

2. 大量实例应用

为了更好地帮助读者将理论知识转化为实际能力，本书在每个章节都提供了丰富的实例应用。通过实例，读者能够直观地了解 HarmonyOS 的开发流程，掌握实际编码技巧，并在不断的实践中逐渐形成自己的编程思维。这些实例从简单到复杂，从基础到高级，覆盖了 HarmonyOS 开发的各个方面，帮助读者逐步深入学习，提高实际应用能力。

3. 系统化的知识结构

本书以系统化的知识结构为读者提供了清晰而有序的学习路径。每一章都按照自然的逻辑顺序展开，从 HarmonyOS 的基础介绍开始，逐步深入讨论开发环境搭建、应用模型、

UI 开发、多媒体开发、网络程序开发等方面的知识，形成了一套紧密联系的知识网络，帮助读者更好地理解 HarmonyOS 应用程序开发的全貌。

4. 实用性强的案例分析

本书以实际项目开发为导向，引用大量实用性强的案例，帮助读者将学到的理论知识应用到实际开发中。每章都附有真实的案例，使读者能深入了解每个知识点的实际应用场景，从而更好地应对日常的开发挑战。

5. 深入浅出的解释和示范

为了确保读者能够深入理解复杂的技术概念，本书在每章都以深入浅出的方式解释关键概念，并通过大量示范代码进行演示。这种教学方法有助于读者更轻松地理解抽象的编程概念，从而更轻松地应用这些知识。

6. 覆盖全面的开发主题

本书涵盖了 HarmonyOS 应用程序开发的各个方面，包括 UI 开发、多媒体开发、网络程序开发、数据管理等主题。这样的涵盖既适合初学者对 HarmonyOS 有全面的认识，也能满足有经验的开发者深入学习和查阅的需求。

本书的读者对象

- 初学者和零基础者：本书以零基础读者为目标受众，适合那些对 HarmonyOS 应用程序开发感兴趣但没有相关经验的读者。通过清晰的教学方式和简单易懂的语言，帮助初学者轻松入门。
- 移动应用开发者：针对已经有一定移动应用开发经验的读者，本书提供了深入 HarmonyOS 开发的机会，使他们能够将已有的开发经验迁移到 HarmonyOS 平台，拓宽技术栈，提高应用开发的全面性。
- 跨平台开发者：对于有跨平台开发需求的读者，本书提供了对 HarmonyOS 的全面介绍，使他们能够了解如何使用 HarmonyOS 框架进行开发，实现在多个平台上的无缝协同。
- HarmonyOS 开发者和爱好者：针对已经在 HarmonyOS 领域有一定基础的开发者，本书通过大量的实例应用和深入的知识讨论，为他们提供进一步拓展技术广度和提升实际项目经验的途径。
- 希望深入了解分布式架构和多设备支持的开发者：本书对 HarmonyOS 特性进行了深入剖析，适合那些对分布式系统和多设备支持有浓厚兴趣的开发者。

总的来说，无论是初学者还是有一定经验的开发者，都能在本书中找到适合自己的学习路径和实践案例，从而更好地应对 HarmonyOS 应用开发的挑战。

致谢

本书在编写过程中，得到了清华大学出版社多位编辑的大力支持，正是他们的求实、耐心和效率，才使本书如期出版。另外，也十分感谢我的家人给予的大力支持。由于编者水平有限，书中难免存在疏漏之处，恳请读者提出宝贵的意见或建议，以便修订并使之完善。

编　者

目　录
CONTENTS

第 1 章　HarmonyOS 开发基础 ⋯⋯1
1.1　智能手机系统介绍 ⋯⋯⋯⋯⋯⋯ 2
1.1.1　智能手机系统的特点 ⋯⋯⋯⋯ 2
1.1.2　Android 系统介绍 ⋯⋯⋯⋯⋯ 2
1.1.3　iOS 系统介绍 ⋯⋯⋯⋯⋯⋯⋯ 3
1.2　HarmonyOS 介绍 ⋯⋯⋯⋯⋯⋯⋯ 4
1.2.1　HarmonyOS 的发展历程 ⋯⋯⋯ 4
1.2.2　HarmonyOS、OpenHarmony、鸿蒙生态的区别与联系 ⋯⋯⋯⋯⋯ 5
1.3　HarmonyOS 的优点 ⋯⋯⋯⋯⋯⋯ 6
1.3.1　分布式架构 ⋯⋯⋯⋯⋯⋯⋯⋯ 6
1.3.2　多设备支持 ⋯⋯⋯⋯⋯⋯⋯⋯ 6
1.3.3　开发者友好 ⋯⋯⋯⋯⋯⋯⋯⋯ 7
1.4　HarmonyOS 架构分析 ⋯⋯⋯⋯⋯ 8
1.4.1　整体架构图 ⋯⋯⋯⋯⋯⋯⋯⋯ 8
1.4.2　内核层 ⋯⋯⋯⋯⋯⋯⋯⋯⋯⋯ 8
1.4.3　系统服务层 ⋯⋯⋯⋯⋯⋯⋯⋯ 9
1.4.4　框架层 ⋯⋯⋯⋯⋯⋯⋯⋯⋯ 10
1.4.5　应用层 ⋯⋯⋯⋯⋯⋯⋯⋯⋯ 10

第 2 章　搭建开发环境 ⋯⋯⋯⋯⋯⋯ 12
2.1　HarmonyOS 应用开发介绍 ⋯⋯⋯ 13
2.1.1　HarmonyOS 应用 / 服务的开发流程 ⋯ 13
2.1.2　配置开发环境的流程 ⋯⋯⋯⋯ 13
2.2　搭建 DevEco Studio 开发环境 ⋯ 13
2.2.1　DevEco Studio 的特点 ⋯⋯⋯ 13
2.2.2　环境要求 ⋯⋯⋯⋯⋯⋯⋯⋯ 14
2.2.3　下载并安装 DevEco Studio ⋯ 14
2.3　配置 DevEco Studio 开发环境 ⋯ 16
2.3.1　下载 SDK 及工具链 ⋯⋯⋯⋯ 16
2.3.2　配置向导 ⋯⋯⋯⋯⋯⋯⋯⋯ 17
2.4　使用 DevEco Studio 开发第一个鸿蒙应用程序 ⋯⋯⋯⋯⋯⋯⋯⋯⋯ 20
2.4.1　创建工程 ⋯⋯⋯⋯⋯⋯⋯⋯ 21
2.4.2　DevEco Studio 界面介绍 ⋯⋯ 23
2.4.3　在模拟器中运行程序 ⋯⋯⋯ 27
2.4.4　在本地真机中运行程序 ⋯⋯ 28

第 3 章　HarmonyOS 应用模型 ⋯ 30
3.1　HarmonyOS 应用模型介绍 ⋯⋯⋯ 31
3.1.1　应用模型的组成元素 ⋯⋯⋯⋯ 31
3.1.2　应用模型的发展进程 ⋯⋯⋯⋯ 31
3.1.3　FA 模型与 Stage 模型的对比 ⋯ 32
3.2　Stage 模型开发基础 ⋯⋯⋯⋯⋯ 33
3.2.1　Stage 模型的知识体系 ⋯⋯⋯ 33
3.2.2　Stage 应用 / 组件级配置 ⋯⋯ 34
3.3　UIAbility 组件 ⋯⋯⋯⋯⋯⋯⋯ 37
3.3.1　UIAbility 组件生命周期 ⋯⋯ 38
3.3.2　UIAbility 组件的启动模式 ⋯ 41
3.3.3　UIAbility 组件的基本用法 ⋯ 44
3.3.4　UIAbility 组件与 UI 的数据同步 ⋯ 46
3.3.5　UIAbility 实战：页面跳转 ⋯ 48
3.4　服务卡片 ⋯⋯⋯⋯⋯⋯⋯⋯⋯⋯ 52
3.4.1　服务卡片架构 ⋯⋯⋯⋯⋯⋯ 52
3.4.2　ArkTS 卡片开发 ⋯⋯⋯⋯⋯ 53
3.4.3　开发基于 JS UI 的卡片 ⋯⋯ 54
3.4.4　卡片实战：多设备自适应服务卡片 · 55

第 4 章　Java UI 开发 ⋯⋯⋯⋯⋯⋯ 61
4.1　Ability 框架 ⋯⋯⋯⋯⋯⋯⋯⋯⋯ 62
4.1.1　Page Ability ⋯⋯⋯⋯⋯⋯⋯ 62
4.1.2　Ability 实战：使用 PageAbility 实现页面跳转 ⋯⋯⋯⋯⋯⋯⋯⋯⋯ 64
4.2　UI 布局 ⋯⋯⋯⋯⋯⋯⋯⋯⋯⋯⋯ 67
4.2.1　代码布局 ⋯⋯⋯⋯⋯⋯⋯⋯ 67
4.2.2　XML 布局 ⋯⋯⋯⋯⋯⋯⋯⋯ 69
4.2.3　Java 布局类 ⋯⋯⋯⋯⋯⋯⋯ 71
4.3　常用组件开发 ⋯⋯⋯⋯⋯⋯⋯⋯ 78
4.3.1　Text 和 Button 组件 ⋯⋯⋯⋯ 78
4.3.2　Image 组件 ⋯⋯⋯⋯⋯⋯⋯ 80
4.3.3　TabList 和 Tab 组件 ⋯⋯⋯⋯ 81

	4.3.4	Picker 和 DatePicker 组件 ·············	82
	4.3.5	TimePicker 组件 ····················	84
	4.3.6	Switch 组件 ·······················	85
	4.3.7	RadioButton 和 Checkbox 组件 ······	86
	4.3.8	ProgressBar、RoundProgressBar 和 Slider 组件 ··························	88
	4.3.9	ToastDialog、PopupDialog 和 CommonDialog 组件 ·····················	93

第 5 章　Ark UI 开发 ·············· 99

- 5.1 方舟开发框架概述 ·············· 100
  - 5.1.1 框架说明 ·················· 100
  - 5.1.2 基本语法 ·················· 101
  - 5.1.3 创建自定义组件 ············ 104
- 5.2 UI 布局 ························ 108
  - 5.2.1 布局结构 ·················· 108
  - 5.2.2 线性布局 ·················· 109
  - 5.2.3 层叠布局 ·················· 112
  - 5.2.4 弹性布局 ·················· 116
  - 5.2.5 相对布局 ·················· 119
  - 5.2.6 栅格布局 ·················· 123
  - 5.2.7 列表布局 ·················· 129
  - 5.2.8 网格布局 ·················· 131
- 5.3 基本组件 ······················ 134
  - 5.3.1 按钮组件 ·················· 134
  - 5.3.2 单选框组件 ················ 138
  - 5.3.3 进度条组件 ················ 140
  - 5.3.4 切换按钮组件 ·············· 142
  - 5.3.5 文本显示组件 ·············· 144
  - 5.3.6 文本输入框 ················ 147
  - 5.3.7 视频播放组件 ·············· 150
  - 5.3.8 气泡提示 ·················· 155
  - 5.3.9 菜单 ······················ 157

第 6 章　图形、图像开发 ········· 159

- 6.1 显示图片 ······················ 160
  - 6.1.1 Image 组件介绍 ············ 160
  - 6.1.2 Image 组件实战：手机相册系统··· 162
- 6.2 绘制几何图形 ·················· 172

	6.2.1	Shape 基础 ·····················	172
	6.2.2	Shape 实战：绘制各种各样的图形 ···	175
6.3	画布 ··························		177
	6.3.1	Canvas 绘制自定义图形 ·········	178
	6.3.2	Canvas 的常用绘图方法 ·········	179
6.4	动画 ··························		182
	6.4.1	ArkUI 动画的分类 ··············	182
	6.4.2	布局更新动画 ················	183
	6.4.3	组件内转场动画 ··············	186

第 7 章　多媒体开发 ·············· 191

- 7.1 HarmonyOS 多媒体开发架构 ········ 192
- 7.2 AVPlayer 和 AVRecorder ··············· 192
  - 7.2.1 AVPlayer ·················· 193
  - 7.2.2 AVRecorder ················ 194
- 7.3 音频播放 ······················ 195
  - 7.3.1 使用 AVPlayer 开发音频播放程序 ··· 195
  - 7.3.2 使用 AudioRenderer 开发音频播放程序·· 198
  - 7.3.3 使用 OpenSL ES 开发音频播放程序 ··· 203
  - 7.3.4 音频播放实战：多功能音乐播放器··· 205
- 7.4 开发音频录制程序 ·············· 212
  - 7.4.1 使用 AVRecorder 开发音频录制程序 ··· 213
  - 7.4.2 使用 AudioCapturer 开发音频录制程序·· 216
  - 7.4.3 使用 OpenSL ES 开发音频录制程序 ··· 218
  - 7.4.4 管理麦克风 ················ 220
- 7.5 音频通话 ······················ 222
  - 7.5.1 音频通话基础 ·············· 222
  - 7.5.2 开发音频通话功能 ·········· 223
- 7.6 视频播放 ······················ 229

第 8 章　相机开发 ················ 233

- 8.1 相机开发概述 ·················· 234
- 8.2 开发相机程序 ·················· 234
  - 8.2.1 相机接口 ·················· 234
  - 8.2.2 创建相机设备 ·············· 235
  - 8.2.3 配置相机设备 ·············· 238
  - 8.2.4 拍照 ······················ 240
- 8.3 相机实战：多功能拍照程序 ······ 246
  - 8.3.1 配置文件 ·················· 246

  8.3.2 布局文件 ················ 246
  8.3.3 主界面逻辑 ·············· 247
  8.3.4 拍照逻辑 ················ 249
  8.3.5 录制视频逻辑 ············ 254

第 9 章 网络程序开发 ············ 259

9.1 网络管理开发 ················ 260
  9.1.1 HTTP 数据请求 ············ 260
  9.1.2 WebSocket 连接 ··········· 264
  9.1.3 Socket 连接 ············· 270
9.2 IPC 与 RPC 通信 ·············· 273
  9.2.1 IPC 与 RPC 的基本概念 ····· 273
  9.2.2 开发 IPC 与 RPC 通信程序 ··· 274

第 10 章 数据管理 ················ 277

10.1 HarmonyOS 数据管理介绍 ········ 278
10.2 应用数据持久化 ·············· 279
  10.2.1 使用用户首选项存储数据 ···· 279
  10.2.2 使用键值型数据库存储数据 ··· 286
  10.2.3 使用关系型数据库存储数据 ··· 290

第 11 章 电话和短信服务 ········ 301

11.1 电话服务开发概述 ············ 302
11.2 跳转拨号界面 ··············· 302
  11.2.1 拨号接口 ··············· 302
  11.2.2 开发一个拨号程序 ········· 303
11.3 获取当前蜂窝网络信号信息 ····· 307
11.4 短信服务 ··················· 308
  11.4.1 sms 模块介绍 ············· 308
  11.4.2 sms 实战：发送指定内容的短信 ·· 311

第 12 章 设备管理 ················ 313

12.1 USB 开发 ··················· 314
  12.1.1 HarmonyOS USB API 介绍 ······ 314
  12.1.2 开发 HarmonyOS USB 程序 ······· 316
12.2 位置服务 ··················· 319
  12.2.1 位置开发概述 ············· 319
  12.2.2 获取设备的位置信息 ········ 320
  12.2.3 地理编码转化 ············· 325
12.3 传感器 ····················· 327
  12.3.1 HarmonyOS 系统传感器介绍 ···· 327
  12.3.2 开发传感器应用程序 ········ 329
12.4 综合实战：健身计步器 ········· 332
  12.4.1 系统配置 ··············· 332
  12.4.2 UI 视图 ················ 334
  12.4.3 项目主界面 ·············· 341

第 13 章 综合实战：新闻客户端（Node.js 服务端 + HarmonyOS 客户端）··· 347

13.1 背景介绍 ··················· 348
13.2 项目介绍 ··················· 348
  13.2.1 主要特点 ··············· 348
  13.2.2 项目结构 ··············· 349
13.3 系统架构 ··················· 349
13.4 服务器端 ··················· 349
  13.4.1 系统配置 ··············· 350
  13.4.2 Model 模块 ·············· 351
  13.4.3 控制器 ················ 353
  13.4.4 视图组件 ··············· 355
13.5 客户端 ····················· 355
  13.5.1 系统配置 ··············· 355
  13.5.2 通用模块 ··············· 357
  13.5.3 数据交互 ··············· 366
  13.5.4 视图界面 ··············· 370
  13.5.5 入口界面 ··············· 378
13.6 调试运行 ··················· 378

第 1 章 HarmonyOS 开发基础

鸿蒙操作系统（HarmonyOS）是华为公司开发的一款分布式操作系统，旨在实现各种设备之间的智能互联和协同工作。HarmonyOS 提供了强大的开发工具和框架，以支持跨设备的应用程序开发。随着时间的推移，HarmonyOS 的生态系统将继续完善，为开发者和用户提供更多的选择和便利。本章将详细讲解 HarmonyOS 系统的基础知识，帮助读者了解该系统的特点和优势。

扫码下载全书
案例源代码

1.1 智能手机系统介绍

智能手机系统是运行在智能手机上的操作系统，它控制着手机的各种功能和应用程序。每个操作系统都有其特点和各自的生态系统，用户和开发者可以根据其需求和偏好选择合适的智能手机系统。

1.1.1 智能手机系统的特点

智能手机系统是现代移动通信和计算的核心设备，其主要特点如下。

- 多功能性：智能手机系统具备多种功能，包括通信、互联网浏览、社交媒体、娱乐、摄影、导航、健康跟踪等，用户可以在一个设备上执行多种任务。
- 应用程序生态系统：智能手机系统拥有丰富的应用程序，用户可以通过应用商店下载各种应用程序，涵盖游戏、社交媒体、工作生产力工具等各个领域。
- 可定制性：用户可以根据自己的需求和喜好自定义智能手机系统的外观和功能，包括更换主题、设置壁纸、安装小部件和应用程序等。
- 联网能力：智能手机系统具有强大的互联网连接能力，可以通过 Wi-Fi、蜂窝数据网络、蓝牙等方式连接到互联网和其他设备。
- 传感器技术：智能手机系统配备了多种传感器，如 GPS、加速度计、陀螺仪、光线传感器、指纹识别等，以支持各种应用程序和功能。
- 移动支付：许多智能手机系统支持移动支付，用户可以使用手机进行付款，而无须使用实体信用卡或现金。
- 云集成：智能手机系统通常与云存储和云服务集成，使用户能够在不同设备之间同步和分享数据。
- 安全性：智能手机系统提供了多层次的安全性，包括屏幕锁定、指纹识别、面部识别和数据加密，以确保用户数据的安全性和私密性。
- 多媒体功能：智能手机系统通常配备高分辨率摄像头和音频功能，用户可以拍摄照片、录制视频、播放音乐和视频等。
- 语音助手：许多智能手机系统内置了语音助手，如苹果（Siri）、谷歌（Google Assistant）、亚马逊（Alexa），以便用户通过语音进行操作和查询。
- 固件和系统更新：智能手机制造商定期提供操作系统和安全性固件更新，以改进性能、修复漏洞和增强功能。

这些特点使智能手机系统成为一个强大的、多用途的移动计算平台，满足了用户的多种需求，从通信到娱乐，从工作到健康跟踪，为用户的生活提供了便利。

1.1.2 Android 系统介绍

Android 是一种流行的移动操作系统，由谷歌开发并维护。它是开源的操作系统，因此可以在各种手机制造商的设备上找到它，包括三星、华为、小米、LG 等。Android 系统的主要特点和功能如下。

- 开源代码：Android 的开源性意味着任何人都可以查看、修改和定制操作系统的代码。这使开发者和制造商可以自由地为其设备创建定制版本。
- 应用生态系统：Android 拥有庞大而多样化的应用商店，称为 Google Play 商店。用户可以从数百万个应用程序中选择，涵盖了各种领域，包括社交媒体、游戏、生产力工具、娱乐和教育应用。
- 多样性的硬件支持：Android 能够适应各种不同规格和功能的手机硬件。这意味着用户可以根据自己的需求选择不同品牌和型号的手机，以满足其需求。
- 定制性：用户可以根据自己的喜好自定义 Android 界面，如更换主题、图标、小部件等。此外，开发者可以创建自定义固件（ROM），以改变 Android 的外观和功能。
- 多用户支持：Android 支持多用户配置，允许多个用户在同一台设备上创建自己的账户，各自拥有自己的应用和数据。
- 通知系统：Android 提供了强大的通知系统，允许应用发送各种通知，包括消息、提醒、更新等。用户可以轻松地管理通知，并根据自己的偏好进行定制。
- Google 集成：Android 与谷歌的生态系统紧密集成，包括 Gmail、Google Drive、Google Maps、Google Photos 等应用。这些应用可以轻松地在 Android 设备上使用。
- 多任务处理：Android 允许用户在多个应用之间轻松切换，并支持多任务处理，允许同时运行多个应用。

总的来说，Android 是一个灵活、定制化和功能丰富的移动操作系统，适用于各种不同类型的用户和场景。它的开放性质使开发者和制造商可以根据市场需求进行创新和定制，因此在全球范围内广受欢迎。

1.1.3 iOS 系统介绍

iOS（原名 iPhone OS）是由苹果公司开发的移动操作系统，专为苹果的移动设备设计，包括 iPhone、iPad 和 iPod Touch。iOS 系统的主要特点和功能如下。

- 直观的用户界面：iOS 以其直观、流畅的用户界面而闻名，采用了触摸屏界面，具有直观的手势控制，包括滑动、轻扫和捏合等操作。
- App Store：iOS 拥有一个庞大的应用商店，称为 App Store。用户可以从中下载各种应用程序，包括社交媒体、游戏、生产力工具、娱乐和教育应用。
- 安全性：iOS 被广泛认为是一个相对安全的移动操作系统，采用了多层次的安全措施。包括应用沙盒、面部识别（Face ID）和指纹识别（Touch ID）等技术，以保护用户的数据和隐私。
- 更新和支持：苹果公司定期发布 iOS 更新，为用户提供新功能、性能改进和安全更新。iOS 设备通常可以获得多年的软件支持，这意味着较旧的设备也可以获得最新版本的 iOS。
- iCloud 集成：iOS 集成了 iCloud，这是苹果的云存储和同步服务。它允许用户轻松备份、共享和同步其照片、联系人、日历、备忘录和其他数据。
- Siri：iOS 引入了 Siri，Siri 是苹果的语音助手。用户可以使用语音命令来执行各种任

- 务，如发送短信、查找信息、设置提醒和控制智能家居设备。
- 家庭与健康：iOS 支持 HomeKit 和 HealthKit，这些是苹果的家庭自动化和健康管理平台。用户可以使用 iOS 设备来控制智能家居设备，同时监测健康和健身数据。
- 多任务处理：iOS 支持多任务处理，允许用户在不同应用之间切换，并实时多任务处理，以便在后台运行应用程序。

总的来说，iOS 是一个稳定、安全且对用户友好的移动操作系统，为用户提供了广泛的应用程序和生态系统集成，并且与苹果的硬件紧密配合，以提供一流的移动体验。iOS 设备通常受到持续的软件更新和支持，使其在长期内保持高度竞争力。

1.2 HarmonyOS 介绍

除了前文介绍的 Android 系统和 iOS 系统外，在智能手机领域还有一款令大家耳熟能详的强大操作系统，这便是本书的主角：HarmonyOS。这不仅是一款智能手机操作系统，还是一款能够实现万物互联的操作系统，手机、平板、电脑、智能家居、汽车等设备都可以互联。本节将详细讲解这款系统的特点，初步了解这款系统的强大之处。

1.2.1 HarmonyOS 的发展历程

HarmonyOS，也称为鸿蒙操作系统，是华为公司开发的多平台分布式操作系统，能够为智能手机、平板电脑、电视、智能家居设备等各种设备提供无缝的互联和协作体验。HarmonyOS 的主要发展历程如下。

2012—2016 年：华为开始研究和开发自己的操作系统。这个过程从最初的设想到技术原型的创建一直持续了几年。

2017 年：鸿蒙项目正式启动。华为宣布计划开发一种统一的操作系统，可以在多种设备上运行，以实现设备之间的无缝互联。

2019 年 8 月：华为首次发布了 HarmonyOS 的开发者预览版本。这个版本的目标是吸引开发者开始构建支持 HarmonyOS 的应用程序。

2020 年 9 月：华为发布了 HarmonyOS 2.0。这个版本是一个重大的里程碑，它将 HarmonyOS 扩展到了更多的设备类型，包括智能手机、平板电脑和智能电视。

2021 年 6 月：华为正式发布了 HarmonyOS 2.0 操作系统，首批支持该系统的智能手机在中国上市。此举标志着华为开始将其智能手机从之前使用的 Android 操作系统转向 HarmonyOS 操作系统。

2021 年 8 月：华为宣布计划将 HarmonyOS 逐渐推广到更多的设备，包括智能家居设备、汽车和智能穿戴设备等。

2022 年：华为继续在全球范围内推广 HarmonyOS，努力将其打造成一个多平台、多设备的操作系统。

2022 年 7 月 27 日：在华为全场景新品发布会正式发布 HarmonyOS 3 版本。

2023 年 8 月 4 日：在华为开发者大会 2023（HDC.Together）大会上，HarmonyOS 4 正式发布。

总的来说，HarmonyOS 的发展历程表明华为致力于构建一个统一的、分布式的操作系统，以满足不同类型设备之间互联互通的需求。虽然该系统仍在不断发展中，但华为将其视为未来智能生态系统的核心组成部分，并在不同领域推广其应用。

1.2.2　HarmonyOS、OpenHarmony、鸿蒙生态的区别与联系

HarmonyOS、OpenHarmony 和鸿蒙生态是在华为的鸿蒙（HarmonyOS）项目下的不同组成部分，它们之间存在着区别与联系。

1. HarmonyOS

HarmonyOS 是华为公司开发的多平台分布式操作系统，旨在实现不同类型设备之间的无缝互联和协作。

- 用途：HarmonyOS 是操作系统本身，可以安装在智能手机、平板电脑、智能电视、智能家居设备等多种设备上，以提供统一的用户体验。
- 特点：HarmonyOS 强调分布式架构，允许不同设备之间实现资源共享、协同工作和互联。它还注重性能、安全性和开发者友好性。

2. OpenHarmony

OpenHarmony 是 HarmonyOS 项目的一部分，是一个开源项目，旨在提供一个开放的、自由可用的分布式操作系统。它是 HarmonyOS 的开源版本。

- 用途：OpenHarmony 是为开发者和制造商提供的一个开放的、自由的操作系统，可用于在各种设备上构建自定义的分布式应用。
- 特点：OpenHarmony 提供了一套开放的代码库和工具，以帮助开发者构建支持分布式架构的应用。它促进了 HarmonyOS 生态系统的扩展。

3. 鸿蒙生态

鸿蒙生态是指在 HarmonyOS 基础上构建的生态系统，包括支持 HarmonyOS 的设备、应用程序、服务和开发者社区等。

- 用途：鸿蒙生态系统的目标是实现不同类型设备之间的互联互通，为用户提供统一的体验，同时为开发者提供构建应用和服务的机会。
- 特点：鸿蒙生态系统的特点是多设备支持，开发者友好性，以及在不同领域扩展，如智能手机、平板电脑、电视、汽车、智能家居等。

通过上面的描述，可以总结出 HarmonyOS、OpenHarmony 和鸿蒙生态的联系如下。

- HarmonyOS 是整个项目的核心操作系统，OpenHarmony 是它的开源版本，用于扩展和促进 HarmonyOS 生态系统的发展。
- 鸿蒙生态是在 HarmonyOS 的基础上建立的，它包括了所有支持 HarmonyOS 的设备和应用程序，以及与 HarmonyOS 相关的服务和社区。HarmonyOS 和 OpenHarmony 共同构成了鸿蒙生态系统的基础。

总的来说，HarmonyOS 是一个操作系统，OpenHarmony 是其开源版本，鸿蒙生态是基于 HarmonyOS 建立的生态系统，共同构成了华为的分布式计算和互联互通战略的关键组成部分。这些组件共同助力华为实现不同设备之间的协同工作和资源共享，为用户提供更强大的体验。

1.3 HarmonyOS 的优点

HarmonyOS 作为华为开发的多平台分布式操作系统，具有多个突出的优点，使其能在竞争激烈的移动操作系统市场中脱颖而出。

1.3.1 分布式架构

分布式架构是 HarmonyOS 的关键优点之一，它为该操作系统带来了多个重要优势，具体内容如下。

- 无缝互联：HarmonyOS 允许不同类型的设备之间实现无缝互联。这意味着可以轻松地在智能手机、平板电脑、电视和其他智能设备之间切换，而不会丧失应用程序的当前状态。这种无缝互联提供了更一致的用户体验。
- 资源共享：HarmonyOS 可以将不同设备上的资源进行共享，包括计算资源、存储空间、传感器和外部设备。这意味着可以利用其他设备上的资源来加速任务，例如，在电视上播放智能手机上的媒体文件。
- 分布式数据管理：HarmonyOS 允许数据在不同设备之间同步和共享。无论是在手机、电视、平板电脑还是其他设备上，都使用户可以随时访问其数据。分布式数据管理提供了出色的数据可用性和灵活性。
- 多任务处理：分布式架构使 HarmonyOS 可以更好地支持多任务处理。用户可以同时在不同设备上运行多个应用程序，从而提高了生产力和效率。
- 设备协同工作：HarmonyOS 允许设备之间协同工作，以执行复杂的任务。例如，智能家居设备可以与手机协同工作，以创建智能家居控制中心。设备协同工作提供了更多的智能化和自动化功能。
- 设备独立性：HarmonyOS 的分布式架构允许不同设备独立运行应用程序和服务。这意味着开发者可以为不同设备定制应用程序，而不必担心兼容性问题。
- 安全性：HarmonyOS 强调安全性，并提供了分布式安全措施，以保护用户数据和隐私。设备独立性对于多设备互联非常关键。

总的来说，HarmonyOS 的分布式架构使不同类型的设备可以更好地协同工作，为用户提供了更出色的用户体验、资源共享、多任务处理和安全性，这些优点有望在未来推动设备互联的发展。

1.3.2 多设备支持

多设备支持是 HarmonyOS 的重要优点之一，这意味着该操作系统可以运行在多种类型的设备上，提供了许多显著的优势。

- 一致的用户体验：无论是在智能手机、平板电脑、智能电视、智能家居设备还是其他类型的设备上，HarmonyOS 都为用户提供了一致的用户界面和操作方式。这使用户能够轻松切换设备而无须重新学习如何操作。
- 多设备协同工作：HarmonyOS 可以使不同类型的设备协同工作。例如，你可以在智能手机上启动一个任务，然后将其无缝切换到智能电视或平板电脑上继续进行。这种多设备协同工作提高了用户的生产力和便利性。

- 统一的应用生态系统：HarmonyOS 通过统一的应用生态系统，允许开发者构建一次应用程序并在多种设备上运行。这简化了应用程序开发和维护的流程，同时扩展了应用的覆盖范围。
- 多端数据同步：HarmonyOS 支持多端数据同步，这意味着可以在不同设备上访问和编辑相同的数据，如照片、联系人、日历和备忘录等。这提供了出色的数据一致性和可用性。
- 扩展性和适应性：HarmonyOS 可以根据不同设备的需求进行定制和优化。它具有扩展性，可以适应各种屏幕尺寸、输入方法和硬件规格。
- 设备无关性：开发者可以编写与设备无关的应用程序，这意味着应用程序可以适应不同设备上的不同屏幕大小和分辨率，从而提供更好的用户体验。
- 统一的开发工具：HarmonyOS 提供了统一的开发工具和 API，使开发者能够更容易地构建支持多设备的应用程序。

总的来说，HarmonyOS 的多设备支持是其核心优势之一，它提供了一种无缝的设备互联和用户体验，为用户和开发者提供了更大的灵活性和便利性。这有助于推动设备之间的互联和协同工作的开展。

1.3.3 开发者友好

HarmonyOS 的开发者友好性不仅在于提供统一的开发框架和开发工具，还包括完整的开发文档和资源，以帮助开发者更轻松地构建应用程序。具体来说，HarmonyOS 对开发者的友好主要体现在以下几个方面。

- 统一开发框架：HarmonyOS 提供了统一的开发框架，使开发者可以编写一次应用程序序代码，并在多个设备上运行，无须进行大规模的修改。
- 多端开发支持：HarmonyOS 支持多端开发，开发者可以构建适用于多种设备类型的应用程序，包括智能手机、平板电脑、电视、智能家居设备等。
- 统一的开发工具：HarmonyOS 提供一套一致的开发工具，包括开发集成环境（IDE）、调试器和模拟器，以提高开发效率和代码质量。
- 开放的开发生态系统：华为鼓励开发者积极参与到 HarmonyOS 生态系统中。它提供了开发者社区、开发者支持和资源，以促进应用程序的创建和发布。
- 开源部分代码：HarmonyOS 的一部分被开源，称为 OpenHarmony，鼓励开发者和社区的参与，使更多人可以为该系统贡献代码和功能。
- 多语言支持：HarmonyOS 支持多种编程语言，包括 C、C++、Java 和 Kotlin，使开发者可以使用他们熟悉的语言来构建应用程序。
- 持续更新和支持：华为承诺为 HarmonyOS 提供持续的软件更新和技术支持，以确保开发者能够访问最新的工具和功能。
- 完整的开发文档：HarmonyOS 提供了详细的开发文档，包括 API 文档、示例代码、教程和开发者指南。这些文档帮助开发者了解如何使用 HarmonyOS 的各种功能和服务。
- 在线支持和社区：开发者可以通过在线支持渠道和社区论坛获取帮助和解答问题。这有助于解决开发中遇到的问题和疑虑。

总之，HarmonyOS 为开发者提供了一系列的开发资源和文档，以帮助他们更轻松地构建应用程序，并确保这些应用程序在不同类型的设备上顺畅运行。这些资源和文档有助于提高开发者的生产力，同时也有助于推动 HarmonyOS 生态系统的发展。

1.4 HarmonyOS 架构分析

HarmonyOS 的架构旨在实现分布式计算和多设备互联，它采用微内核设计，强调安全性和模块化，同时提供了多种关键组件和工具，以支持开发者构建应用程序并在多种设备上运行。这种架构使 HarmonyOS 成为一个适用于各种设备和场景的多平台操作系统。

1.4.1 整体架构图

HarmonyOS 整体遵从分层设计，从下向上依次为：内核层、系统服务层、框架层和应用层。系统功能按照"系统 > 子系统 > 功能 / 模块"逐级展开，在多设备部署场景下，支持根据实际需求裁剪某些非必要的子系统或功能 / 模块。HarmonyOS 技术架构如图 1-1 所示，本架构图来源于华为官网。

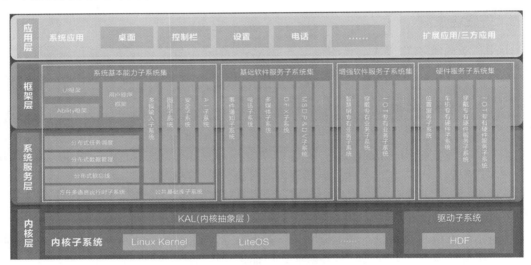

图 1-1　HarmonyOS 技术架构

1.4.2 内核层

HarmonyOS 的内核层是系统的核心部分，它负责管理硬件资源、进程管理、调度任务以及提供基本的系统服务。HarmonyOS 的内核层采用微内核架构，将核心功能分解成小模块，每个模块运行在独立的地址空间中，这有助于提高系统的稳定性和安全性。以下是 HarmonyOS 内核层的主要组成部分。

- 轻量级内核（Lite Kernel）：Lite Kernel 是 HarmonyOS 的核心微内核，它提供了基本的硬件抽象和管理，包括处理器管理、内存管理、中断处理和线程管理。Lite Kernel 是 HarmonyOS 的运行时引擎，负责执行系统任务和应用程序。
- 进程管理模块：负责创建、销毁和管理应用程序进程。它确保不同的应用程序在系

- 调度器模块：负责任务的调度和管理。它决定了哪个任务在何时运行，并且根据任务的优先级和资源需求来分配处理器时间。
- 驱动程序和硬件抽象层（HAL）模块：允许操作系统与硬件设备进行通信。每个设备都有相应的驱动程序和硬件抽象层，以便操作系统可以与硬件设备交互，包括传感器、显示屏、声音设备等。
- 进程间通信（IPC）模块：负责不同进程之间的通信。它允许应用程序在不同进程之间共享数据和信息。
- 文件系统模块：负责文件的管理和存储。它允许应用程序创建、读取、写入和删除文件。
- 安全子系统模块：负责管理系统的安全性，包括身份验证、数据加密和权限管理等。它确保用户的数据和隐私得到保护。
- 内存管理模块：负责分配和释放系统内存。它确保不同任务和进程之间的内存隔离。

这些组成部分共同构成了 HarmonyOS 的内核层，它们协同工作，以确保系统的稳定性、性能和安全性。这个微内核架构使 HarmonyOS 更加灵活，可以轻松适应不同类型的设备和应用场景。通过将核心功能分解成小模块，也有助于简化系统的开发和维护。

1.4.3 系统服务层

HarmonyOS 的系统服务层是该操作系统的一个重要组成部分，它提供了各种系统级服务以支持应用程序和用户体验。系统服务层包括许多服务模块，重要的服务模块如下。

- 应用管理服务模块：负责应用程序的安装、卸载、更新和管理。它还包括应用启动管理，确保应用程序以有效的方式启动和运行。
- 图形用户界面（GUI）服务模块：处理用户界面的渲染、事件处理和用户交互。它允许应用程序创建图形界面并与用户进行互动。
- 音频和多媒体服务模块：提供音频播放、视频播放和多媒体处理的功能。它支持多媒体应用程序的开发，包括音乐播放器、视频播放器等。
- 网络服务模块：负责网络连接、通信和数据传输。它支持各种网络协议和通信方式，包括 Wi-Fi、蓝牙、蜂窝数据和互联网连接。
- 位置服务模块：提供位置定位和地理信息服务。它支持应用程序获取设备的位置信息，用于导航、地图和位置感知的应用。
- 传感器服务模块：允许应用程序访问设备上的传感器数据，如加速度计、陀螺仪、环境传感器等。
- 文件系统服务模块：提供文件访问和管理的功能。它允许应用程序读取、写入和管理设备上的文件和数据。
- 通知服务模块：负责管理系统和应用程序的通知。它确保用户及时收到重要信息和通知。
- 安全和权限服务模块：负责管理应用程序的权限和数据安全。它确保用户数据得到保护，只有授权的应用程序可以访问敏感信息。
- 云服务模块：允许应用程序与云端服务器进行通信和数据同步，支持云存储、云备份和云计算等功能。

这些系统服务模块共同构成了 HarmonyOS 的系统服务层，它们为应用程序提供了必要的支持和功能，同时也为用户提供了丰富的功能和体验。这些服务模块使 HarmonyOS 适用于多种设备类型，包括智能手机、平板电脑、电视、智能家居设备等，并支持分布式应用和多设备互联。

1.4.4 框架层

HarmonyOS 的框架层是构建在系统服务层之上的一组框架和工具，它们提供了开发应用程序所需的核心功能和抽象层。框架层的目标是简化应用程序开发，使开发者能够更轻松地构建跨多设备平台的应用程序。HarmonyOS 框架层的主要组成部分如下。

- 应用框架：提供了应用程序的基本结构和生命周期管理。它包括应用启动、界面创建、活动管理、应用间通信等功能。开发者可以使用应用框架来构建应用程序并定义其行为。
- 界面框架：包括 UI 组件库、布局管理器和绘图引擎等，用于创建应用程序的用户界面。它提供了各种界面元素，如按钮、文本框、图像视图等，以及用于排列和显示这些元素的工具。
- 多媒体框架：支持音频和视频的播放和处理。它包括音频和视频编解码器、媒体播放器、相机接口等，使开发者能够创建多媒体应用程序。
- 数据管理框架：帮助开发者管理应用程序的数据，包括本地数据存储、数据库访问、数据同步和数据共享等功能。
- 网络框架：允许应用程序进行网络通信和数据传输。它支持各种网络协议和通信方式，包括 HTTP、WebSocket 等。
- 安全框架：提供了应用程序权限管理、数据加密和身份验证等安全功能，以保护用户数据和隐私。
- 分布式能力框架：支持分布式应用程序的开发，允许应用程序在不同设备之间协同工作和通信。
- 设备能力框架：提供了访问设备硬件和传感器的接口，使应用程序能够利用设备的各种功能。
- 跨平台支持：HarmonyOS 的框架层支持跨多个设备平台的开发，允许开发者编写一次代码并在不同操作系统（如 Android、iOS）上运行。
- 开发工具：HarmonyOS 提供了一套开发工具，包括开发集成环境（IDE）、调试器和模拟器，以帮助开发者创建、测试和部署应用程序。

这些框架和工具构成了 HarmonyOS 的框架层，为开发者提供了丰富的资源和抽象层，以简化应用程序的开发过程。这有助于开发者更轻松地构建多设备互联的应用程序，并提供一致的用户体验。框架层的设计和功能使 HarmonyOS 成为一个多平台、多设备操作系统，有助于推动设备互联和跨设备应用的发展。

1.4.5 应用层

HarmonyOS 的应用层是操作系统的顶层，包括各种应用程序和用户界面。这一层主要由应用程序、系统应用和用户界面组成，它们为用户提供了各种功能和体验。HarmonyOS 应用层的主要组成部分如下。

- 应用程序：用户直接与操作系统交互的部分。它们包括各种类型的应用，如社交媒体应用、游戏、办公套件、娱乐应用、生产力工具等。这些应用程序由开发者创建，并可以通过应用商店或其他渠道安装到设备上。
- 系统应用：由操作系统提供的应用程序，用于管理设备的核心功能和服务。这些应用程序通常包括设置、日历、联系人、邮件、短信等，它们是操作系统的一部分，为用户提供基本的功能。
- 用户界面（UI）：用户与设备交互的界面，包括主屏幕、应用程序图标、通知中心、设置菜单、锁屏界面等。HarmonyOS 的用户界面通常采用现代、直观的设计，以提供良好的用户体验。
- 多窗口管理：HarmonyOS 支持多窗口管理，允许用户在同一设备上同时运行多个应用程序并进行分屏或多窗口操作。这提高了多任务处理的效率。
- 分布式应用：HarmonyOS 的应用层支持分布式应用程序，允许应用程序在不同设备之间协同工作和通信。例如，用户可以在智能手机上启动一个任务，然后在智能电视上继续操作。
- 应用商店：用户获取和安装应用程序的渠道。HarmonyOS 包括应用商店，如华为的 AppGallery，用户可以从中浏览、搜索和下载应用程序。
- 通知系统：允许应用程序向用户发送通知和消息。用户可以从通知中心查看和管理通知，以保持对重要信息的关注。
- 虚拟助手：HarmonyOS 可能包括虚拟助手，如华为的 Celia，用于语音识别和语音控制，以帮助用户完成各种任务。
- 用户设置和个性化：用户可以在系统中进行各种设置和个性化配置，包括主题、壁纸、语言设置、隐私设置等，以满足他们的需求和偏好。

HarmonyOS 的应用层是用户与操作系统互动的前沿，它提供了各种应用和工具，以满足用户的需求，并为用户提供一致的、流畅的体验。应用层还支持分布式应用和多设备互联，使用户能够无缝切换设备并协同工作。这一层的设计和功能对于提高用户满意度和系统的多用途性至关重要。

第 2 章
搭建开发环境

搭建开发环境是指为开发特定的应用程序或平台配置所需的工具、软件和资源，以便开发者可以进行编码、测试和调试工作。对于 HarmonyOS 应用开发来说，同样需要搭建开发环境。在进行 HarmonyOS 应用开发之前，需要配置一些基本的开发工具和资源，以确保开发者顺利进行开发工作。本章将详细讲解搭建 HarmonyOS 应用开发环境的知识，为读者学习本书后续章节的知识打下基础。

2.1 HarmonyOS 应用开发介绍

HUAWEI DevEco Studio 是基于 IntelliJ IDEA Community 开源版本打造的，面向全场景多设备，提供一站式的分布式应用开发平台。它支持分布式多端开发、分布式多端调测、多端模拟仿真，以及提供全方位的质量与安全保障。

2.1.1 HarmonyOS 应用 / 服务的开发流程

DevEco Studio 是 HarmonyOS 应用开发的必备工具，可以使用 DevEco Studio 开发并上架一个 HarmonyOS 应用程序到华为应用市场。华为官网中给出了 HarmonyOS 应用 / 服务的开发流程，如图 2-1 所示。

图 2-1　HarmonyOS 应用 / 服务的开发流程

2.1.2 配置开发环境的流程

DevEco Studio 同时支持 Windows 系统和 macOS 系统。在开发 HarmonyOS 应用 / 服务前，需要配置 HarmonyOS 应用 / 服务的开发环境。配置开发环境的基本流程，如图 2-2 所示。

图 2-2　配置开发环境的基本流程

2.2 搭建 DevEco Studio 开发环境

前文已经介绍，DevEco Studio 是 HarmonyOS 应用开发的必备工具。本节将详细讲解搭建 DevEco Studio 开发环境的知识。

2.2.1 DevEco Studio 的特点

作为一款专业的开发工具，DevEco Studio 除了具有基本的代码开发、编译构建及调测等功能外，还具有如下特点。

- 高效智能代码编辑：支持 eTS、JavaScript、C/C++ 等语言的代码高亮显示、代码智能补齐、代码错误检查、代码自动跳转、代码格式化、代码查找等功能，提升代码编写效率。更多详细信息，请参考编辑器使用技巧。
- 低代码可视化开发：丰富的 UI 界面编辑能力，支持自由拖曳组件和可视化数据绑定，可快速预览效果，所见即所得；同时支持卡片的零代码开发，降低开发门槛和提升界面开发效率。更多详细信息，请参考使用低代码开发应用/服务。
- 多端双向实时预览：支持 UI 界面代码的双向预览、实时预览、动态预览、组件预览及多端设备预览，便于快速查看代码运行效果。更多详细信息，请参考使用预览器预览应用/服务界面效果。
- 多端设备模拟仿真：提供 HarmonyOS 本地模拟器、远程模拟器、超级终端模拟器，支持手机、智慧屏、智能穿戴等多端设备的模拟仿真，便捷获取调试环境。更多详细信息，请参考使用模拟器运行应用/服务。

2.2.2 环境要求

1. Windows 运行环境要求

为保证 DevEco Studio 正常运行，建议电脑配置满足如下要求。
- 操作系统：Windows 10，64 位。
- 内存：8GB 及以上。
- 硬盘：100GB 及以上。
- 分辨率：1280 像素 ×800 像素及以上。

2. macOS 运行环境要求

为保证 DevEco Studio 正常运行，建议电脑配置满足如下要求。
- 操作系统：macOS 10.15/11.x/12.x。
- 芯片类型：Intel 系列。
- 内存：8GB 及以上。
- 硬盘：100GB 及以上。
- 分辨率：1280 像素 ×800 像素及以上。

2.2.3 下载并安装 DevEco Studio

下面以 Windows 系统为例，介绍下载并安装 DevEco Studio 的流程。

（1）登录华为开发者网站：https://developer.harmonyos.com/，依次单击网页顶部的"开发"及 DevEco Studio 链接，如图 2-3 所示。

图 2-3　依次单击"开发"及 DevEco Studio 链接

（2）在弹出的新页面中单击"立即下载"按钮，如图 2-4 所示。

（3）在新弹出的页面中显示下载列表信息，展示了不同操作系统的下载版本，如图 2-5 所示。大家需要根据计算机的操作系统选择对应的版本下载，本书将以 64 位操作系统为例进行讲解，单击"Windows(64-bit)"后面的图标↓开始下载。

图 2-4　单击"立即下载"按钮　　　　　　　图 2-5　下载列表

（4）下载完成后会得到一个 deveco-studio-3.xxx.exe 格式的安装文件，用鼠标双击这个文件开始安装。在安装时首先弹出一个 Welcome 欢迎界面，如图 2-6 所示。

图 2-6　Welcome 欢迎界面

（5）单击 Next 按钮，选择安装位置界面，并选择安装 DevEco Studio 的位置，如图 2-7 所示。

（6）单击 Next 按钮，打开安装选项界面，勾选 3 个复选框，如图 2-8 所示。

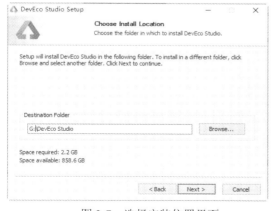
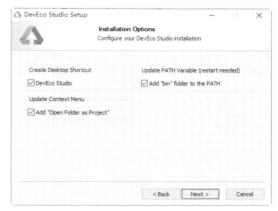

图 2-7　选择安装位置界面　　　　　　　　图 2-8　安装选项界面

（7）单击 Next 按钮，打开选择启动菜单界面，此处使用默认选择，如图 2-9 所示。
（8）单击 Installt 按钮，开始安装 DevEco Studio 并显示安装进度条，如图 2-10 所示。

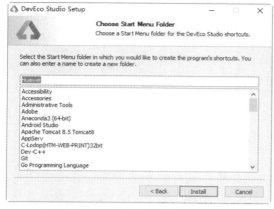
图 2-9　选择启动菜单界面

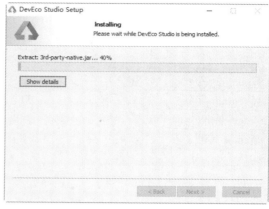
图 2-10　安装进度条界面

（9）单击 Next 按钮弹出安装完成界面，如图 2-11 所示，单击 Finish 按钮完成安装。

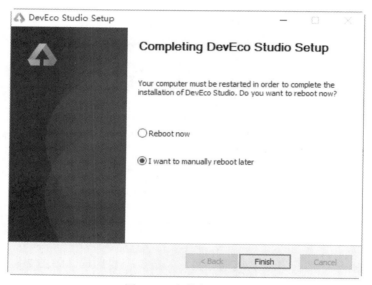
图 2-11　安装完成界面

2.3　配置 DevEco Studio 开发环境

在开发 HarmonyOS 应用 / 服务前，请提前下载 SDK 及配套的工具链，并配置 HDC 工具环境变量。在安装 DevEco Studio 后，第一次打开 DevEco Studio 时会提示系统配置信息。

2.3.1　下载 SDK 及工具链

DevEco Studio 提供 SDK Manager 统一管理 SDK 及工具组件，包括如表 2-1 所示的组件包。

表 2-1　SDK 及配套的工具链说明

组件包名	说　　明
Native	C/C++ 语言 SDK 包
eTS	eTS（Extended TypeScript）SDK 包
JS	JS 语言 SDK 包
Java	Java 语言 SDK 包
System-image-phone	本地模拟器 Phone 设备镜像文件，仅支持 API Version 6
System-image-tv	本地模拟器 TV 设备镜像文件，仅支持 API Version 6
System-image-wearable	本地模拟器 Wearable 设备镜像文件，仅支持 API Version 6
EmulatorX86	本地模拟器工具包
Toolchains	SDK 工具链，HarmonyOS 应用 / 服务开发必备工具集，包括编译、打包、签名、数据库管理等工具的集合
Previewer	HarmonyOS 应用 / 服务预览器，在开发过程中可以动态预览 Phone、TV、Wearable、LiteWearable 等设备的应用 / 服务效果，支持 JS、eTS 和 Java 应用 / 服务预览

2.3.2　配置向导

HarmonyOS 应用 / 服务支持 API Version 4 至 API Version 8，如果是第一次使用 DevEco Studio，其配置向导会引导下载 SDK 及工具链。配置向导默认下载 API Version 8 的 SDK 及工具链。如需下载 API Version 4 至 7，可以在工程配置完成后，进入 HarmonyOS SDK 界面手动下载，方法如下。

（1）在 DevEco Studio 欢迎界面，依次单击 Configure（或图标）> Settings > HarmonyOS SDK 进入 SDK Manager 界面（macOS 系统为依次单击 Configure > Preferences > HarmonyOS SDK）。

（2）在 DevEco Studio 打开工程的情况下，单击 Tools > SDK Manager 界面进入；或者单击 Files > Settings > HarmonyOS SDK 进入（macOS 系统为 DevEco Studio > Preferences > HarmonyOS SDK）。

如果是第一次启动 DevEco Studio，使用配置向导的流程如下。

（1）运行已安装的 DevEco Studio，弹出如图 2-12 所示的欢迎界面，在此单击 Agree 按钮。

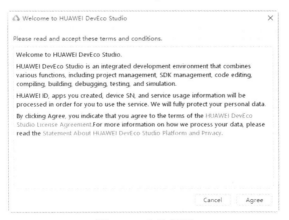

图 2-12　欢迎界面

（2）如果是第一次启动 DevEco Studio，在单击 Agree 按钮后会弹出配置向导界面。在此选中 Do not import settings 单选按钮，然后单击"OK"按钮，如图 2-13 所示。

图 2-13　勾选"Do not import settings"选项

（3）进入 DevEco Studio 操作向导界面，修改 npm registry，DevEco Studio 已预置对应仓库（默认的 npm 仓库，可能出现部分开发者无法访问或访问速度缓慢的情况），直接单击 Start using DevEco Studio 按钮进行下一步。

（4）在弹出的 Basic Setup 界面中设置 Node.js 和 Ohpm 信息，可以指定本地已安装的 Node.js（Node.js 版本要求为 v14.19.1 及以上，且低于 v15.0.0；对应的 npm 版本要求为 6.14.16 及以上，且低于 7.0.0 版本）。建议大家选择 Install 选项，在线下载新的 Node.js 和 Ohpm。在设置好存储路径后，单击 Next 按钮进行下一步，如图 2-14 所示。

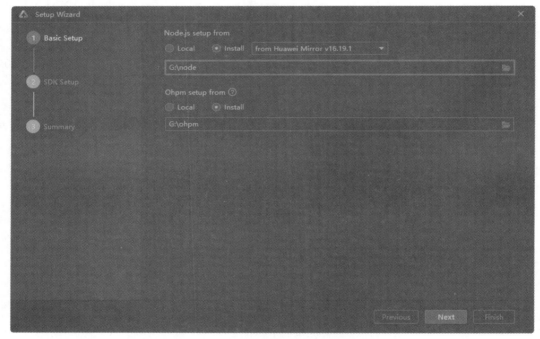

图 2-14　Basic Setup 界面

（5）在弹出的 SDK Setup 界面中设置保存 HarmonyOS SDK 的位置，在此界面还会显示当前 SDK 的信息，包括版本信息和大小信息等，如图 2-15 所示。单击 Next 按钮进入下一步。

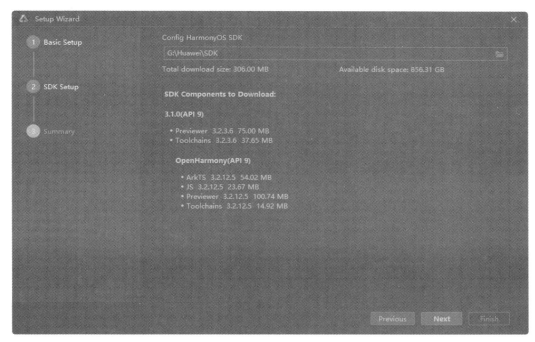

图 2-15　SDK Setup 界面

（6）在弹出的 Summary 界面中显示安装信息，在此选中 Accept 单选按钮，然后单击 Next 按钮进行下一步，如图 2-16 所示。

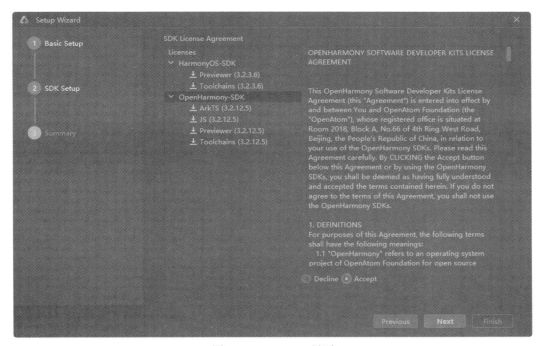

图 2-16　Summary 界面

（7）在弹出的新界面中显示安装位置和所占用空间大小的信息，如图 2-17 所示。在此单击 Next 按钮进行下一步。

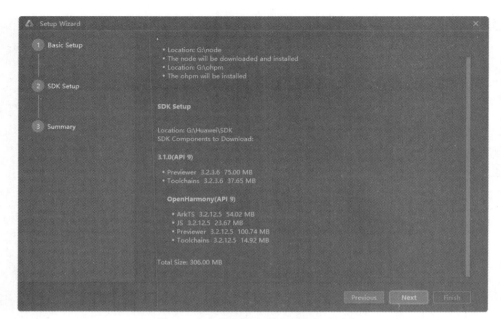

图 2-17　安装位置和所占用空间大小

（8）在弹出的界面中显示安装进度，如图 2-18 所示。进度条到达 100% 后单击 Finish 按钮完成安装。

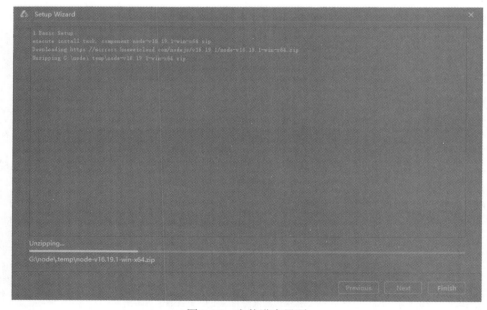

图 2-18　安装进度界面

2.4　使用 DevEco Studio 开发第一个鸿蒙应用程序

DevEco Studio 开发环境配置工作完成后，可以通过运行 Hello World 工程来验证环境设置是否正确。

扫码看视频

2.4.1 创建工程

（1）打开 DevEco Studio，在欢迎界面的左侧导航栏中单击 Create Project 按钮，创建一个新工程，如图 2-19 所示。

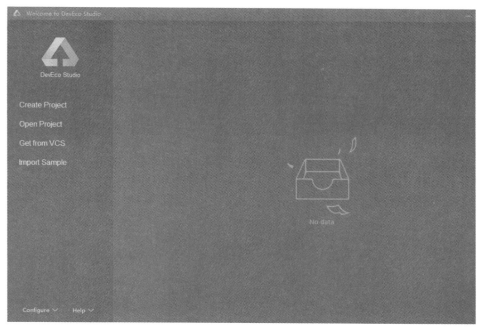

图 2-19　单击 Create Project 按钮

（2）此时，弹出"创建工程"界面，如图 2-20 所示，在此选择 Empty Ability 模板，然后单击 Next 按钮。

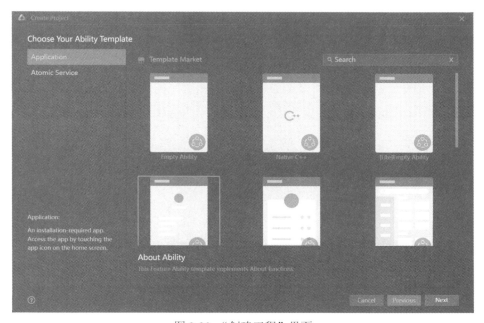

图 2-20　"创建工程"界面

（3）在弹出的"设置工程"界面中填写相关信息，如图 2-21 所示。在此可以设置工程的如下信息。

- Project name：工程的名称，可以自定义，由大小写字母、数字和下划线组成。
- Bundle name：软件包名称，默认情况下，应用/服务 ID 也会使用该名称，应用/服务发布时，应用/服务 ID 需要唯一。如果 Project type 选择了 Atomic service，则 Bundle name 的后缀名必须是 .hmservice。
- Save location：保存当前工程文件的路径，由大小写字母、数字和下划线等组成，不能包含中文字符。
- Compile SDK：应用/服务的目标 API Version。在编译构建时，DevEco Studio 会根据指定的 Compile API 版本进行编译打包。
- Model：应用支持的模式。API Version 4 至 8 只支持 FA 模式，建议选择 Stage 模型。因为 Stage 是 OpenHarmony API 9 开始新增的模型，是目前主推且会长期演进的模型。该模型提供了 AbilityStage、WindowStage 等作为应用组件和 Window 窗口的"舞台"，因此称这种应用模型为 Stage 模型。
- Enable Super Visual：支持低代码开发模式，部分模板支持低代码开发，可选择打开该开关。
- Language：开发语言。
- Compatible SDK：兼容的最低 API Version。
- Device type：该工程模板支持的设备类型。
- Show in service center：是否在服务中心展示。如果 Project type 为 Atomic service，则会同步创建一个 2×2 宫格的服务卡片模板；如果 Project type 为 Application，则会同步创建一个 2×2 宫格的服务卡片模板，同时还会创建入口卡片。

设置完成后单击 Finish 按钮。

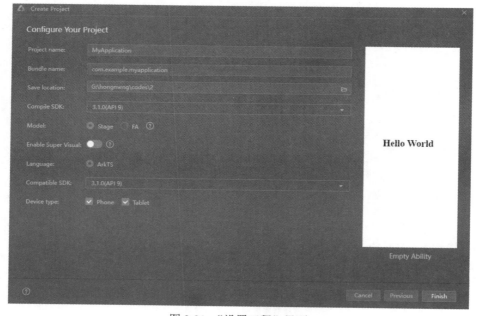

图 2-21 "设置工程"界面

（4）工程创建完成后，DevEco Studio 会自动进行工程的同步，自动生成示例代码和相关资源，如图 2-22 所示。

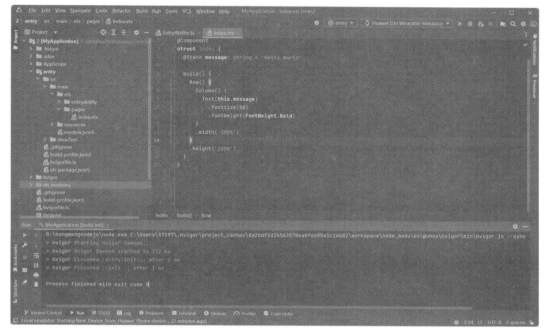

图 2-22 创建的工程和自动生成的代码

2.4.2 DevEco Studio 界面介绍

DevEco Studio 界面大致可以分为四个部分，分别为代码编辑区、通知栏、工程目录区和预览区，如图 2-23 所示。

图 2-23 IDE 界面

1. 代码编辑区

DevEco Studio 界面中间部分是代码编辑区，开发者在这里编写代码、修改代码，以及切换显示的文件。通过按住 Ctrl 键加鼠标滚轮，可以实现界面的放大与缩小。

2. 通知栏

在 DevEco Studio 界面下端是通知栏，在编辑器底部有一行工具栏，主要包括常用信息栏，其中 Run 是项目运行时的信息栏，Problems 是当前工程错误与提醒信息栏，Terminal 是命令行终端，在这里执行命令行操作，PreviewerLog 是预览器日志输出栏，Log 是模拟器和真机运行时的日志输出栏，如图 2-24 所示。

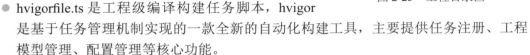

图 2-24　通知栏

3. 工程目录区

DevEco Studio 界面左侧为工程目录区，在此列出了当前项目工程的所有程序文件，如图 2-25 所示。各个工程目录的具体说明如下。

- AppScope 目录中存放的是应用全局所需要的资源文件。
- entry 目录是应用的主模块，用于存放 HarmonyOS 应用的代码、资源等。
- oh_modules 文件夹是工程的依赖包，用于存放工程依赖的源文件。
- build-profile.json5 是工程级配置文件，在里面包含了签名、产品配置等信息。

图 2-25　工程目录区

- hvigorfile.ts 是工程级编译构建任务脚本，hvigor 是基于任务管理机制实现的一款全新的自动化构建工具，主要提供任务注册、工程模型管理、配置管理等核心功能。
- oh-package.json5 是工程级依赖配置文件，用于记录引入包的配置信息。

1）AppScope 目录

在 AppScope 目录下有 resources 文件夹和配置文件 app.json5，在 AppScope>resources>base 目录中包含了 element 和 media 两个文件夹，如图 2-26 所示。其中 element 目录用于存放公共的字符串、布局文件等资源，media 目录中存放了全局的、公共的多媒体资源文件。

2）entry 目录

在 entry>src 目录中主要包含了 main 目录，单元测试目录 ohosTest，以及模块级的配置文件。在 main 目录中，ets 目录用于存放 ets 代码，resources 目录用于存放模块内的多媒体及布局文件等，文件 module.json5 为模块的配置文件，如图 2-27 所示。

各个子目录的具体说明如下。
- ohosTest 是单元测试目录，保存和测试相关的程序文件。
- build-profile.json5 是模块级配置信息，包括编译构建配置项。
- hvigorfile.ts 文件是模块级构建脚本。
- oh-package.json5 是模块级依赖配置信息文件。

在 src>main>ets 目录中包含 entryability 和 pages 两个文件夹，如图 2-28 所示。
- entryability 存放 ability 文件，用于当前 ability 应用逻辑和生命周期管理。
- pages 存放 UI 界面相关代码文件，初始会生成一个 Index 页面。

图 2-26　AppScope 目录　　　图 2-27　entry 目录　　　图 2-28　src>main>ets 目录

4. 文件 app.json5

AppScope>app.json5 是应用全局的配置文件，用于存放应用公共的配置信息，如图 2-29 所示。

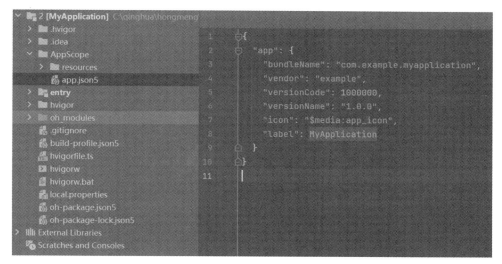

图 2-29　文件 app.json5

各个配置信息的具体说明如下。
- bundleName：包名。
- vendor：应用程序供应商。
- versionCode：用于区分应用版本。

- versionName：版本号。
- icon：应用的显示图标。
- label：应用名称。

5. 文件 module.json5

文件 entry>src>main>module.json5 是模块的配置文件，包含当前模块的配置信息，如图 2-30 所示。

图 2-30　文件 module.json5

其中 module 对应的是模块的配置信息，一个模块对应一个打包后的 hap 包，hap 包全称是 HarmonyOS Ability Package，其中包含了 ability、第三方库、资源和配置文件。模块配置信息的属性及其描述，如表 2-2 所示。

表 2-2　模块配置信息的属性及其描述

属　性	描　述
name	该标签标识当前 module 的名字，module 打包成 hap 后，表示 hap 的名称，标签值采用字符串表示（最大长度 31 个字节），该名称在整个应用中具有唯一性
type	表示模块的类型有三种，分别是 entry、feature 和 har
srcEntry	当前模块的入口文件路径
description	当前模块的描述信息
mainElement	该标签标识 hap 的入口 ability 名称或者 extension 名称。只有配置为 mainElement 的 ability 或者 extension 才允许在服务中心露出
deviceTypes	该标签标识 hap 可以运行在哪类设备上，标签值采用字符串数组表示
deliveryWithInstall	标识当前 Module 是否在用户主动安装的时候安装，表示该 Module 对应的 HAP 是否跟随应用一起安装。- true：主动安装时安装。- false：主动安装时不安装
installationFree	标识当前 Module 是否支持免安装特性。- true：表示支持免安装特性，且符合免安装约束。- false：表示不支持免安装特性
pages	对应的是 main_pages.json 文件，用于配置 ability 中用到的 page 信息
abilities	是一个数组，存放当前模块中所有的 ability 元能力的配置信息，其中可以有多个 ability

6. 文件 main_pages.json

文件 src/main/resources/base/profile/main_pages.json 保存的是页面 page 的路径配置信息，所有需要进行路由跳转的 page 页面都要在这里进行配置，如图 2-31 所示。

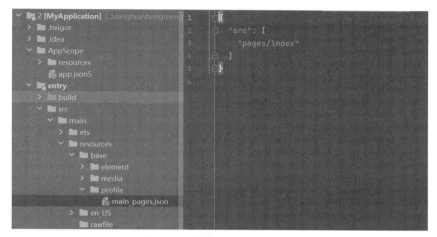

图 2-31　文件 main_pages.json

2.4.3　在模拟器中运行程序

我们可以在模拟器中运行前文创建的 Hello World 工程，接下来展示 Hello World 工程在模拟器中运行的过程。

（1）在 DevEco Studio 菜单栏中选择 Tools > Device Manager 命令。

（2）在 Remote Emulator 界面中单击 Sign In 按钮，在浏览器中弹出华为开发者联盟账号登录界面，请输入已实名认证的华为开发者联盟账号的用户名和密码进行登录。

（3）登录后，单击界面的"允许"按钮进行授权，如图 2-32 所示，授权完成后，切换到 Device Manager 界面。

图 2-32　单击"允许"按钮，进行授权

（4）授权完成后，在弹出的"设备列表"界面中选择一个 Phone 设备，如选择 P50，然后，单击后面的"▶"按钮运行模拟器，如图 2-33 所示。

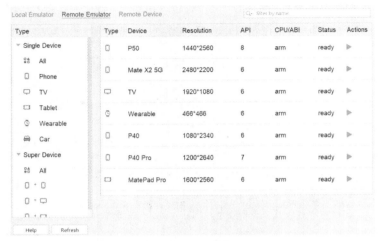

图 2-33　运行模拟器

（5）单击 DevEco Studio 工具栏中的"▶"按钮运行当前工程（Hello World），或使用默认 Shift+F10 快捷键（macOS 为 Control+R）运行工程，如图 2-34 所示。

（6）DevEco Studio 会启动应用/服务的编译构建工作，构建完成后即可在模拟器上成功运行前面创建的 Hello World 工程，如图 2-35 所示。

图 2-34　单击 DevEco Studio 工具栏中的"▶"按钮　　图 2-35　在模拟器中的运行 hello World 工程效果

2.4.4　在本地真机中运行程序

在 Phone 和 Tablet 真机中运行 HarmonyOS 应用/服务的操作方法一致，可以采用 USB 连接方式或者 IP Connection 连接方式。采用 IP Connection 连接方式要求 Phone/Tablet 和 PC 端在同一个网段，建议将 Phone/Tablet 和 PC 连接到同一个 WLAN 网络。

1. 使用 USB 连接方式

（1）在 Phone 或 Tablet 中，打开"开发者模式"。打开方法如下：依次进入"设置">"关于手机"/"关于平板"，然后连续多次点击"版本号"，直到提示"您正处于开发者模式"。然后在"设置"的"系统与更新">"开发人员选项"中，打开"USB 调试"开关，如图 2-36 所示。

（2）使用 USB 线将 Phone 或 Tablet 与 PC 连接。

（3）在 Phone 或 Tablet 中，USB 连接方式选择"传输文件"。

（4）在 Phone 或 Tablet 中，会弹出"是否允许 USB 调试"的提示框，点击"确定"按钮，如图 2-37 所示。

图 2-36　打开"USB 调试"开关　　图 2-37　"是否允许 USB 调试"提示框

（5）在 DevEco Studio 的菜单栏中，依次选择 Run>Run' 模块名称 ' 命令，或使用默认快捷键 Shift+F10（macOS 为"Control+R"）运行应用 / 服务，如图 2-38 所示。

图 2-38　选择 Run>Run' 模块名称 ' 命令

（6）DevEco Studio 启动 HAP 的编译构建和安装。在真机设备中安装应用程序成功后，Phone 或 Tablet 会自动运行安装的 HarmonyOS 应用 / 服务。

2. 使用 IP Connection 连接方式

（1）将 Phone/Tablet 和 PC 连接到同一 WLAN 网络。

（2）确保 Phone/Tablet 上的 5555 端口为打开状态。默认是关闭状态，可以通过使用 USB 连接方式连接上设备后，执行如下命令打开：

```
hdc tmode port 5555
```

在 Phone/Tablet 中运行应用 / 服务，需要根据为应用 / 服务进行签名的章节，提前对应用 / 服务进行签名。

（3）打开 DevEco Studio，选择菜单栏中的 Tools>IP Connection 命令，输入连接设备的 IP 地址，点击运行"▶"按钮，连接正常后，设备状态应显示为 online，如图 2-39 所示。

（4）在 DevEco Studio 菜单栏中，选择 Run>Run' 模块名称 ' 命令，或使用默认快捷键 "Shift+F10"（macOS 为 Control+R）运行应用 / 服务，如图 2-40 所示。

图 2-39　连接设备　　　　　　图 2-40　选择 Run>Run' 模块名称 ' 命令

第 3 章 HarmonyOS 应用模型

应用模型是 HarmonyOS 为开发者提供的对应用程序所需能力的抽象提炼,它提供了应用程序必备的组件和运行机制。有了应用模型,开发者可以基于一套统一的模型进行应用开发,使应用开发更简单、更高效。本章将详细讲解 HarmonyOS 应用模型的知识。

3.1 HarmonyOS 应用模型介绍

HarmonyOS 中的应用模型十分重要，其提供了应用程序必备的组件和运行机制，可以基于一套统一的模型进行应用开发。

3.1.1 应用模型的组成元素

作为 HarmonyOS 系统中的一个重要概念，应用模型的组成元素包括如下内容。

1）应用组件

应用组件是应用的基本组成单位，也是应用的运行入口。用户在启动、使用和退出应用的过程中，应用组件会在不同状态间切换，这些状态被称为应用组件的生命周期。应用组件提供生命周期的回调函数，开发者可以通过应用组件的生命周期回调感知应用的状态变化。开发者在编写应用时，首先需要编写应用组件，并编写生命周期回调函数，在应用配置文件中配置相关信息。操作系统在运行期间通过配置文件创建应用组件的实例，并调度其生命周期回调函数，执行开发者的代码。

2）应用进程模型

应用进程模型定义应用进程的创建和销毁方式，以及进程间的通信方式。

3）应用线程模型

应用线程模型定义了应用进程内线程的创建和销毁方式、主线程和 UI 线程的创建方式，以及线程间的通信方式。

4）应用任务管理模型

应用任务管理模型定义了任务（Mission）的创建和销毁方式，以及任务与组件间的关系。HarmonyOS 的应用任务管理由系统应用负责，第三方应用无须关注，下文不作具体介绍。

5）应用配置文件

应用配置文件包含应用配置信息、应用组件信息、权限信息、开发者自定义信息等，这些信息在编译构建、分发和运行阶段分别提供给编译工具、应用市场和操作系统使用。

3.1.2 应用模型的发展进程

随着 HarmonyOS 系统的发展，先后提供了以下两种应用模型。
- FA（Feature Ability）模型：HarmonyOS 早期版本支持的模型，现在已不再主推。
- Stage 模型：HarmonyOS 3.1 Developer Preview 版本新增的模型，是目前主推且会长期发展的模型。该模型由于提供了 AbilityStage、WindowStage 等作为应用组件和 Window 窗口的"舞台"，因此称为 Stage 模型。

Stage 模型之所以成为主推模型，源于其设计思想，Stage 模型的设计出发点如下。

（1）为复杂应用而设计。

- 多个应用组件共享同一个 ArkTS 引擎（运行 ArkTS 语言的虚拟机）实例，应用组件之间可以方便地共享对象和状态，同时减少复杂应用运行对内存的占用。
- 采用面向对象的开发方式，使复杂应用代码可读性高、易维护性好、可扩展性强。

（2）支持多设备和多窗口形态，应用组件管理和窗口管理在架构层面解耦。
- 便于系统对应用组件进行裁剪（例如，无屏设备可裁剪窗口）。
- 便于系统扩展窗口形态。
- 在多设备（如桌面设备和移动设备）上，应用组件可使用同一套生命周期。

（3）平衡应用能力和系统管控成本：Stage 模型重新定义了应用能力的边界，平衡了应用能力和系统管控成本。
- 提供特定场景（如卡片、输入法）的应用组件，以便满足更多的使用场景。
- 规范化后台进程管理：为保障用户体验，Stage 模型对后台应用进程进行了有序治理，应用程序不能随意驻留在后台，同时应用的后台行为受到严格管理，防止恶意应用行为。

3.1.3　FA 模型与 Stage 模型的对比

HarmonyOS 系统中，Stage 模型与 FA 模型的最大区别是：在 FA 模型中，每个应用组件独享一个 ArkTS 引擎实例。在 Stage 模型中，多个应用组件共享同一个 ArkTS 引擎实例。应用组件之间可以方便地共享对象和状态，同时减少复杂应用运行对内存的占用。当下，Stage 模型作为华为主推的应用模型，开发者通过它能够更加便利地开发出分布式场景下的复杂应用。FA 模型与 Stage 模型的对比信息，如表 3-1 所示。

表 3-1　FA 模型与 Stage 模型的对比

项目	FA 模型	Stage 模型
应用组件	1. 组件分类 FA Model：PageAbility、ServiceAbility、DataAbility - PageAbility 组件：包含 UI 界面，提供展示 UI 的能力。 - ServiceAbility 组件：提供后台服务的能力，无 UI 界面。 - DataAbility 组件：提供数据分享的能力，无 UI 界面。 2. 开发方式 通过导出匿名对象、固定入口文件的方式指定应用组件。开发者无法进行派生，不利于扩展能力	1. 组件分类 Stage Model：UIAbility、ExtensionAbility、ServiceExtensionAbility…… - UIAbility 组件：包含 UI 界面，提供展示 UI 的能力，主要用于和用户交互。 - ExtensionAbility 组件：提供特定场景（如卡片、输入法）的扩展能力，满足更多的使用场景。 2. 开发方式 采用面向对象的方式，将应用组件以类接口的形式开放给开发者，可以进行派生，利于扩展能力
进程模型	有两类进程： 1. 主进程 2. 渲染进程	有三类进程： 1. 主进程 2. ExtensionAbility 进程 3. 渲染进程

续表

项目	FA 模型	Stage 模型
线程模型	1. ArkTS 引擎实例的创建 一个进程可以运行多个应用组件实例，每个应用组件实例运行在一个单独的 ArkTS 引擎实例中。 2. 线程模型 每个 ArkTS 引擎实例都在一个单独线程（非主线程）上创建，主线程没有 ArkTS 引擎实例。 3. 进程内对象共享：不支持	1. ArkTS 引擎实例的创建 一个进程可以运行多个应用组件实例，所有应用组件实例共享一个 ArkTS 引擎实例。 2. 线程模型 ArkTS 引擎实例在主线程上创建。 3. 进程内对象共享：支持
应用配置文件	使用 config.json 描述应用信息、HAP 信息和应用组件信息	使用 app.json5 描述应用信息，module.json5 描述 HAP 信息、应用组件信息

3.2 Stage 模型开发基础

从现在开始，华为公司将主推并长期发展 Stage 模型。因此，本书将重点讲解 Stage 应用模型的知识，使读者尽快地掌握并使用这项技术。本节将首先介绍 Stage 模型开发的基础知识。

3.2.1 Stage 模型的知识体系

Stage 模型的知识体系架构，如图 3-1 所示。

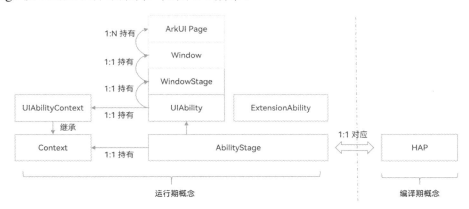

图 3-1　Stage 模型的知识体系架构

在图 3-1 所示的架构图中，各个知识体系的具体说明如下。

（1）UIAbility 组件和 ExtensionAbility 组件：Stage 模型提供了 UIAbility 和 ExtensionAbility 两种类型的组件，这两种组件都有具体的类承载，支持面向对象的开发方式。

- UIAbility 组件是一种包含 UI 界面的应用组件，主要用于用户交互。例如，图库类应用可以在 UIAbility 组件中展示图片瀑布流。当用户选择某个图片后，可以在新的页面中展示图片的详细内容。同时用户可以通过返回键返回到瀑布流页面。UIAbility 的生命周期只包含创建、销毁、前台、后台等状态，与显示相关的状态通过 WindowStage 的事件暴露给开发者。
- ExtensionAbility 组件是一种面向特定场景的应用组件。

（2）WindowStage：每个 UIAbility 类实例都会与一个 WindowStage 类实例绑定，该类提供了应用进程内窗口管理器的作用。它包含一个主窗口，也就是说 UIAbility 通过 WindowStage 持有了一个窗口，该窗口为 ArkUI 提供了绘制区域。

（3）Context：在 Stage 模型上，Context 及其派生类向开发者提供在运行期可以调用的各种能力。这种设计模式遵循了面向对象编程中的继承原则，其中基类 Context 定义了一组基本的功能和属性，而派生类则根据它们所属的组件（如 UIAbility 组件和各种 ExtensionAbility 派生类）提供额外的、特定的功能。

（4）AbilityStage：每个 Entry 类型或者 Feature 类型的 HAP 在运行期都有一个 AbilityStage 类实例。当 HAP 中的代码首次被加载到进程中时，系统会先创建 AbilityStage 实例。每个在该 HAP 中定义的 UIAbility 类，在实例化后都会与该实例产生关联。开发者可以使用 AbilityStage 获取该 HAP 中 UIAbility 实例的运行时信息。

3.2.2 Stage 应用 / 组件级配置

在开发 HarmonyOS 应用程序时，需要配置应用程序的一些标签，如应用的包名、图标等标识特征属性。在配置时，通常将图标和标签一起配置，可以分为应用图标、应用标签和入口图标、入口标签。这些配置信息分别对应于配置文件 app.json5 和配置文件 module.json5 中的 icon 和 label 标签。应用图标和标签在设置应用中使用，例如，在设置应用中的应用列表中显示。入口图标是应用安装完成后在设备桌面上显示出来的，如图 3-2 所示。入口图标是以 UIAbility 为粒度，支持同一个应用存在多个入口图标和标签，单击后进入对应的 UIAbility 界面。

图 3-2　应用图标和标签及入口图标和标签

1. 配置应用程序的包名

在应用程序工程的 AppScope 目录下打开配置文件 app.json5，找到里面的 bundleName

标签并进行配置,该标签用于标识应用程序的唯一性。推荐采用反域名形式命名(例如 com.example.demo,建议第一级为域名后缀 com,第二级为厂商 / 个人名,第三级为应用名,也可以多级)。

2. 应用图标和标签配置

使用 Stage 模型创建应用程序后,需要配置应用程序的图标和标签。在设置应用中使用应用图标和标签,例如,在设置应用中的应用列表中会显示出对应的图标和标签。

要设置应用程序的图标,需要打开工程的 AppScope 目录下的配置文件 app.json5,然后配置里面的 icon 标签。应用图标需配置为图片的资源索引,配置完成后,该图片即为应用的图标。

要设置应用程序的标签,需要打开工程的 AppScope 目录下的配置文件 app.json5 中,然后配置里面的 label 标签。该标签标识了应用在用户界面中显示的名称,需要将 label 配置为字符串资源的索引。

3. 入口图标和标签配置

Stage 模型支持对组件配置入口图标和入口标签,这些入口图标和入口标签会显示在桌面上。需要在配置文件 module.json5 中配置入口图标,在 abilities 标签下面有 icon 标签。例如,希望在桌面上显示该 UIAbility 的图标,则需要在 skills 标签下的 entities 中添加 entity.system.home,在 actions 中添加 action.system.home。如果同一个应用有多个 UIAbility 配置上述字段时,桌面上会显示出多个图标,分别对应各自的 UIAbility。

4. 应用版本声明配置

应用版本声明需要在工程的 AppScope 目录下的配置文件 app.json5 中配置 versionCode 标签和 versionName 标签。versionCode 用于标识应用的版本号,该标签的值为 32 位非负整数。此数字仅用于确定某个版本是否比另一个版本更新,数值越大表示版本越高。versionName 标签标识版本号的文字描述。

5. Module 支持的设备类型配置

Module 支持的设备类型需要在配置文件 module.json5 中配置 deviceTypes 标签,如果 deviceTypes 标签中添加了某种设备,则表明当前的 Module 支持在该设备上运行。

6. Module 权限配置

如果要配置 Module 访问系统或其他应用受保护部分所需的权限信息,需要在配置文件 module.json5 中配置 requestPermission 标签。该标签用于声明需要申请权限的名称、申请权限的原因以及权限使用的场景。

例如,下面是一个基本的鸿蒙应用程序配置文件 app.json5 的代码,位于 AppScope 目录下。

```
{
  "app": {
    "bundleName": "com.application.myapplication",
    "vendor": "example",
    "versionCode": 1000000,
    "versionName": "1.0.0",
    "icon": "$media:app_icon",
    "label": "$string:app_name",
```

```
      "description": "$string:description_application",
      "minAPIVersion": 9,
      "targetAPIVersion": 9,
      "apiReleaseType": "Release",
      "debug": false,
      "car": {
        "minAPIVersion": 8,
      }
    },
  }
```

在配置文件 app.json5 中，各个配置标签的具体说明如下。

（1）bundleName：标识应用的 Bundle 名称，用于标识应用的唯一性，该标签不可缺省。命名规则如下。

- 字符串由字母、数字、下划线和符号"."组成。
- 必须以字母开头。
- 最小长度为 7 个字节，最大长度为 127 个字节。

推荐采用反域名形式命名（如 com.example.demo），建议第一级为域名后缀 com，第二级为厂商 / 个人名，第三级为应用名，也可以采用多级结构。

（2）bundleType：标识应用的 Bundle 类型，用于区分应用或者原子化服务。该标签可选值为 app 和 atomicService，具体说明如下。

- app：表示当前 Bundle 为普通应用。
- atomicService：表示当前 Bundle 为元服务。

（3）debug：标识应用是否可调试，该标签由 IDE 在编译构建时生成，具体取值说明如下。

- true：表示可调试。
- false：表示不可调试。

（4）icon：标识应用的图标，标签值为图标资源文件的索引。

（5）label：标识应用的名称，标签值为字符串资源的索引。

（6）description：标识应用的描述信息，标签值为字符串类型（最大 255 个字节），或对描述内容的字符串资源索引。

（7）vendor：标识应用开发厂商的描述。该标签的值是字符串类型（最大 255 个字节）。

（8）versionCode：标识应用的版本号，该标签值为 32 位非负整数。此数字仅用于确定某个版本是否比另一个版本更新，数值越大表示版本更新。开发者可以将该值设置为任何正整数，但必须确保应用的新版本都使用比旧版本更大的值。该标签不可缺省，versionCode 值应小于 2^{31}。

（9）versionName：标识应用版本号的文字描述，用于向用户展示。该标签由数字和点构成，推荐采用"A.B.C.D"四段式形式，各段的含义如下。

- 第一段：主版本号 /Major，范围为 0—99，重大修改的版本，如实现新功能或重大变化。
- 第二段：次版本号 /Minor，范围为 0—99，表示新增添加功能或重要问题修复。
- 第三段：特性版本号 /Feature，范围为 0—99，标识新版本的特性。
- 第四段：修订版本号 /Patch，范围为 0—999，表示维护版本，用于修复 bug。

（10）minCompatibleVersionCode：标识应用能够兼容的最低历史版本号，用于跨设备兼容性判断。

（11）minAPIVersion：标识应用运行需要的 SDK 的 API 最小版本。

（12）targetAPIVersion：标识应用运行需要的 API 目标版本。

（13）apiReleaseTyp：标识应用运行需要的 API 目标版本的类型，采用字符串类型表示。取值为 CanaryN、BetaN 或 Release，其中 N 代表大于零的整数具体如下。

- Canary：受限发布的版本。
- Beta：公开发布的测试版本。
- Release：公开发布的正式版本。

（14）multiProjects：标识当前工程是否支持多个工程的联合开发，具体取值说明如下。

- true：表示当前工程支持多个工程的联合开发。
- false：表示当前工程不支持多个工程的联合开发。

（15）tablet：标识对 tablet 设备做的特殊配置，可以配置的属性字段有前文提到的：minAPIVersion、distributedNotificationEnabled。如果使用该属性对 tablet 设备做了特殊配置，则应用在 tablet 设备中会采用此处配置的属性值，并忽略在 app.json5 公共区域配置的属性值。

（16）tv：标识对 tv 设备做的特殊配置，可以配置的属性字段有前文提到的：minAPIVersion、distributedNotificationEnabled。如果使用该属性对 tv 设备做了特殊配置，则应用在 tv 设备中会采用此处配置的属性值，并忽略在 app.json5 公共区域配置的属性值。

（17）wearable：标识对 wearable 设备做的特殊配置，可以配置的属性字段有前文提到的：minAPIVersion、distributedNotificationEnabled。如果使用该属性对 wearable 设备做特殊配置，则应用在 wearable 设备中会采用此处配置的属性值，并忽略在 app.json5 公共区域配置的属性值。

（18）car：标识对 car 设备做的特殊配置，可以配置的属性字段有前文提到的：minAPIVersion、distributedNotificationEnabled。如果使用该属性对 car 设备做特殊配置，则应用在 car 设备中会采用此处配置的属性值，并忽略在 app.json5 公共区域配置的属性值。

（19）phone：标识对 phone 设备做的特殊配置，可以配置的属性字段有前文提到的：minAPIVersion、distributedNotificationEnabled。如果使用该属性对 phone 设备做特殊配置，则应用在 phone 设备中会采用此处配置的属性值，并忽略在 app.json5 公共区域配置的属性值。

3.3 UIAbility 组件

在 HarmonyOS 系统中，UIAbility 组件是一种包含 UI 界面的应用组件，主要用于和用户交互。UIAbility 组件是系统调度的基本单元，为应用提供绘制界面的窗口。一个 UIAbility 组件可以通过多个页面来实现一个功能模块，每一个 UIAbility 组件实例，都对应于一个最近任务列表中的任务。为了使 HarmonyOS 应用程序能够正常使用 UIAbility，需要在配置文件 module.json5 的 abilities 标签中声明 UIAbility 的名称、入口、标签等相关信息。例如，下面的配置文件配置了一个名为 EntryAbility 的 UI 组件，这是应用程序中的一个可视化界面的入口点，定义了它的各种属性，包括代码路径、描述、图标等。

```json
{
  "module": {
    // ...
    "abilities": [
      {
        "name": "EntryAbility", // UIAbility 组件的名称
        "srcEntrance": "./ets/entryability/EntryAbility.ts", // UIAbility 组件的代码路径
        "description": "$string:EntryAbility_desc", // UIAbility 组件的描述信息
        "icon": "$media:icon", // UIAbility 组件的图标
        "label": "$string:EntryAbility_label", // UIAbility 组件的标签
        "startWindowIcon": "$media:icon", // UIAbility 组件启动页面图标资源文件的索引
        "startWindowBackground": "$color:start_window_background", // UIAbility 组件启动页面背景颜色资源文件的索引
        // ...
      }
    ]
  }
}
```

3.3.1 UIAbility 组件生命周期

当用户打开、切换和返回到 HarmonyOS 应用程序时，应用中的 UIAbility 实例会在其生命周期的不同状态之间转换。UIAbility 提供了一系列回调，通过这些回调可以知道当前 UIAbility 实例的某个状态发生改变，会经过 UIAbility 实例的创建和销毁，或者 UIAbility 实例发生了前后台的状态切换。UIAbility 的生命周期包括 Create、Foreground、Background、Destroy 四个状态，如图 3-3 所示。

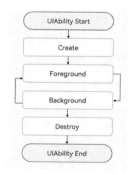

图 3-3 UIAbility 生命周期的状态

1. Create 状态

Create 状态在应用加载过程中，为 UIAbility 实例创建完成时触发，系统会调用 onCreate() 回调。可以在该回调中进行应用初始化操作，例如，变量定义、资源加载等，用于后续的 UI 界面展示。

```typescript
import UIAbility from '@ohos.app.ability.UIAbility';
import Window from '@ohos.window';
export default class EntryAbility extends UIAbility {
    onCreate(want, launchParam) {
        // 应用初始化
    }
    // ...
}
```

2. WindowStageCreate 和 WindowStageDestroy 状态

在创建 UIAbility 实例完成之后，在进入 Foreground 之前，系统会创建一个 WindowStage。WindowStage 创建完成后会进入 onWindowStageCreate() 回调，可以在该回

调中设置 UI 界面加载、设置 WindowStage 的事件订阅，如图 3-4 所示。

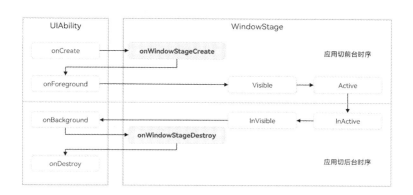

图 3-4　WindowStageCreate 和 WindowStageDestroy 的状态

例如，在下面的 onWindowStageCreate() 回调中，通过 loadContent() 方法设置应用要加载的页面并根据需要订阅 WindowStage 的事件（获焦 / 失焦、可见 / 不可见）。

```
import UIAbility from '@ohos.app.ability.UIAbility';
import Window from '@ohos.window';

export default class EntryAbility extends UIAbility {
    onWindowStageCreate(windowStage: Window.WindowStage) {
        // 设置 WindowStage 的事件订阅（获焦 / 失焦、可见 / 不可见）

        // 设置 UI 界面加载
        windowStage.loadContent('pages/Index', (err, data) => {
            // ...
        });
    }
}
```

下面的代码对应于 onWindowStageCreate() 回调，在 UIAbility 实例销毁之前会先进入 onWindowStageDestroy() 回调，可以在该回调中释放 UI 界面资源。例如，在 onWindowStageDestroy() 中注销获焦 / 失焦等 WindowStage 事件。

```
import UIAbility from '@ohos.app.ability.UIAbility';
import Window from '@ohos.window';

export default class EntryAbility extends UIAbility {
    // ...

    onWindowStageDestroy() {
        // 释放 UI 界面资源
    }
}
```

3. Foreground 状态和 Background 状态

Foreground 状态和 Background 状态分别在 UIAbility 实例切换至前台和切换至后台时触发，对应于 onForeground() 回调和 onBackground() 回调。

- onForeground() 回调：在 UIAbility 的 UI 界面可见之前，如 UIAbility 切换至前台时触发。可以在 onForeground() 回调中申请系统需要的资源，或者重新申请在 onBackground() 中释放的资源。
- onBackground() 回调：在 UIAbility 的 UI 界面完全不可见之后，如 UIAbility 切换至后台时候触发。可以在 onBackground() 回调中释放 UI 界面不可见时无用的资源，或者在此回调中执行较为耗时的操作，如状态保存等。

例如，在使用应用程序的过程中，需要使用用户定位，假设应用程序已经获得用户的定位权限授权，那么在显示 UI 界面之前，可以在 onForeground() 回调中开启定位功能，从而获取到当前的位置信息。当应用程序切换到后台状态时，可以在 onBackground() 回调中停止定位功能，以节省系统的资源消耗。例如，下面的代码展示了生命周期方法 onForeground 和 onBackground 的用法。

```
import UIAbility from '@ohos.app.ability.UIAbility';
export default class EntryAbility extends UIAbility {
    onForeground() {
        // 申请系统需要的资源，或者重新申请在 onBackground 中释放的资源
    }
    onBackground() {
        // 释放 UI 界面不可见时无用的资源，或者在此回调中执行较为耗时的操作
        // 如状态保存等
    }
}
```

- onForeground 方法：当 UI 能力组件处于前台时，即可见时，该方法会被调用。通常在这里执行申请系统需要的资源的操作，或者重新申请在 onBackground 方法中释放的资源。
- onBackground 方法：当 UI 能力组件处于后台时，即不可见时，该方法会被调用。通常在这里执行释放 UI 界面不可见时无用的资源的操作，或者执行一些在后台操作的任务，如状态保存等。

这两个方法提供了对 UI 能力组件生命周期的管理，允许在前台和后台状态之间进行必要的资源管理和操作。可以根据具体的业务需求在这两个方法中添加相应的逻辑。

4. Destroy 状态

Destroy 状态在 UIAbility 实例销毁时被触发，可以在 onDestroy() 回调中进行系统资源的释放、数据的保存等操作。例如，在下面的代码中，调用 terminateSelf() 方法停止当前的 UIAbility 实例，从而完成 UIAbility 实例的销毁工作；或者用户使用最近任务列表关闭该 UIAbility 实例，完成 UIAbility 实例的销毁。

```
import UIAbility from '@ohos.app.ability.UIAbility';
import Window from '@ohos.window';

export default class EntryAbility extends UIAbility {
    onDestroy() {
        // 系统资源的释放、数据的保存等
    }
}
```

3.3.2 UIAbility 组件的启动模式

UIAbility 的启动模式指的是 UIAbility 实例在启动时的不同呈现状态。根据不同的业务场景，系统提供了三种启动模式：singleton 模式（单实例模式）、multiton 模式（多实例模式）和 specified 模式（指定实例模式）。

1. singleton 模式

singleton 模式是单实例模式，也是默认情况下的启动模式。每次调用 startAbility() 方法时，如果应用进程中该类型的 UIAbility 实例已经存在，则复用系统中的 UIAbility 实例。系统中只存在唯一的一个该 UIAbility 实例，即在最近任务列表中只存在一个该类型的 UIAbility 实例。单实例模式的演示效果，如图 3-5 所示。

如果需要使用 singleton 模式启动应用程序，需要将配置文件 module.json5 中的 "launchType" 字段配置为 "singleton"，代码如下。

```
{
  "module": {
    // ...
    "abilities": [
      {
        "launchType": "singleton",
        // ...
      }
    ]
  }
}
```

2. multiton 模式

multiton 模式是多实例模式。当调用 startAbility() 方法时，都会在应用进程中创建一个新的该类型的 UIAbility 实例。也就是说，在最近任务列表中可以看到有多个该类型的 UIAbility 实例。多实例模式的演示效果，如图 3-6 所示。

图 3-5　单实例模式的演示效果　　图 3-6　多实例模式的演示效果

如果需要使用 multiton 模式启动应用程序，需要将配置文件 module.json5 中的 "launchType" 字段配置为 "multiton"，代码如下。

```
{
  "module": {
```

```
    // ...
    "abilities": [
      {
        "launchType": "multiton",
        // ...
      }
    ]
  }
}
```

3. specified 模式

specified 模式是指定实例模式，能够被一些特殊场景使用，例如，在文档应用中，每次新建文档都希望能新建一个文档实例，重复打开一个已保存的文档都希望打开的是同一个文档实例。

在创建 UIAbility 实例之前，允许开发者为该实例创建一个唯一的字符串 Key。当将创建的 UIAbility 实例绑定 Key 之后，每当后续调用 startAbility() 方法时，都会询问应用使用哪个 Key 对应的 UIAbility 实例来响应 startAbility() 请求。在运行时由 UIAbility 内部业务决定是否创建多实例。如果匹配有该 UIAbility 实例的 Key，则直接拉起与之绑定的 UIAbility 实例，否则创建一个新的 UIAbility 实例。指定实例模式的演示效果，如图 3-7 所示。

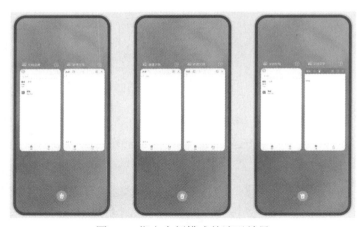

图 3-7 指定实例模式的演示效果

注意：在创建应用程序的 UIAbility 实例后，如果该 UIAbility 配置为指定实例模式，再次调用 startAbility() 方法启动该 UIAbility 实例，且 AbilityStage 的 onAcceptWant() 回调匹配到一个已创建的 UIAbility 实例。此时，再次启动该 UIAbility 时，只会进入该 UIAbility 的 onNewWant() 回调，不会进入其 onCreate() 和 onWindowStageCreate() 生命周期回调。

例如，有两个 UIAbility：EntryAbility 和 FuncAbility，将 FuncAbility 配置为 specified 启动模式，则需要在 EntryAbility 的页面中启动 FuncAbility。具体实现流程如下。

（1）在 FuncAbility 中，将 module.json5 配置文件的"launchType"字段配置为"specified"。

```
{
  "module": {
    // ...
```

```json
    "abilities": [
      {
        "launchType": "specified",
        // ...
      }
    ]
  }
}
```

（2）当在 EntryAbility 中调用 startAbility() 方法时，在参数 want 中增加一个自定义参数来区别 UIAbility 实例，例如，增加一个"instanceKey"自定义参数。

```
// 在启动指定实例模式的 UIAbility 时，给每一个 UIAbility 实例配置一个独立的 Key 标识
// 例如在文档使用场景中，可以用文档路径作为 Key 标识
function getInstance() {
    // ...
}

let want = {
    deviceId: '', // deviceId 为空表示本设备
    bundleName: 'com.example.myapplication',
    abilityName: 'FuncAbility',
    moduleName: 'module1', // moduleName 非必选
    parameters: { // 自定义信息
        instanceKey: getInstance(),
    },
}
// context 为调用方 UIAbility 的 AbilityContext
this.context.startAbility(want).then(() => {
    // ...
}).catch((err) => {
    // ...
})
```

（3）由于 FuncAbility 的启动模式配置了指定实例启动模式，所以在启动 FuncAbility 之前，会先进入其对应的 AbilityStage 的 onAcceptWant() 生命周期回调中，解析传入的 want 参数，获取"instanceKey"自定义参数。根据业务需要通过 AbilityStage 的 onAcceptWant() 生命周期回调返回一个字符串 Key 标识。如果返回的 Key 对应一个已启动的 UIAbility，则会将之前的 UIAbility 拉回前台并获焦，而不创建新的实例，否则创建新的实例并启动。

```
import AbilityStage from '@ohos.app.ability.AbilityStage';

export default class MyAbilityStage extends AbilityStage {
    onAcceptWant(want): string {
        // 在被调用方的 AbilityStage 中，针对启动模式为 specified 的 UIAbility 返回一个 UIAbility 实例对应的一个 Key 值
        // 当前示例指的是 module1 Module 的 FuncAbility
        if (want.abilityName === 'FuncAbility') {
            // 返回的字符串 Key 标识为自定义拼接的字符串内容
            return `ControlModule_EntryAbilityInstance_${want.parameters.instanceKey}`;
```

```
        }
        return";
    }
}
```

例如，在文档类应用程序中，可以对不同的文档实例内容绑定不同的 Key 值。当每次新建文档时，可以传入不同的新 Key 值（例如，可以将文件的路径作为一个 Key 标识），此时在 AbilityStage 中启动 UIAbility 时都会创建一个新的 UIAbility 实例。当新建的文档保存后，回到桌面，或者新打开一个已保存的文档，再次回到桌面，此时再次打开该已保存的文档，在 AbilityStage 中再次启动该 UIAbility 时，打开的仍然是之前已保存的文档界面。具体步骤如下。

- 打开文件 A，对应启动一个新的 UIAbility 实例，例如，启动 UIAbility 实例 1。
- 在最近任务列表中关闭文件 A 的进程，此时 UIAbility 实例 1 被销毁，回到桌面，再次打开文件 A，此时对应启动一个新的 UIAbility 实例，例如，启动 UIAbility 实例 2。
- 回到桌面，打开文件 B，此时对应启动一个新的 UIAbility 实例，例如，启动 UIAbility 实例 3。
- 回到桌面，再次打开文件 A，此时对应启动的还是之前的 UIAbility 实例 2。

3.3.3 UIAbility 组件的基本用法

在接下来的内容中将详细讲解 UIAbility 组件的基本用法，主要包括指定 UIAbility 的启动页面以及获取 UIAbility 的上下文 UIAbilityContext。

1. 指定 UIAbility 的启动页面

在 HarmonyOS 应用程序中，在启动 UIAbility 的过程中需要指定启动页面，否则应用启动后会没有默认加载页面导致白屏。可以在 UIAbility 的 onWindowStageCreate() 生命周期回调中，通过 WindowStage 对象的 loadContent() 方法设置启动页面。

```
import UIAbility from '@ohos.app.ability.UIAbility';
import Window from '@ohos.window';

export default class EntryAbility extends UIAbility {
    onWindowStageCreate(windowStage: Window.WindowStage) {
        // Main window is created, set main page for this ability
        windowStage.loadContent('pages/Index', (err, data) => {
            // ...
        });
    }
    // ...
}
```

注意：当使用 DevEco Studio 创建 HarmonyOS 应用程序后，在创建的 UIAbility 中，该 UIAbility 实例会默认加载 Index 页面，根据需要将 Index 页面路径替换为所需的页面路径即可。

2. 获取 UIAbility 的上下文信息

在 HarmonyOS 应用程序中，UIAbility 类拥有自身的上下文信息，该信息为 UIAbilityContext 类的实例。UIAbilityContext 类拥有 abilityInfo、currentHapModuleInfo 等属性。通过 UIAbilityContext

可以获取 UIAbility 的相关配置信息，如包代码路径、Bundle 名称、Ability 名称和应用程序需要的环境状态等属性信息，以及可以获取操作 UIAbility 实例的方法；如 startAbility()、connectServiceExtensionAbility()、terminateSelf() 等。

- 在 UIAbility 中，可以通过 this.context 获取 UIAbility 实例的上下文信息，代码如下。

```
import UIAbility from '@ohos.app.ability.UIAbility';

export default class EntryAbility extends UIAbility {
    onCreate(want, launchParam) {
        // 获取 UIAbility 实例的上下文
        let context = this.context;

        // ...
    }
}
```

- 可以在页面中获取 UIAbility 实例的上下文信息，包括导入依赖资源 context 模块和在组件中定义一个 context 变量两个部分。代码如下。

```
import common from '@ohos.app.ability.common';

@Entry
@Component
struct Index {
  private context = getContext(this) as common.UIAbilityContext;

  startAbilityTest() {
    let want = {
      // Want 参数信息
    };
    this.context.startAbility(want);
  }

  // 页面展示
  build() {
    // ...
  }
}
```

也可以在导入依赖的 context 模块资源后，在具体使用 UIAbilityContext 前定义变量。代码如下。

```
import common from '@ohos.app.ability.common';

@Entry
@Component
struct Index {

  startAbilityTest() {
    let context = getContext(this) as common.UIAbilityContext;
    let want = {
```

```
      // Want 参数信息
    };
    context.startAbility(want);
  }

  // 页面展示
  build() {
    // ...
  }
}
```

3.3.4　UIAbility 组件与 UI 的数据同步

在 HarmonyOS 系统的应用模型中，可以通过以下两种方式来实现 UIAbility 组件与 UI 之间的数据同步。
- EventHub：基于发布订阅模式来实现，事件需要先订阅后发布，订阅者收到消息后进行处理。
- globalThis：ArkTS 引擎实例内部的一个全局对象，在 ArkTS 引擎实例内部都能访问。

1. 使用 EventHub 进行数据通信

EventHub 提供了 UIAbility 组件 /ExtensionAbility 组件级别的事件机制，以 UIAbility 组件 /ExtensionAbility 组件为中心，提供了订阅、取消订阅和触发事件的数据通信能力。在使用 EventHub 进行数据通信之前，首先需要通过 Context 基类获取到 EventHub 实例。本节将重点介绍如何通过 EventHub 在 UIAbility 组件和 UI 之间实现有效的数据通信。
- 在 UIAbility 中调用 eventHub.on() 方法注册一个自定义事件"event1"，有如下两种调用 eventHub.on() 的方式，我们只需使用其中一种即可。

```
import UIAbility from '@ohos.app.ability.UIAbility';
const TAG: string = '[Example].[Entry].[EntryAbility]';

export default class EntryAbility extends UIAbility {
    func1(...data) {
        // 触发事件，完成相应的业务操作
        console.info(TAG, '1. ' + JSON.stringify(data));
    }

    onCreate(want, launch) {
        // 获取 eventHub
        let eventhub = this.context.eventHub;
        // 执行订阅操作
        eventhub.on('event1', this.func1);
        eventhub.on('event1', (...data) => {
            // 触发事件，完成相应的业务操作
            console.info(TAG, '2. ' + JSON.stringify(data));
        });
    }
}
```

- 在 UI 界面中通过 eventHub.emit() 方法触发该事件,在触发事件的同时,根据需要传入参数信息。

```
import common from '@ohos.app.ability.common';

@Entry
@Component
struct Index {
  private context = getContext(this) as common.UIAbilityContext;

  eventHubFunc() {
    // 不带参数触发自定义 "event1" 事件
    this.context.eventHub.emit('event1');
    // 带 1 个参数触发自定义 "event1" 事件
    this.context.eventHub.emit('event1', 1);
    // 带 2 个参数触发自定义 "event1" 事件
    this.context.eventHub.emit('event1', 2, 'test');
    // 开发者可以根据实际的业务场景设计事件传递的参数
  }

  // 页面展示
  build() {
    // ...
  }
}
```

- 在 UIAbility 的注册事件回调中可以得到对应的触发事件结果,运行日志结果如下。

```
[]
[1]
[2,'test']
```

- 在自定义事件"event1"使用完成后,可以根据需要调用 eventHub.off() 方法取消该事件的订阅。

```
// context 为 UIAbility 实例的 AbilityContext
this.context.eventHub.off('event1');
```

2. 使用 globalThis 进行数据同步

globalThis 是 ArkTS 引擎实例内部的一个全局对象,引擎内部的 UIAbility/ExtensionAbility/Page 都可以使用,因此可以使用 globalThis 全局对象进行数据同步。同步过程如图 3-8 所示。

1)UIAbility 和 Page 之间使用 globalThis

globalThis 是 ArkTS 引擎实例下的全局对象,可以通过 globalThis 绑定属性/方法来进行 UIAbility 组件与 UI 的数据同步。例如,在 UIAbility 组件中绑定 want 参数,即可在 UIAbility 对应的 UI 界面上使用 want 参数信息。

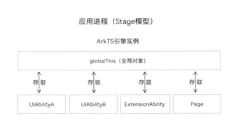

图 3-8 使用 globalThis 进行数据同步

2）UIAbility 和 UIAbility 之间使用 globalThis

同一个应用中 UIAbility 和 UIAbility 之间的数据传递，可以通过将数据绑定到全局变量 globalThis 上进行同步，如在 UIAbilityA 中将数据保存在 globalThis，然后跳转到 UIAbilityB 中取得该数据。

3.3.5 UIAbility 实战：页面跳转

接下来，将展示一个使用 UIAbility 的例子，实例 3-1 基于 Stage 模型下的 UIAbility 开发，实现了 UIAbility 内页面之间的跳转功能。跳转功能是通过 HarmonyOS 系统中的 router.pushUrl 模块实现的。模块 router.pushUrl 提供了通过不同的 URL 访问不同页面的功能，包括跳转到应用内的指定页面、用应用内的某个页面替换当前页面、返回上一页面或来到指定的页面等。

实例 3-1：实现不同页面之间的跳转（源码路径 :codes\3\PagesRouter）

（1）编写文件 AppScope/app.json5：这是一个鸿蒙应用程序配置文件，包含了应用的基本信息。

扫码看视频

```
{
  "app": {
    "bundleName": "com.example.pagesrouter",
    "vendor": "example",
    "versionCode": 1000000,
    "versionName": "1.0.0",
    "icon": "$media:app_icon",
    "label": "$string:app_name"
  }
}
```

这个配置文件描述了一个具有包名"com.example.pagesrouter"的应用程序，版本号为"1.0.0"，开发者是"example"。应用图标和名称是通过相应的资源引用获取的。对上述代码的具体说明如下。

- bundleName：应用的包名，用于唯一标识应用。
- vendor：应用的开发者或供应商。
- versionCode：应用的版本号，用于在更新时比较版本。
- versionName：应用的版本名称，人类可读的版本标识。
- icon：应用的图标路径，可能是媒体文件（这里使用了 $media:app_icon，可能是相对于媒体资源的路径）。
- label：应用的显示名称，可能是从字符串资源中获取的。

（2）编写文件 src/main/resources/base/profile/main_pages.json：这是鸿蒙应用程序中关于页面（pages）配置的一部分，这个配置文件指定了两个页面的相对路径，分别是"IndexPage"和"SecondPage"。这两个页面是应用程序中的视图或界面，用于显示用户界面的不同部分。

```
{
  "src": [
    "pages/IndexPage",
```

```
      "pages/SecondPage"
    ]
}
```

（3）编写文件 src/main/resources/base/element/string.json：这段 JSON 配置文件是鸿蒙应用程序中关于字符串资源（string）的配置，用于定义应用程序中使用的字符串值。

```
{
  "string": [
    {
      "name": "module_desc",
      "value": "description"
    },
    {
      "name": "EntryAbility_desc",
      "value": "description"
    },
    {
      "name": "EntryAbility_label",
      "value": "pageRouter"
    },
    {
      "name": "next",
      "value": "下个页面"
    },
    {
      "name": "back",
      "value": "后退"
    }
  ]
}
```

这个配置文件定义了一些字符串资源，其中包括模块描述、EntryAbility 描述、EntryAbility 标签、"下个页面"和"后退"等。在应用程序中，可以通过引用这些名称来获取相应的字符串值，使应用程序中的文本内容更容易维护和本地化。

（4）编写文件 src/main/ets/entryability/EntryAbility.ets，使用鸿蒙应用框架的 TypeScript 类创建一个名为 EntryAbility 的 UI 界面。这个 UI 界面的功能主要包括在创建和窗口创建阶段执行一些操作，如记录日志、存储数据及加载页面内容等。

```
import UIAbility from '@ohos.app.ability.UIAbility';
import Want from '@ohos.app.ability.Want';
import AbilityConstant from '@ohos.app.ability.AbilityConstant';
import window from '@ohos.window';
import Logger from '../common/utils/Logger';

const TAG = '[EntryAbility]';

export default class EntryAbility extends UIAbility {
  onCreate(want: Want, launchParam: AbilityConstant.LaunchParam) {
    Logger.info(TAG, 'onCreate');
    AppStorage.SetOrCreate('abilityWant', want);
```

```
    }

    onWindowStageCreate(windowStage: window.WindowStage) {
      // Main window is created, set the main page for this ability
      Logger.info(TAG, 'onWindowStageCreate');
      windowStage.loadContent('pages/IndexPage', (err, data) => {
        if (err.code) {
          Logger.info(TAG, 'Failed to load the content. Cause:' + JSON.stringify(err));
          return;
        }
        Logger.info(TAG, 'Succeeded in loading the content. Data: ' + JSON.stringify(data));
      });
    }
  };
```

（5）编写文件 src/main/ets/pages/IndexPage.ets，功能是创建一个名为 IndexPage 的页面组件，这个页面组件主要用于应用程序的入口，包含了一个简单的 UI 布局和页面导航逻辑。

```
import router from '@ohos.router';
import CommonConstants from '../common/constants/CommonConstants';
import Logger from '../common/utils/Logger';

const TAG = '[IndexPage]';

@Entry
@Component
struct IndexPage {
  @State message: string = CommonConstants.INDEX_MESSAGE;

  build() {
    Row() {
      Column() {
        Text(this.message)
          .fontSize(CommonConstants.FONT_SIZE)
          .fontWeight(FontWeight.Bold)
        Blank()
        Button($r('app.string.next'))
          .fontSize(CommonConstants.BUTTON_FONT_SIZE)
          .width(CommonConstants.BUTTON_WIDTH)
          .height(CommonConstants.BUTTON_HEIGHT)
          .backgroundColor($r('app.color.button_bg'))
          .onClick(() => {
            router.pushUrl({
              url: CommonConstants.SECOND_URL,
              params: {
                src: CommonConstants.SECOND_SRC_MSG
              }
            }).catch((error: Error) => {
              Logger.info(TAG, 'IndexPage push error' + JSON.stringify(error));
```

```
        }));
      })
    }
    .width(CommonConstants.FULL_WIDTH)
    .height(CommonConstants.LAYOUT_HEIGHT)
  }
  .height(CommonConstants.FULL_HEIGHT)
  .backgroundColor($r('app.color.page_bg'))
}
```

（6）编写文件 src/main/ets/pages/SecondPage.ets，功能是创建一个鸿蒙应用程序的页面组件 SecondPage。这个页面组件主要用于显示消息和导航参数，并提供一个返回按钮以进行页面返回。

```
import router from '@ohos.router';
import CommonConstants from '../common/constants/CommonConstants';

@Entry
@Component
struct SecondPage {
  @State message: string = CommonConstants.SECOND_MESSAGE;
  @State src: string = (router.getParams() as Record<string, string>)
[CommonConstants.SECOND_SRC_PARAM];

  build() {
    Row() {
      Column() {
        Text(this.message)
          .fontSize(CommonConstants.FONT_SIZE)
          .fontWeight(FontWeight.Bold)
        Text(this.src)
          .fontSize(CommonConstants.PARAMS_FONT_SIZE)
          .opacity(CommonConstants.PARAMS_OPACITY)
        Blank()
        Button($r('app.string.back'))
          .fontSize(CommonConstants.BUTTON_FONT_SIZE)
          .width(CommonConstants.BUTTON_WIDTH)
          .height(CommonConstants.BUTTON_HEIGHT)
          .backgroundColor($r('app.color.button_bg'))
          .onClick(() => {
            router.back();
          })
      }
      .width(CommonConstants.FULL_WIDTH)
      .height(CommonConstants.LAYOUT_HEIGHT)
    }
    .height(CommonConstants.FULL_HEIGHT)
    .backgroundColor($r('app.color.page_bg'))
  }
}
```

执行效果如图 3-9 所示。

主界面　　　　　　　　　第二个页面

图 3-9　执行效果

3.4　服务卡片

服务卡片（以下简称"卡片"）是一种界面展示形式，可以将应用的重要信息或操作前置到卡片上，以达到服务直达、减少体验层级的目的。卡片常用于嵌入其他应用中（当前卡片使用方只支持系统应用，如桌面），作为其界面显示的一部分，并支持拉起页面、发送消息等基础的交互功能。

3.4.1　服务卡片架构

在 HarmonyOS 系统的 Stage 模型中，服务卡片的架构，如图 3-10 所示。

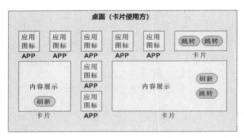
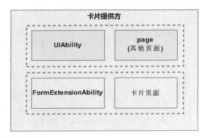

图 3-10　服务卡片的架构

在图 3-10 所示的架构中，服务卡片的相关概念如下。
- 卡片使用方：如图 3-10 的桌面，是显示卡片内容的宿主应用，控制卡片在宿主中展示的位置。
- 应用图标：应用入口图标，点击后可拉起应用进程，图标内容不支持交互。
- 卡片：具备不同规格大小的界面展示，卡片的内容可以进行交互，例如，实现按钮进行界面的刷新、应用的跳转等。
- 卡片提供方：包含卡片的应用，提供卡片的显示内容、控件布局以及控件点击处理逻辑。
- FormExtensionAbility：卡片业务逻辑模块，提供卡片创建、销毁、刷新等生命周期回调。
- 卡片页面：卡片 UI 模块，包含页面控件、布局、事件等显示和交互信息。

使用服务卡片的基本步骤如下，整个过程如图 3-11 所示。

（1）长按"桌面图标"，弹出操作菜单。

（2）选择"服务卡片"选项，进入卡片预览界面。

（3）点击"添加到桌面"按钮，即可在桌面上看到新添加的卡片。

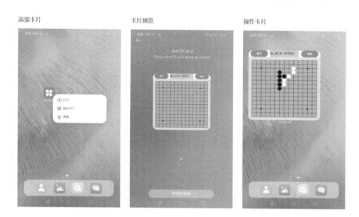

图 3-11 使用服务卡片的步骤

3.4.2 ArkTS 卡片开发

在 HarmonyOS 系统中，实现 ArkTS 卡片的基本原理，如图 3-12 所示。

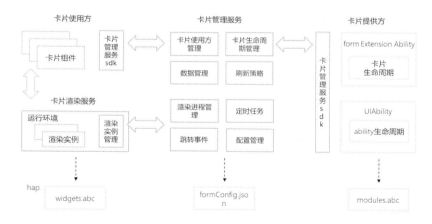

图 3-12 实现 ArkTS 卡片的基本原理

- 卡片使用方：显示卡片内容的宿主应用，控制卡片在宿主中展示的位置，当前仅系统应用可以作为卡片使用方。
- 卡片提供方：提供卡片显示内容的应用，控制卡片的显示内容、控件布局以及控件单击事件。
- 卡片管理服务：用于管理系统中所添加卡片的常驻代理服务，提供 formProvider 接口能力，同时提供卡片对象的管理与使用以及卡片周期性刷新等能力。
- 卡片渲染服务：用于管理卡片渲染实例，渲染实例与卡片使用方上的卡片组件一一绑定。卡片渲染服务运行卡片页面代码 widgets.abc 进行渲染，并将渲染后的数据发送至卡片使用方对应的卡片组件。

在 HarmonyOS 系统中，ArkTS 卡片渲染服务运行原理，如图 3-13 所示。

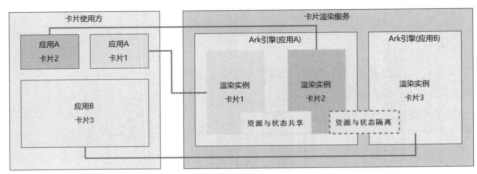

图 3-13　ArkTS 卡片渲染服务运行原理

3.4.3　开发基于 JS UI 的卡片

在 HarmonyOS 系统中，基于 JS UI 的卡片框架的运作机制，如图 3-14 所示。

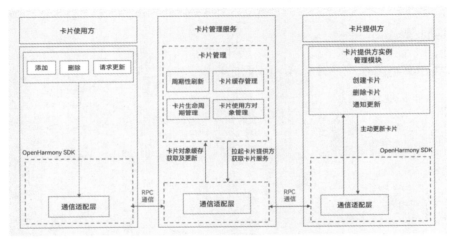

图 3-14　基于 JS UI 的卡片框架的运作机制

在图 3-14 中，卡片使用方包含以下模块。
- 卡片使用：包含卡片的添加、删除、请求更新等操作。
- 通信适配层：由 OpenHarmony SDK 提供，负责与卡片管理服务通信，用于将卡片的相关操作请求转发给卡片管理服务。

卡片管理服务包含以下模块。
- 周期性刷新：在卡片添加后，根据卡片的刷新策略启动定时任务周期性触发卡片的刷新。
- 卡片缓存管理：在卡片添加到卡片管理服务后，对卡片的视图信息进行缓存，以便下次获取卡片时可以直接返回缓存数据，降低时延。
- 卡片生命周期管理：对于卡片切换到后台或者被遮挡时，暂停卡片的刷新；以及卡片的升级 / 卸载场景下对卡片数据的更新和清理。
- 卡片使用方对象管理：对卡片使用方的 RPC 对象进行管理，用于使用方请求进行校验及对卡片更新后的回调处理。
- 通信适配层：负责与卡片使用方和提供方进行 RPC 通信。

卡片提供方包含以下模块。
- 创建卡片、更新卡片以及删除卡片：这些卡片服务由卡片提供方开发者实现，提供相应的卡片服务。
- 卡片提供方实例管理模块：由卡片提供方开发者实现，负责对卡片管理服务分配的卡片实例进行持久化管理。
- 通信适配层：由 OpenHarmony SDK 提供，负责与卡片管理服务通信，用于将卡片的更新数据主动推送到卡片管理服务。

3.4.4　卡片实战：多设备自适应服务卡片

本小节将通过一个具体实例来展示使用卡片的过程。实例 3-2 通过沉浸式卡片、图文卡片、宫格卡片、纯文本卡片、多维度信息卡片五种类型，分别展示了卡片在不同尺寸设备上的自适应能力。当卡片在极宽和极高的情况下，通过拉伸、缩放、隐藏、折行、均分等自适应能力将卡片展示出来。

实例 3-2：多设备自适应服务卡片（源码路径 :codes\3\JsAdaptiveServiceWidget）

（1）编写文件 src/main/config.json，这是一个鸿蒙应用程序的配置文件，这个配置文件包含了应用程序的基本信息、模块配置、能力组件配置以及页面和表单的配置。具体实现代码如下。

扫码看视频

```json
{
  "app": {
    "bundleName": "ohos.samples.adaptiveservicewidget",
    "version": {
      "code": 1000000,
      "name": "1.0.0"
    }
  },
  "deviceConfig": {},
  "module": {
    "package": "ohos.samples.adaptiveservicewidget",
    "name": ".MyApplication",
    "mainAbility": "ohos.samples.adaptiveservicewidget.MainAbility",
    "deviceType": [
      "phone",
      "tablet"
    ],
    "distro": {
      "deliveryWithInstall": true,
      "moduleName": "entry",
      "moduleType": "entry",
      "installationFree": false
    },
    "abilities": [
      {
        "skills": [
```

```json
        {
          "entities": [
            "entity.system.home"
          ],
          "actions": [
            "action.system.home"
          ]
        }
      ],
      "name": "ohos.samples.adaptiveservicewidget.MainAbility",
      "icon": "$media:icon",
      "description": "$string:mainability_description",
      "label": "$string:entry_MainAbility",
      "type": "page",
      "launchType": "standard",
      "formsEnabled": true,
      "forms": [
        {
          "jsComponentName": "complex",
          "isDefault": true,
          "scheduledUpdateTime": "10:30",
          "defaultDimension": "4*4",
          "name": "complex",
          "description": "This is a service widget",
          "colorMode": "auto",
          "type": "JS",
          "supportDimensions": [
            "4*4"
          ],
          "updateEnabled": true,
          "updateDuration": 1
        },
        {
          "jsComponentName": "text",
          "isDefault": false,
          "scheduledUpdateTime": "10:30",
          "defaultDimension": "2*2",
          "name": "text",
          "description": "This is a service widget",
          "colorMode": "auto",
          "type": "JS",
          "supportDimensions": [
            "2*2",
            "4*4"
          ],
          "updateEnabled": true,
          "updateDuration": 1
        },
        {
          "jsComponentName": "immersive",
          "isDefault": false,
          "scheduledUpdateTime": "10:30",
```

```json
          "defaultDimension": "2*2",
          "name": "immersive",
          "description": "This is a service widget",
          "colorMode": "auto",
          "type": "JS",
          "supportDimensions": [
            "2*2"
          ],
          "updateEnabled": true,
          "updateDuration": 1
        },
        {
          "jsComponentName": "grid",
          "isDefault": false,
          "scheduledUpdateTime": "10:30",
          "defaultDimension": "2*2",
          "name": "grid",
          "description": "This is a service widget",
          "colorMode": "auto",
          "type": "JS",
          "supportDimensions": [
            "2*2"
          ],
          "updateEnabled": true,
          "updateDuration": 1
        },
        {
          "jsComponentName": "imgText",
          "isDefault": false,
          "scheduledUpdateTime": "10:30",
          "defaultDimension": "2*4",
          "name": "imgText",
          "description": "This is a service widget",
          "colorMode": "auto",
          "type": "JS",
          "supportDimensions": [
            "2*4"
          ],
          "updateEnabled": true,
          "updateDuration": 1
        }
      ]
    }
  ],
  "js": [
    {
      "pages": [
        "pages/index/index"
      ],
      "name": "default",
      "window": {
        "designWidth": 720,
```

```json
      "autoDesignWidth": true
    }
  },
  {
    "pages": [
      "pages/index/index"
    ],
    "name": "complex",
    "window": {
      "designWidth": 720,
      "autoDesignWidth": true
    },
    "type": "form"
  },
  {
    "pages": [
      "pages/index/index"
    ],
    "name": "text",
    "window": {
      "designWidth": 720,
      "autoDesignWidth": true
    },
    "type": "form"
  },
  {
    "pages": [
      "pages/index/index"
    ],
    "name": "immersive",
    "window": {
      "designWidth": 720,
      "autoDesignWidth": true
    },
    "type": "form"
  },
  {
    "pages": [
      "pages/index/index"
    ],
    "name": "grid",
    "window": {
      "designWidth": 720,
      "autoDesignWidth": true
    },
    "type": "form"
  },
  {
    "pages": [
      "pages/index/index"
    ],
    "name": "imgText",
```

```
        "window": {
          "designWidth": 720,
          "autoDesignWidth": true
        },
        "type": "form"
      }
    ]
  }
}
```

在上述代码中，jsComponentName 是指在鸿蒙应用程序的配置中用于定义页面或表单的 JS 组件名称，其用于指定在页面或表单中使用的具体的 JavaScript 组件。在上述配置文件中有五种不同的卡片类型，每种卡片类型都使用了不同的 jsComponentName，具体说明如下。

- "complex" 卡片：复杂（多维度）类型卡片，包含多个组件和交互元素。
- "text" 卡片：文本类型卡片，用于显示文本信息。
- "immersive" 卡片：沉浸式类型卡片，用于提供全屏交互体验。
- "grid" 卡片：网格（宫格）类型卡片，用于以表格或网格形式显示信息。
- "imgText" 卡片：图文混排类型卡片，包含图像和文本的组合。

这些不同类型的卡片允许我们在应用程序中使用不同的布局和交互元素，以满足不同的需求和设计风格。

（2）为了节省本书篇幅，接下来只介绍一种卡片类型的实现过程：text 文本类型卡片，用于显示文本信息。编写文件 src/main/js/text/pages/index/index.hml，使用 HTML 模板代码在页面中渲染动态数据。在这个模板中，{{ array }} 是一个表示循环的指令，它会遍历一个名为 array 的数组，然后使用 $item 来引用数组中的每个元素。

```
<div class="container">
    <!-- 使用循环迭代数组中的元素 -->
    <div class="main-div" for="{{ array }}" tid="id">
        <!-- 每个数组元素对应的主要 div -->
        <div class="display-div">
            <!-- 显示名称 -->
            <text class="exponent-div">{{ $item.name }}</text>
            <!-- 显示指数 -->
            <text class="exponent">{{ $item.exponent }}</text>
        </div>
        <div class="display-div">
            <!-- 显示编码 -->
            <text class="coded-div">{{ $item.coded }}</text>
            <!-- 显示百分比 -->
            <text class="per-div">{{ $item.per }}</text>
        </div>
    </div>
</div>
```

编写文件 src/main/js/text/pages/index/index.json，这个文件作为上面文件 index.hml 中的 HTML 模板数据，这些数据是一个 JSON 对象，其中包含了一个名为 data 的属性，其值是一个包含多个对象的数组。每个对象代表了一条数据记录，包含了 id、name、exponent、coded 和 per 等属性。

```
{
    "data": {
        "array": [
            {
                "id": 1,
                "name": " 上证指数 ",
                "exponent": "3599.54",
                "coded": "000001",
                "per": "0.21%"
            },
            {
                "id": 2,
                "name": " 上证指数 ",
                "exponent": "3599.54",
                "coded": "000001",
                "per": "0.21%"
            },
            {
                "id": 3,
                "name": " 上证指数 ",
                "exponent": "3599.54",
                "coded": "000001",
                "per": "0.21%"
            }
        ]
    },
    "actions": {
        "routerEvent": {
            "action": "router",
            "bundleName": "ohos.samples.adaptiveservicewidget",
            "abilityName": "ohos.samples.adaptiveservicewidget.MainAbility",
            "params": {
                "message": "add detail"
            }
        }
    }
}
```

在前面的文件 src/main/js/text/pages/index/index.hml 中的 HTML 模板中，使用了 {{ array }}，这表明它会迭代 data 对象中的 array 数组，并在每次迭代中使用 $item 来引用数组中的每个对象。这些数据来源于上述的 src/main/js/text/pages/index/index.json。

注意：本实例需要在真机中运行，有如下两种使用方法。
- 方法 1：上滑本示例应用图标，选择需要的卡片尺寸添加到屏幕。
- 方法 2：长按应用，等待出现 "服务卡片" 字样，点击后选择需要的卡片尺寸，添加到屏幕。

第 4 章
Java UI 开发

鸿蒙系统支持 Java 语言，因此可以使用 Java 语言来编写应用程序的业务逻辑和用户界面。通过 Java 语言，可以为鸿蒙系统开发出具有分布式能力的应用程序，使应用程序能够在不同设备之间共享数据和服务。本章将详细讲解使用 Java 开发鸿蒙应用程序的知识。

4.1 Ability 框架

Ability 是应用所具备能力的抽象，也是应用程序的重要组成部分。一个应用可以具备多种能力（即可以包含多个 Ability），HarmonyOS 支持应用以 Ability 为单位进行部署。Ability 可以分为 FA（Feature Ability）和 PA（Particle Ability）两种类型，每种类型为开发者提供不同的模板，以便实现不同的业务功能。

（1）FA 支持 Page Ability：Page 模板是 FA 唯一支持的模板，用于提供与用户交互的能力。一个 Page 实例可以包含一组相关页面，每个页面用一个 AbilitySlice 实例表示。

（2）PA 支持 Service Ability 和 Data Ability。

- Service 模板：用于提供后台运行任务的能力。
- Data 模板：用于对外部提供统一的数据访问抽象。

4.1.1 Page Ability

Page 模板（以下简称"Page"）是 FA 唯一支持的模板，用于提供与用户交互的能力。一个 Page 可以由一个或多个 AbilitySlice 构成，AbilitySlice 是指应用的单个页面及其控制逻辑的总和。当一个 Page 由多个 AbilitySlice 共同构成时，这些 AbilitySlice 页面提供的业务能力应具有高度相关性。例如，新闻浏览功能可以通过一个 Page 来实现，其中包含了两个 AbilitySlice：一个 AbilitySlice 用于展示新闻列表，另一个 AbilitySlice 用于展示新闻详情。Page 和 AbilitySlice 的关系，如图 4-1 所示。

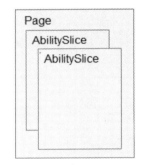

图 4-1 Page 与 AbilitySlice

与桌面场景相比，移动场景下应用之间的交互更为频繁。通常，单个应用专注于某个方面的能力开发，当它需要其他能力辅助时，会调用其他应用提供的能力。例如，外卖应用提供了联系商家的业务功能入口，当用户在使用该功能时，会跳转到通话应用的拨号页面。与此类似，HarmonyOS 支持不同 Page 之间的跳转，并可以指定跳转到目标 Page 中的某个具体的 AbilitySlice。

虽然一个 Page 可以包含多个 AbilitySlice，但是 Page 进入前台时界面默认只展示一个 AbilitySlice。默认展示的 AbilitySlice 是通过 setMainRoute() 方法来指定的。如果需要更改默认展示的 AbilitySlice，可以通过 addActionRoute() 方法为此 AbilitySlice 配置一条路由规则。此时，当其他 Page 实例期望导航到此 AbilitySlice 时，可以在 Intent 中指定 Action。

使用 setMainRoute() 方法与使用 addActionRoute() 方法的示例如下：

```
public class MyAbility extends Ability {
    @Override
    public void onStart(Intent intent) {
```

```
        super.onStart(intent);
        // set the main route
        setMainRoute(MainSlice.class.getName());

        // set the action route
        addActionRoute("action.pay", PaySlice.class.getName());
        addActionRoute("action.scan", ScanSlice.class.getName());
    }
}
```

1. Page Ability 的生命周期

Page Ability 生命周期的不同状态转换及其对应的回调,如图 4-2 所示。

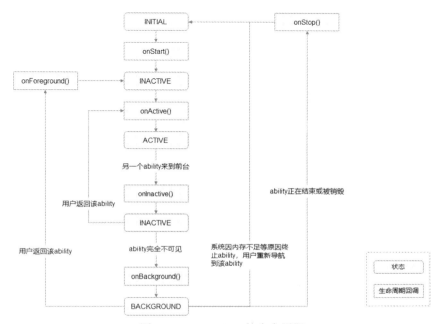

图 4-2　Page Ability 的生命周期

1) onStart()

当系统首次创建 Page 实例时,触发该回调。对于一个 Page 实例,该回调在其生命周期过程中仅触发一次,Page 在该逻辑后将进入 INACTIVE 状态。开发者必须重写该方法,并在此配置默认展示的 AbilitySlice。

2) onActive()

Page 会进入 INACTIVE 状态后来到前台,然后系统调用此回调。Page 在此之后进入 ACTIVE 状态,该状态是应用与用户交互的状态。Page 将保持在此状态,除非某类事件发生导致 Page 失去焦点,比如用户点击返回键或导航到其他 Page。当此类事件发生时,会触发 Page 回到 INACTIVE 状态,系统将调用 onInactive() 回调。此后,Page 可能重新回到 ACTIVE 状态,系统将再次调用 onActive() 回调。因此,开发者通常需要成对实现 onActive() 和 onInactive(),并在 onActive() 中获取在 onInactive() 中被释放的资源。

3) onInactive()

当 Page 失去焦点时,系统将调用此回调,此后 Page 进入 INACTIVE 状态。开发者可以在此回调中实现 Page 失去焦点时应表现的恰当行为。

4) onBackground()

如果 Page 不再对用户可见，系统将调用此回调通知开发者进行相应的资源释放，此后 Page 进入 BACKGROUND 状态。开发者应该在此回调中释放 Page 不可见时无用的资源，或在此回调中执行较为耗时的状态保存操作。

5) onForeground()

处于 BACKGROUND 状态的 Page 仍然驻留在内存中，当重新回到前台时（比如用户重新导航到此 Page），系统将先调用 onForeground() 回调通知开发者，而后 Page 的生命周期状态回到 INACTIVE 状态。开发者应当在此回调中重新申请在 onBackground() 中释放的资源，最后 Page 的生命周期状态进一步回到 ACTIVE 状态，系统将通过 onActive() 回调通知开发者。

6) onStop()

系统销毁 Page 时，将会触发此回调函数，通知用户进行系统资源的释放。销毁 Page 的原因包括以下几个方面。

- 用户通过系统管理能力关闭指定 Page，例如，使用任务管理器关闭 Page。
- 用户行为触发 Page 的 terminateAbility() 方法调用，例如，使用应用的退出功能。
- 配置变更导致系统暂时销毁 Page 并重建。
- 系统出于资源管理目的，自动触发对处于 BACKGROUND 状态 Page 的销毁。

2. AbilitySlice 的生命周期

AbilitySlice 作为 Page 的组成单元，其生命周期是依托于其所属 Page 生命周期的。AbilitySlice 和 Page 具有相同的生命周期状态和同名的回调，当 Page 生命周期发生变化时，它的 AbilitySlice 也会发生相同的生命周期变化。此外，AbilitySlice 还具有独立于 Page 的生命周期变化，这发生在同一 Page 中的 AbilitySlice 之间导航时，此时 Page 的生命周期状态不会改变。

AbilitySlice 生命周期回调与 Page 的相应回调类似，在此不再赘述。由于 AbilitySlice 承载具体的页面，开发者必须重写 AbilitySlice 的 onStart() 回调，并在此方法中通过 setUIContent() 方法设置页面，具体内容如下。

```
@Override
protected void onStart(Intent intent) {
    super.onStart(intent);

    setUIContent(ResourceTable.Layout_main_layout);
}
```

AbilitySlice 实例的创建和管理通常由应用负责，系统仅在特定情况下会创建 AbilitySlice 实例。例如，通过导航启动某个 AbilitySlice 时，是由系统负责实例化；但是在同一个 Page 中不同的 AbilitySlice 间导航时则由应用负责实例化。

4.1.2 Ability 实战：使用 PageAbility 实现页面跳转

实例 4-1，演示了基于 Page 模板的 Ability，实现与用户进行交互的过程。本实例展示了一个 Page 可以由一个或多个 AbilitySlice 构成，AbilitySlice 是应用的

扫码看视频

单个页面及其控制逻辑的总和的用法。

实例 4-1：使用 PageAbility 实现页面跳转（源码路径 :codes\4\PageAbility）

（1）主布局文件 src/main/resources/base/layout/main_ability_slice.xml 包含两个按钮（Button），按钮的内容和图标等属性通过 XML 的方式定义。这是一个垂直方向的 DirectionalLayout，即子元素（按钮）会垂直排列。具体实现代码如下。

```xml
<DirectionalLayout
    xmlns:ohos="http://schemas.huawei.com/res/ohos"
    ohos:height="match_parent"
    ohos:width="match_parent"
    ohos:orientation="vertical">

    <Button
        ohos:id="$+id:navigation_button"
        ohos:height="match_content"
        ohos:width="match_parent"
        ohos:element_right="$media:ic_arrow_right"
        ohos:start_padding="14vp"
        ohos:end_padding="0vp"
        ohos:padding="10vp"
        ohos:text="$string:navigation_button"
        ohos:text_alignment="left"
        ohos:text_size="16vp"/>

    <Button
        ohos:id="$+id:continuation_button"
        ohos:height="match_content"
        ohos:width="match_parent"
        ohos:element_right="$media:ic_arrow_right"
        ohos:start_padding="14vp"
        ohos:end_padding="0vp"
        ohos:padding="10vp"
        ohos:text="$string:continuation_button"
        ohos:text_alignment="left"
        ohos:text_size="16vp"/>
</DirectionalLayout>
```

（2）文件 src/main/resources/base/layout/first_ability_main_slice.xml 也是一个 XML 布局文件，描述了一个垂直方向的 DirectionalLayout，其中包含了一些文本（Text）和按钮（`Button）。

（3）文件 src/main/resources/base/layout/first_ability_second_slice.xml 也是一个 XML 布局文件，描述了一个垂直方向的 DirectionalLayout，其中包含了一个文本元素和一个按钮元素。

（4）文件 src/main/java/ohos/samples/pageability/MainAbility.java 是一个鸿蒙（HarmonyOS）应用程序的主要入口 Ability 的代码，其中包含了一个 MainAbility 类。在鸿蒙系统中，Ability 通常表示一个应用程序的主要组成部分，类似于 Android 中的 Activity。

```
package ohos.samples.pageability;
```

```
import ohos.samples.pageability.slice.MainAbilitySlice;

import ohos.aafwk.ability.Ability;
import ohos.aafwk.content.Intent;

/**
 * MainAbility
 */
public class MainAbility extends Ability {
    @Override
    public void onStart(Intent intent) {
        super.onStart(intent);
        super.setMainRoute(MainAbilitySlice.class.getName());
    }
}
```

上述代码是一个简单的鸿蒙应用程序的主要 Ability，它通过 MainAbilitySlice 提供用户界面，并在应用程序启动时设置这个 Slice 为主要界面。

（5）文件 src/main/java/ohos/samples/pageability/FirstAbility.java 是一个鸿蒙（HarmonyOS）应用程序的一个 Ability 类，称为 FirstAbility。

```
package ohos.samples.pageability;

import ohos.samples.pageability.slice.FirstAbilityMainSlice;

import ohos.aafwk.ability.Ability;
import ohos.aafwk.content.Intent;

/**
 * FirstAbility
 */
public class FirstAbility extends Ability {
    @Override
    public void onStart(Intent intent) {
        super.onStart(intent);
        super.setMainRoute(FirstAbilityMainSlice.class.getName());
    }
}
```

（6）文件 src/main/java/ohos/samples/pageability/SecondAbility.java 是一个鸿蒙（HarmonyOS）应用程序的一个 Ability 类，称为 SecondAbility。这个类包含两个 Slice，分别是 SecondAbilityMainSlice 和 SecondAbilitySecondSlice。

```
package ohos.samples.pageability;

import ohos.samples.pageability.slice.SecondAbilityMainSlice;
import ohos.samples.pageability.slice.SecondAbilitySecondSlice;

import ohos.aafwk.ability.Ability;
import ohos.aafwk.content.Intent;
```

```
/**
 * SecondAbility
 */
public class SecondAbility extends Ability {
    /**
     * Route action for SecondAbilitySecondSlice
     */
    public static final String ACTION = "SECOND_ABILITY_SECOND_SLICE_ACTION";

    @Override
    public void onStart(Intent intent) {
        super.onStart(intent);
        super.setMainRoute(SecondAbilityMainSlice.class.getName());
        addActionRoute(ACTION, SecondAbilitySecondSlice.class.
    }
}
```

执行本实例后先显示主界面（main_ability_slice.xml 布局），效果如图 4-3 所示，具体描述如下。

图 4-3　主界面

- 单击 "Ability Slice Navigation" 按钮会启动 FirstAbility。
- 单击 "First Ability Second Slice" 按钮，跳转至 FirstAbility 的 Second Slice。
- 单击 "Second Ability Second Slice" 按钮，启动 SecondAbility 的 Second Slice。
- 单击 "Second Ability" 按钮，启动 SecondAbility。
- 单击 "Back" 按钮，返回上一个界面。

4.2　UI 布局

HarmonyOS 提供了 Ability 和 AbilitySlice 两个基础类。一个有界面的 Ability 可以由一个或多个 AbilitySlice 构成。AbilitySlice 主要用于承载单个页面的具体逻辑实现和界面 UI，是应用显示、运行和跳转的最小单元。AbilitySlice 通过 setUIContent 为界面设置布局。

4.2.1　代码布局

在使用代码方式布局时，需要先创建 Component 和 ComponentContainer 对象，为这些对象设置合适的布局参数和属性值，并将 Component 添加到 ComponentContainer 中，从而创建出完整的界面。

下面的代码，创建了一个鸿蒙（HarmonyOS）应用程序类 AbilitySlice，这个类负责创建用户界面，包含一个 DirectionalLayout，在其中添加了一个 Text 组件和一个 Button 组件。

```
import ohos.aafwk.ability.AbilitySlice;
import ohos.aafwk.content.Intent;
import ohos.agp.colors.RgbColor;
import ohos.agp.components.*;
import ohos.agp.components.element.ShapeElement;
import ohos.agp.utils.LayoutAlignment;
```

```java
public class MainAbilitySlice extends AbilitySlice {
    @Override
    public void onStart(Intent intent) {
        super.onStart(intent);
        // 声明布局
        DirectionalLayout directionalLayout = new DirectionalLayout(getContext());
        // 设置布局大小
        directionalLayout.setWidth(ComponentContainer.LayoutConfig.MATCH_PARENT);
        directionalLayout.setHeight(ComponentContainer.LayoutConfig.MATCH_PARENT);
        // 设置布局属性
        directionalLayout.setOrientation(Component.VERTICAL);
        directionalLayout.setPadding(32, 32, 32, 32);

        Text text = new Text(getContext());
        text.setText("My name is Text.");
        text.setTextSize(50);
        text.setId(100);
        // 为组件添加对应布局的布局属性
        DirectionalLayout.LayoutConfig layoutConfig = new DirectionalLayout.LayoutConfig(ComponentContainer.LayoutConfig.MATCH_CONTENT, ComponentContainer.LayoutConfig.MATCH_CONTENT);
        layoutConfig.alignment = LayoutAlignment.HORIZONTAL_CENTER;
        text.setLayoutConfig(layoutConfig);

        // 将 Text 添加到布局中
        directionalLayout.addComponent(text);

        // 类似的添加一个 Button
        Button button = new Button(getContext());
        layoutConfig.setMargins(0, 50, 0, 0);
        button.setLayoutConfig(layoutConfig);
        button.setText("My name is Button.");
        button.setTextSize(50);
        ShapeElement background = new ShapeElement();
        background.setRgbColor(new RgbColor(0, 125, 255));
        background.setCornerRadius(25);
        button.setBackground(background);
        button.setPadding(10, 10, 10, 10);
        button.setClickedListener(new Component.ClickedListener() {
            @Override
            // 在组件中增加对单击事件的检测
            public void onClick(Component component) {
                // 此处添加按钮被单击需要执行的操作
            }
        });
        directionalLayout.addComponent(button);

        // 将布局作为根布局添加到视图树中
```

```
        super.setUIContent(directionalLayout);
    }
}
```

上述代码创建了一个简单的用户界面，其中包含一个文本（Text）和一个按钮（Button）。上述代码的布局功能是通过布局类 DirectionalLayout 实现的，并通过 setLayoutConfig 为组件设置布局属性，最后将布局作为根布局添加到视图树中。并且，为界面中的按钮添加了一个单击事件监听器，可以在监听器中定义按钮被单击时的操作。

4.2.2　XML 布局

通过使用 XML 代码，可以按层级结构描述 Component 和 ComponentContainer 的关系，给组件节点设定合适的布局参数和属性值，直接加载生成此布局。

在 Java 中，建议使用 XML 方式布局鸿蒙 UI，因为 XML 声明布局的方式更加简便直观。每一个 Component 和 ComponentContainer 对象的大部分属性都支持在 XML 中进行设置，它们都有各自的 XML 属性列表。某些属性仅适用于特定的组件，例如，只有 Text 支持"text_color"属性。不支持该属性的组件如果添加了该属性，该属性将被忽略。具有继承关系的组件子类将继承父类的属性列表，Component 作为组件的基类，拥有各个组件常用的属性，如 ID、布局参数等。

创建 XML 布局文件的流程如下。

（1）在 DevEco Studio 的 Project 窗口，依次打开"entry > src > main > resources > base"，右键单击"layout"文件夹，选择"New > Layout Resource File"，命名为"first_layout"，如图 4-4 所示。

图 4-4　给布局文件命名

（2）打开上面新创建的布局文件 first_layout.xml，修改其中的内容，对布局和组件的属性和层级进行描述。

```
<?xml version="1.0" encoding="utf-8"?>
<DirectionalLayout
    xmlns:ohos="http://schemas.huawei.com/res/ohos"
    ohos:width="match_parent"
    ohos:height="match_parent"
    ohos:orientation="vertical"
    ohos:padding="32">
    <Text
        ohos:id="$+id:text"
```

```xml
        ohos:width="match_content"
        ohos:height="match_content"
        ohos:layout_alignment="horizontal_center"
        ohos:text="My name is Text."
        ohos:text_size="25fp"/>
    <Button
        ohos:id="$+id:button"
        ohos:margin="50"
        ohos:width="match_content"
        ohos:height="match_content"
        ohos:layout_alignment="horizontal_center"
        ohos:text="My name is Button."
        ohos:text_size="50"/>
</DirectionalLayout>
```

（3）在 Java 代码中加载上面的布局文件 first_layout.xml，并添加为根布局或作为其他布局的子 Component。

```java
// 请根据实际工程 / 包名情况引入
package com.example.myapplication.slice;

import com.example.myapplication.ResourceTable;
import ohos.aafwk.ability.AbilitySlice;
import ohos.aafwk.content.Intent;
import ohos.agp.colors.RgbColor;
import ohos.agp.components.*;
import ohos.agp.components.element.ShapeElement;

public class ExampleAbilitySlice extends AbilitySlice {
    @Override
    public void onStart(Intent intent) {
        super.onStart(intent);
        // 加载 XML 布局作为根布局
        super.setUIContent(ResourceTable.Layout_first_layout);
        Button button = (Button) findComponentById(ResourceTable.Id_button);
        if (button != null) {
            // 设置组件的属性
            ShapeElement background = new ShapeElement();
            background.setRgbColor(new RgbColor(0, 125, 255));
            background.setCornerRadius(25);
            button.setBackground(background);

            button.setClickedListener(new Component.ClickedListener() {
                @Override
                // 在组件中增加对单击事件的检测
                public void onClick(Component component) {
                    // 此处添加按钮被单击需要执行的操作
                }
            });
        }
    }
}
```

4.2.3 Java 布局类

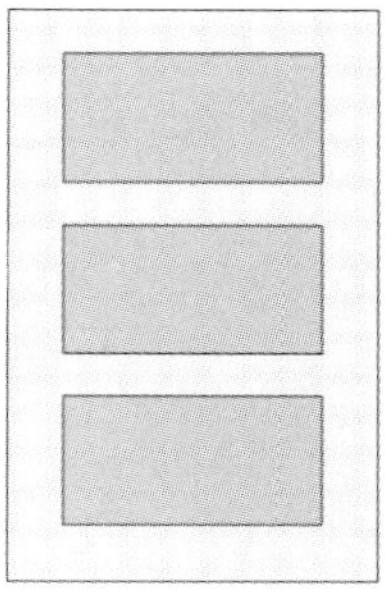

鸿蒙系统中的布局类有 PositionLayout、DirectionalLayout、StackLayout、DependentLayout、TableLayout、AdaptiveBoxLayout，它们提供了不同布局规范的组件容器，如以单一方向排列的 DirectionalLayout、以相对位置排列的 DependentLayout、以确切位置排列的 PositionLayout 等。

1. DirectionalLayout 布局

DirectionalLayout 是 Java UI 中的一种重要组件布局，用于将一组组件（Component）按照水平或垂直方向排布，能够方便地对齐布局内的组件。该布局和其他布局的组合，可以实现更加丰富的布局方式。DirectionalLayout 的布局效果，如图 4-5 所示。

图 4-5 DirectionalLayout 的布局效果

DirectionalLayout 的自有 XML 属性，如表 4-1 所示。

表 4-1 DirectionalLayout 的自有 XML 属性

属性名称	中文描述	取值	取值说明	使用案例
alignment	对齐方式	left	表示左对齐	可以设置取值项如表中所列，也可以使用"|"进行多项组合。 ohos:alignment="top|left" ohos:alignment="left"
		top	表示顶部对齐	
		right	表示右对齐	
		bottom	表示底部对齐	
		horizontal_center	表示水平居中对齐	
		vertical_center	表示垂直居中对齐	
		center	表示居中对齐	
		start	表示靠起始端对齐	
		end	表示靠结束端对齐	
orientation	子布局排列方向	horizontal	表示水平方向布局	ohos:orientation="horizontal"
		vertical	表示垂直方向布局	ohos:orientation="vertical"
total_weight	所有子视图的权重之和	float 类型	可以直接设置浮点数值，也可以引用 float 浮点数资源	ohos:total_weight="2.5" ohos:total_weight="$float:total_weight"

DirectionalLayout 所包含组件可支持的 XML 属性，如表 4-2 所示。

表 4-2 DirectionalLayout 所包含组件可支持的 XML 属性

属性名称	中文描述	取值	取值说明	使用案例
layout_alignment	对齐方式	left	表示左对齐	可以设置取值项如表中所列，也可以使用"|"进行多项组合。 ohos:layout_alignment="top" ohos:layout_alignment="top|left"
		top	表示顶部对齐	
		right	表示右对齐	
		bottom	表示底部对齐	
		horizontal_center	表示水平居中对齐	
		vertical_center	表示垂直居中对齐	
		center	表示居中对齐	
weight	比重	float 类型	可以直接设置浮点数值，也可以引用 float 浮点数资源	ohos:weight="1" ohos:weight="$float:weight"

2. DependentLayout 布局

DependentLayout 是 Java UI 框架里的一种常见布局方式，与 DirectionalLayout 相比，DependentLayout 拥有更多的排布方式，每个组件可以指定相对于其他同级元素的位置，或者指定相对于父组件的位置。DependentLayout 的布局效果，如图 4-6 所示。

DependentLayout 的自有 XML 属性，如表 4-3 所示。

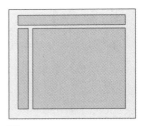

图 4-6　DependentLayout 的布局效果

表 4-3　DependentLayout 的自有 XML 属性

属性名称	中文描述	取值	取值说明	使用案例
alignment	对齐方式	left	表示左对齐	可以设置取值项如表中所列，也可以使用"|"进行多项组合。 ohos:alignment="top|left" ohos:alignment="left"
		top	表示顶部对齐	
		right	表示右对齐	
		bottom	表示底部对齐	
		horizontal_center	表示水平居中对齐	
		vertical_center	表示垂直居中对齐	
		center	表示居中对齐	

DependentLayout 所包含组件可支持的 XML 属性如下。

- left_of：将右边缘与另一个子组件的左边缘对齐。
- right_of：将左边缘与另一个子组件的右边缘对齐。
- start_of：将结束边与另一个子组件的起始边对齐。
- end_of：将起始边与另一个子组件的结束边对齐。
- above：将下边缘与另一个子组件的上边缘对齐。
- below：将上边缘与另一个子组件的下边缘对齐。
- align_baseline：将子组件的基线与另一个子组件的基线对齐。
- align_left：将左边缘与另一个子组件的左边缘对齐。
- align_top：将上边缘与另一个子组件的上边缘对齐。
- align_right：将右边缘与另一个子组件的右边缘对齐。
- align_bottom：将底边与另一个子组件的底边对齐。
- align_start：将起始边与另一个子组件的起始边对齐。
- align_end：将结束边与另一个子组件的结束边对齐。
- align_parent_left：将左边缘与父组件的左边缘对齐。
- align_parent_top：将上边缘与父组件的上边缘对齐。
- align_parent_right：将右边缘与父组件的右边缘对齐。
- align_parent_bottom：将底边与父组件的底边对齐。
- align_parent_start：将起始边与父组件的起始边对齐。
- align_parent_end：将结束边与父组件的结束边对齐。
- center_in_parent：将子组件保持在父组件的中心。
- horizontal_center：将子组件保持在父组件水平方向的中心。
- vertical_center：将子组件保持在父组件垂直方向的中心。

3. StackLayout 布局

StackLayout 直接在屏幕上开辟出一块空白的区域，添加到这个布局中的视图都是以层叠的方式显示，而它会把这些视图默认放到这块区域的左上角，第一个添加到布局中的视图显示在最底层，最后一个被放在最顶层，上一层的视图会覆盖下一层的视图。StackLayout 的布局效果，如图 4-7 所示。

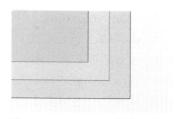

图 4-7 StackLayout 的布局效果

StackLayout 布局可支持的 XML 属性，如表 4-4 所示。

表 4-4 StackLayout 布局支持的 XML 属性

属性名称和功能	取值	取值说明	使用案例
layout_alignment：对齐方式	left	表示左对齐	可以设置取值项如表中所列，也可以使用"|"进行多项组合。 ohos:layout_alignment="top" ohos:layout_alignment="top|left"
	top	表示顶部对齐	
	right	表示右对齐	
	bottom	表示底部对齐	
	horizontal_center	表示水平居中对齐	
	vertical_center	表示垂直居中对齐	
	center	表示居中对齐	

4. TableLayout 布局

TableLayout 使用表格的方式划分子组件。TableLayout 的布局效果，如图 4-8 所示。TableLayout 布局的常用 XML 属性，如表 4-5 所示。

表 4-5 TableLayout 布局的常用 XML 属性

属性名称	中文描述	取值	取值说明	使用案例
alignment_type	对齐方式	align_edges	表示 TableLayout 内的组件按边界对齐	ohos:alignment_type="align_edges"
		align_contents	表示 TableLayout 内的组件按边距对齐	ohos:alignment_type="align_contents"
column_count	列数	integer 类型	可以直接设置整型数值，也可以引用 integer 资源	ohos:column_count="3" ohos:column_count="$integer:count"
row_count	行数	integer 类型	可以直接设置整型数值，也可以引用 integer 资源	ohos:row_count="2" ohos:row_count="$integer:count"
orientation	排列方向	horizontal	表示水平方向布局	ohos:orientation="horizontal"
		vertical	表示垂直方向布局	ohos:orientation="vertical"

5. PositionLayout 布局

在 PositionLayout 中，子组件通过指定准确的 x/y 坐标值在屏幕上显示。(0, 0) 为左上角；当向下或向右移动时，坐标值变大；并且允许组件之间互相重叠。PositionLayout 的布局效果，如图 4-9 所示。

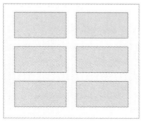
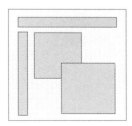

图 4-8 TableLayout 的布局效果　　图 4-9 PositionLayout 的布局效果

6. AdaptiveBoxLayout 布局

AdaptiveBoxLayout 是自适应盒子布局，该布局提供了在不同屏幕尺寸设备上的自适应布局能力，主要用于相同级别的多个组件需要在不同屏幕尺寸设备上自动调整列数的场景。AdaptiveBoxLayout 布局的特点如下。

- 该布局中的每个子组件都用一个单独的"盒子"装起来，子组件设置的布局参数都是以盒子作为父布局生效，不以整个自适应布局为生效范围。
- 该布局中每个盒子的宽度固定为布局总宽度除以自适应得到的列数，高度为 match_content，每一行中的所有盒子按高度最高的进行对齐。
- 该布局水平方向是自动分块，因此水平方向不支持 match_content，布局水平宽度仅支持 match_parent 或固定宽度。
- 自适应仅在水平方向进行了自动分块，纵向没有做限制，因此如果某个子组件的高设置为 match_parent 类型，可能导致后续行无法显示。

AdaptiveBoxLayout 的布局效果，如图 4-10 所示。

AdaptiveBoxLayout 布局常用方法如下。

- addAdaptiveRule(int minWidth, int maxWidth, int columns)：用于添加一个自适应盒子布局规则。
- removeAdaptiveRule(int minWidth, int maxWidth, int columns)：用于移除一个自适应盒子布局规则。
- clearAdaptiveRules()：用于移除所有自适应盒子布局规则。

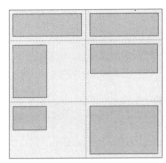

图 4-10 AdaptiveBoxLayout 的布局效果

实例 4-2，演示了使用 DirectionalLayout 和 DependentLayout 两种布局方式的过程。

实例 4-2：使用 DirectionalLayout 和 DependentLayout 两种布局（源码路径 :codes\4\CommonLayout）

（1）编写文件 src/main/resources/base/layout/main_ability_slice.xml，在 HarmonyOS 中定义了一个简单的垂直 DirectionalLayout 布局，其中包含两个 Button 元素。

（2）编写文件 src/main/resources/base/layout/directionalLayout_ability_slice.xml，功能是在 HarmonyOS 中定义了一个垂直 DirectionalLayout 布局，其中包含了多个子布局，每个子布局演示了不同的布局属性和特性。

```xml
<DirectionalLayout
    xmlns:ohos="http://schemas.huawei.com/res/ohos"
    ohos:height="match_parent"
    ohos:width="match_parent">

    <Text
        ohos:id="$+id:vertical_text"
        ohos:height="25vp"
        ohos:width="match_parent"
        ohos:background_element="#0d000000"
        ohos:bottom_margin="10vp"
        ohos:layout_alignment="horizontal_center"
        ohos:start_margin="15vp"
```

```xml
        ohos:end_margin="15vp"
        ohos:text="$string:vertical"
        ohos:text_size="16fp"
        ohos:top_margin="10vp"/>
```

// 省略部分代码

```xml
    <DirectionalLayout
        ohos:height="match_content"
        ohos:width="match_parent"
        ohos:start_margin="15vp"
        ohos:orientation="horizontal"
        ohos:end_margin="15vp">

        <Button
            ohos:height="30vp"
            ohos:width="50vp"
            ohos:background_element="$graphic:color_light_gray_element"
            ohos:text="$string:btn_1"
            ohos:weight="1"
            ohos:text_size="15fp"/>

        <Button
            ohos:height="30vp"
            ohos:width="50vp"
            ohos:background_element="$graphic:color_gray_element"
            ohos:text="$string:btn_2"
            ohos:weight="2"
            ohos:text_size="16fp"/>

        <Button
            ohos:height="30vp"
            ohos:width="50vp"
            ohos:background_element="$graphic:color_light_gray_element"
            ohos:text="$string:btn_3"
            ohos:weight="3"
            ohos:text_size="16fp"/>
    </DirectionalLayout>
</DirectionalLayout>
```

在上述代码中，每个 DirectionalLayout 中的 Button 元素都演示了不同的布局属性，如大小、边距、背景、文本样式、对齐方式和权重等。这个 XML 文件很好地展示了 HarmonyOS 中使用 DirectionalLayout 布局的一些基本用法。

（3）编写文件 src/main/resources/base/layout/dependentlayout_ability_slice.xml，功能是在 HarmonyOS 中定义了一个 DependentLayout 布局，它包含了多个子组件，如 Text 和 Button。

（4）编写文件 src/main/java/ohos/samples/commonlayout/MainAbility.java，功能是在应用程序启动时设置主切片的路由，确保用户看到的是 MainAbilitySlice 的界面。在 HarmonyOS 开发中，能力是应用程序的主要组件，而切片则代表应用程序的各个界面或功能模块。

```java
import ohos.samples.commonlayout.slice.MainAbilitySlice;
```

```java
import ohos.aafwk.ability.Ability;
import ohos.aafwk.content.Intent;

/**
 * CommonLayout MainAbility
 */
public class MainAbility extends Ability {
    @Override
    public void onStart(Intent intent) {
        super.onStart(intent);
        super.setMainRoute(MainAbilitySlice.class.getName());
    }
}
```

（5）编写文件 src/main/java/ohos/samples/commonlayout/slice/MainAbilitySlice.java，主要功能是在应用程序启动时，设置主切片（MainAbilitySlice）的界面内容，并初始化两个按钮组件（showDirectionalLayoutButton 和 showDependentLayoutButton）。这两个按钮分别关联了单击事件监听器，当用户单击它们时，会启动新的切片（DirectionalLayoutAbilitySlice 和 DependentLayoutAbilitySlice）。在具体的实现中，通过 ResourceTable.Layout_main_ability_slice 引用了布局文件，其中包含了两个按钮（Id_directional_layout_button 和 Id_dependent_layout_button）。通过 findComponentById 方法找到这两个按钮组件，然后为它们设置单击事件监听器，以便在用户单击时执行相应的操作。

```java
public class MainAbilitySlice extends AbilitySlice {
    @Override
    public void onStart(Intent intent) {
        super.onStart(intent);
        super.setUIContent(ResourceTable.Layout_main_ability_slice);
        initComponents();
    }

    private void initComponents() {
        // 找到布局中的两个按钮组件
        Component showDirectionalLayoutButton = findComponentById(ResourceTable.Id_directional_layout_button);
        Component showDependentLayoutButton = findComponentById(ResourceTable.Id_dependent_layout_button);

        // 设置按钮的单击事件监听器
        showDirectionalLayoutButton.setClickedListener(
            component -> present(new DirectionalLayoutAbilitySlice(), new Intent()));
        showDependentLayoutButton.setClickedListener(
            component -> present(new DependentLayoutAbilitySlice(), new Intent()));
    }
}
```

总体而言，上述代码负责初始化主切片的界面，并为两个按钮添加单击事件监听器，以便在用户单击时展示不同的切片。

（6）编写文件 src/main/java/ohos/samples/commonlayout/slice/DirectionalLayoutAbilitySlice.java，这是一个 HarmonyOS 应用程序中的切片（DirectionalLayoutAbilitySlice）的实现。其主要功能是在切片启动时设置 UI 内容，使用了布局文件 directionalLayout_ability_slice.xml。

```
public class DirectionalLayoutAbilitySlice extends AbilitySlice {
    @Override
    public void onStart(Intent intent) {
        super.onStart(intent);
            super.setUIContent(ResourceTable.Layout_directionalLayout_ability_slice);
        }
}
```

（7）编写文件 src/main/java/ohos/samples/commonlayout/slice/DependentLayoutAbilitySlice.java，功能是在切片启动时设置 UI 内容，使用了布局文件 dependentlayout_ability_slice.xml。在具体实现中，onStart 方法是在切片启动时调用的，通过调用 super.onStart(intent) 执行父类的逻辑，然后使用 super.setUIContent(ResourceTable.Layout_dependentlayout_ability_slice) 设置切片的 UI 内容，即加载了名为 dependentlayout_ability_slice.xml 的布局文件。

```
public class DependentLayoutAbilitySlice extends AbilitySlice {
    @Override
    public void onStart(Intent intent) {
        super.onStart(intent);
        super.setUIContent(ResourceTable.Layout_dependentlayout_ability_slice);
    }
}
```

执行后，首先显示由 DirectionalLayout 实现的主界面，如图 4-11 所示。

点击主界面中的 DirectionalLayout 按钮，来到子界面，如图 4-12 所示，这个子界面也是由 DirectionalLayout 布局实现的。点击主界面中的 DependentLayout 按钮，来到子界面，如图 4-13 所示，这个子界面是由 DependentLayout 布局实现的。

图 4-11　DirectionalLayout 实现的主界面

图 4-12　DirectionalLayout 布局实现的子界面

图 4-13　DependentLayout 布局实现的子界面

4.3 常用组件开发

在 HarmonyOS 系统中，组件是应用程序界面的构建块，用于构建用户界面和处理用户交互。组件可以是诸如按钮、文本框、图像等用户界面元素，在本书前文的例子中已经多次用到了按钮组件 Button、文本组件 Text。组件也可以是容器，用于组织和布局其他组件。组件在 HarmonyOS 中是视图层次结构的一部分，负责显示和处理用户界面。

4.3.1 Text 和 Button 组件

Text 是用来显示字符串的组件，在界面上显示为一块文本区域。Text 作为一个基本组件，有很多扩展，常见的有按钮组件 Button，文本编辑组件 TextField。

Button 是一种常见的组件，用于显示一个按钮，点击时可以触发对应的操作，通常由文本或图标组成，也可以由图标和文本共同组成。

实例 4-3，演示了使用 Text 和 Button 组件的过程。在这个例子中，点击 Start Text 按钮与 Start Button 按钮会显示对应的文本样式或者按钮样式，再点击文本会展示对应的测试组件布局。

实例 4-3：使用 Text 和 Button 组件（源码路径:codes\4\Components）

（1）编写文件 src/main/resources/base/layout/text_example_slice.xml，使用 DependentLayout 的 XML 布局文件，每个组件都有相应的布局参数，如边距、内边距、位置关系等，以实现期望的界面布局。

扫码看视频

（2）编写文件 src/main/resources/base/element/string.json，设置了要显示的文字内容。

（3）编写文件 src/main/java/ohos/samples/components/slice/TextExampleSlice.java，这是一个 AbilitySlice，用于展示一个与 XML 布局文件 text_example_slice.xml 关联的界面。在 onStart 方法中，它调用了 setUIContent 方法，将布局文件 text_example_slice.xml 设置为界面的内容。

```
public class TextExampleSlice extends AbilitySlice {
    @Override
    protected void onStart(Intent intent) {
        super.onStart(intent);
        setUIContent(ResourceTable.Layout_text_example_slice);
    }
}
```

执行效果如图 4-14 所示。

（4）编写文件 src/main/resources/base/layout/button_example_slice.xml，这是一个使用 DirectionalLayout 组件和 Button 组件创建的 XML 布局文件，模拟了手机拨号键盘界面效果，用户可以通过点击相应的按钮输入数字或字符。

```
<DirectionalLayout
    xmlns:ohos="http://schemas.huawei.com/res/ohos"
    ohos:height="match_parent"
    ohos:width="match_parent"
```

图 4-14 显示 Text 文本内容

```xml
        ohos:background_element="$graphic:color_light_gray_element"
        ohos:orientation="vertical">

    <Text
        ohos:id="$+id:txt_num"
        ohos:height="match_content"
        ohos:width="match_content"
        ohos:background_element="$graphic:green_text_element"
        ohos:layout_alignment="horizontal_center"
        ohos:text="$string:number"
        ohos:text_alignment="center"
        ohos:text_size="20fp"/>

    <DirectionalLayout
        ohos:height="match_content"
        ohos:width="match_parent"
        ohos:alignment="horizontal_center"
        ohos:bottom_margin="5vp"
        ohos:orientation="horizontal"
        ohos:top_margin="5vp">
```

// 省略部分代码

```xml
        <Button
            ohos:id="$+id:btn_pound"
            ohos:height="40vp"
            ohos:width="40vp"
            ohos:background_element="$graphic:green_circle_button_element"
            ohos:text="$string:w"
            ohos:text_alignment="center"
            ohos:text_size="15fp"/>
    </DirectionalLayout>

    <Button
        ohos:id="$+id:btn_call"
        ohos:height="match_content"
        ohos:width="match_content"
        ohos:background_element="$graphic:green_capsule_button_element"
        ohos:bottom_margin="5vp"
        ohos:bottom_padding="2vp"
        ohos:layout_alignment="horizontal_center"
        ohos:start_padding="10vp"
        ohos:end_padding="10vp"
        ohos:text="$string:call"
        ohos:text_alignment="center"
        ohos:text_size="15fp"
        ohos:top_padding="2vp"/>
</DirectionalLayout>
```

执行效果如图 4-15 所示。

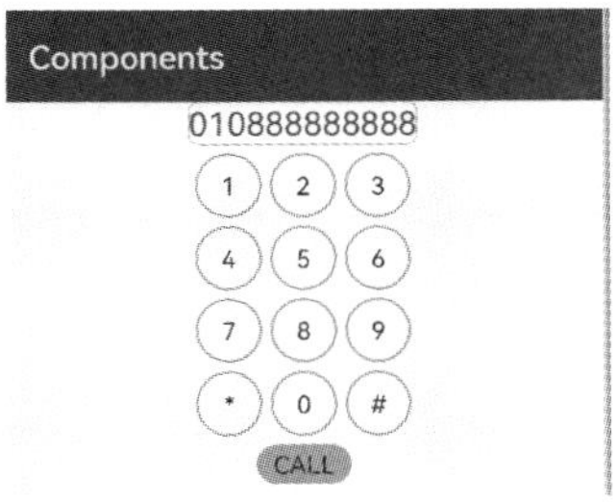

图 4-15 使用 DirectionalLayout 组件和 Button 组件实现模拟手机拨号键盘界面

4.3.2 Image 组件

在鸿蒙系统中，Image 是用来显示图片的组件。Image 组件的自有 XML 属性，如表 4-6 所示。

表 4-6 Image 组件的自有 XML 属性

属性名称	取值	取值说明
clip_alignment：图像裁剪对齐方式	left	表示按左对齐裁剪
	right	表示按右对齐裁剪
	top	表示按顶部对齐裁剪
	bottom	表示按底部对齐裁剪
	center	表示按居中对齐裁剪
image_src：图像	Element 类型	可直接配置色值，也可引用 color 资源或引用 media/graphic 下的图片资源
scale_mode：图像的缩放类型	zoom_center	表示原图按照比例缩放到与 Image 最窄边一致，并居中显示
	zoom_start	表示原图按照比例缩放到与 Image 最窄边一致，并靠起始端显示
	zoom_end	表示原图按照比例缩放到与 Image 最窄边一致，并靠结束端显示
	stretch	表示将原图缩放到与 Image 大小一致
	center	表示不缩放，按 Image 大小显示原图中间部分
	inside	表示将原图按比例缩放到与 Image 相同或更小的尺寸，并居中显示
	clip_center	表示将原图按比例缩放到与 Image 相同或更大的尺寸，并居中显示

以下代码，演示了使用 XML 方式创建 Image 的过程。

```xml
<?xml version="1.0" encoding="utf-8"?>
<DirectionalLayout
    xmlns:ohos="http://schemas.huawei.com/res/ohos"
    ohos:height="match_parent"
    ohos:width="match_parent"
    ohos:orientation="vertical">

    <Image
        ohos:id="$+id:imageComponent"
        ohos:height="200vp"
        ohos:width="200vp"
        ohos:image_src="$media:plant"
        />
</DirectionalLayout>
```

上面的布局代码在垂直方向上放置一张图片，图片的来源是通过 $media:plant 指定的。在这个布局中，图片的大小是 200vp × 200vp。

而下面的代码为演示使用 Java 编程方式创建 Image 的过程。

```java
import ohos.aafwk.ability.AbilitySlice;
import ohos.aafwk.content.Intent;
import ohos.agp.components.DirectionalLayout;
import ohos.agp.components.Image;

public class MainAbilitySlice extends AbilitySlice {
    @Override
    public void onStart(Intent intent) {
```

```
        super.onStart(intent);
        // 创建一个 Image 组件
        Image image = new Image(getContext());
        // 请传入 resources/media 下的图片 id
        image.setPixelMap(ResourceTable.Media_plant);
        // 创建一个布局
        DirectionalLayout layout = new DirectionalLayout(getContext());
        //Image 组件添加到 DirectionalLayout 布局中
        layout.addComponent(image);
        super.setUIContent(layout);
    }
}
```

4.3.3　TabList 和 Tab 组件

Tablist 可以实现多个页签栏的切换，Tab 为某个页签。子页签通常放在内容区上方，展示不同的分类。页签名称应该简洁明了，清晰地描述分类的内容。Tablist 的自有 XML 属性，如表 4-7 所示。

表 4-7　Tablist 的自有 XML 属性

属性名称	中文描述	取值说明
fixed_mode	固定所有页签并同时显示	可以直接设置 true/false，也可以引用 boolean 资源
orientation	页签排列方向	表示水平排列
		表示垂直排列
normal_text_color	未选中的文本颜色	可以直接设置色值，也可以引用 color 资源
selected_text_color	选中的文本颜色	可以直接设置色值，也可以引用 color 资源
selected_tab_indicator_color	选中页签的颜色	可以直接设置色值，也可以引用 color 资源
selected_tab_indicator_height	选中页签的高度	表示尺寸的 float 类型。可以是浮点数值，其默认单位为 px；也可以是带 px/vp/fp 单位的浮点数值；也可以引用 float 资源
tab_indicator_type	页签指示类型	表示选中的页签无指示标记
		表示选中的页签通过底部下划线标记
		表示选中的页签通过左侧分割线标记
		表示选中的页签通过椭圆背景标记
tab_length	页签长度	表示尺寸的 float 类型。可以是浮点数值，其默认单位为 px；也可以是带 px/vp/fp 单位的浮点数值；也可以引用 float 资源
tab_margin	页签间距	表示尺寸的 float 类型。可以是浮点数值，其默认单位为 px；也可以是带 px/vp/fp 单位的浮点数值；也可以引用 float 资源
text_alignment	文本对齐方式	表示文本靠左对齐
		表示文本靠顶部对齐
		表示文本靠右对齐
		表示文本靠底部对齐
		表示文本水平居中对齐
		表示文本垂直居中对齐
		表示文本居中对齐
		表示文本靠起始端对齐
		表示文本靠结尾端对齐

续表

属性名称	中文描述	取值说明
text_size	文本大小	表示尺寸的 float 类型。 可以是浮点数值，其默认单位为 px；也可以是带 px/vp/fp 单位的浮点数值；也可以引用 float 资源

4.3.4 Picker 和 DatePicker 组件

Picker 提供了滑动选择器，允许用户从预定义范围中进行选择。Picker 的自有 XML 属性，如表 4-8 所示。

表 4-8 Picker 的自有 XML 属性

属性名称	中文描述
element_padding	文本和 Element 之间的间距 Element 必须通过 setElementFormatter 接口配置
max_value	最大值
min_value	最小值
value	当前值
normal_text_color	未选中的文本颜色
normal_text_size	未选中的文本大小
selected_text_color	选中的文本颜色
selected_text_size	选中的文本大小
selector_item_num	显示的项目数量
selected_normal_text_margin_ratio	已选文本边距与常规文本边距的比例
shader_color	着色器颜色
top_line_element	选中项的顶行
bottom_line_element	选中项的底线
wheel_mode_enabled	选择轮是否循环显示数据

DatePicker 组件的功能是供用户选择日期，此组件的自有 XML 属性，如表 4-9 所示。

表 4-9 DatePicker 的自有 XML 属性

属性名称	中文描述
date_order	显示格式，年月日
day_fixed	日期是否固定
month_fixed	月份是否固定
year_fixed	年份是否固定
max_date	最大日期
min_date	最小日期
text_size	文本大小
normal_text_size	未选中文本的大小
selected_text_size	选中文本的大小
normal_text_color	未选中文本的颜色
selected_text_color	选中文本的颜色
operated_text_color	操作项的文本颜色

属性名称	中文描述
selected_normal_text_margin_ratio	已选文本边距与常规文本边距的比例
selector_item_num	显示的项目数量
shader_color	着色器颜色
top_line_element	选中项的顶行
bottom_line_element	选中项的底线
wheel_mode_enabled	选择轮是否循环显示数据

实例 4-4，演示了使用 Picker 和 DatePicker 组件的过程。

实例 4-4：使用 Picker 和 DatePicker 组件（源码路径 :codes\4\TabList）

（1）编写文件 src/main/resources/base/layout/main_ability_slice.xml，使用 DirectionalLayout 组件，将 Picker 组件和 DatePicker 组件安置在界面中。

（2）编写文件 src/main/java/ohos/samples/components/slice/MainAbilitySlice.java，功能是在初始化界面中显示 Picker 和 DatePicker 组件，以便用户可以使用它们进行选择。

```java
public class MainAbilitySlice extends AbilitySlice {
    @Override
    public void onStart(Intent intent) {
        super.onStart(intent);
        super.setUIContent(ResourceTable.Layout_main_ability_slice);
        initComponents();
    }

    private void initComponents() {
        Picker picker = (Picker) findComponentById(ResourceTable.Id_test_picker);
        picker.setMinValue(0); // 设置选择器中的最小值
        picker.setMaxValue(6); // 设置选择器中的最大值

        // 获取 DatePicker 实例
        DatePicker datePicker = (DatePicker) findComponentById(ResourceTable.Id_date_pick);
        int day = datePicker.getDayOfMonth();
        int month = datePicker.getMonth();
        int year = datePicker.getYear();
    }
}
```

执行效果如图 4-16 所示。

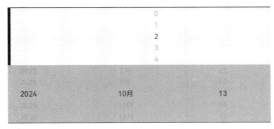

图 4-16　Picker 和 DatePicker 组件的执行效果

4.3.5 TimePicker 组件

TimePicker 组件的作用是供用户选择时间，其自有 XML 属性，如表 4-10 所示。

表 4-10 TimePicker 组件的自有 XML 属性

属性名称	中文描述
am_pm_order	在 12 小时制显示的情况下，控制上午下午排列顺序
mode_24_hour	是否 24 小时制显示
hour	显示小时
minute	显示分钟
second	显示秒
normal_text_color	未选中文本的颜色
selected_text_color	选中文本的颜色
operated_text_color	操作项的文本颜色
normal_text_size	未选中文本的大小
selected_text_size	选中文本的大小
selected_normal_text_margin_ratio	已选文本边距与常规文本边距的比例
selector_item_num	显示的项目数量
shader_color	着色器颜色
text_am	上午文本
text_pm	下午文本
top_line_element	选中项的顶行
bottom_line_element	选中项的底线
wheel_mode_enabled	选择轮是否循环显示数据

实例 4-5，演示了使用 TimePicker 组件的过程。

实例 4-5：使用 TimePicker 组件（源码路径 :codes\4\TimePicker）

（1）编写文件 src/main/resources/base/layout/main_ability_slice.xml，在 XML 布局中添加了一个 TimePicker 组件，设置选择文本的颜色为蓝色，并设置选择文本的大小为 20fp。

```xml
<?xml version="1.0" encoding="utf-8"?>
<DirectionalLayout
    xmlns:ohos="http://schemas.huawei.com/res/ohos"
    ohos:height="match_parent"
    ohos:width="match_parent"
    ohos:background_element="black"
    ohos:orientation="vertical">

    <TimePicker
        ohos:id="$+id:time_picker"
        ohos:height="match_content"
        ohos:selected_text_color="#007DFF"
        ohos:selected_text_size="20fp"
        ohos:width="match_parent" />

</DirectionalLayout>
```

（2）编写文件 src/main/java/ohos/samples/components/slice/MainAbilitySlice.java，首先，通过 findComponentById 方法找到了 XML 布局文件中的 TimePicker 组件，该组件的 ID 为 ResourceTable.Id_time_picker。其次，使用 getHour、getMinute 和 getSecond 方法获取了当前 TimePicker 的小时、分钟和秒。最后，通过 setHour、setMinute 和 setSecond 方法，设置了 TimePicker 的小时为 19，分钟为 18，秒为 12。这样可以在界面初始化时将 TimePicker 的初始时间设置为指定的值。

```
private void initComponents() {
    TimePicker timePicker = (TimePicker) findComponentById(ResourceTable. Id_time_picker);
    int hour = timePicker.getHour();
    int minute = timePicker.getMinute();
    int second = timePicker.getSecond();

    timePicker.setHour(19);
    timePicker.setMinute(18);
    timePicker.setSecond(12);
```

执行效果，如图 4-17 所示。

图 4-17　TimePicker 组件的执行效果

4.3.6　Switch 组件

Switch 是切换单个设置开 / 关两种状态的组件，其自有 XML 属性如下。
- text_state_on：开启时显示的文本。
- text_state_off：关闭时显示的文本。
- track_element：轨迹样式。
- thumb_element：滑块样式。
- marked：当前状态（选中或未选中）。
- check_element：状态标志样式。

实例 4-6，演示了使用 Switch 组件的过程。

实例 4-6：使用 Switch 组件（源码路径 :codes\4\TabList）

（1）编写文件 src/main/resources/base/layout/main_ability_slice.xml，创建 Switch 组件，设置 Switch 在开启和关闭时的文本，并且设置 Switch 的滑块和轨迹的样式。

```
<Switch
    ohos:id="$+id:btn_switch"
    ohos:height="30vp"
    ohos:thumb_element="$graphic:thumb_state_element_bounds"
    ohos:width="60vp"/>
```

（2）编写文件 src/main/java/ohos/samples/components/slice/MainAbilitySlice.java，功能是在初始化界面中设置 Switch 的显示状态。

```
private void initComponents() {
    Switch btnSwitch = (Switch) findComponentById(ResourceTable.Id_btn_switch);
            // 设置 Switch 默认状态
            btnSwitch.setChecked(true);
```

执行效果，如图 4-18 所示。

图 4-18　Switch 组件的执行效果

4.3.7　RadioButton 和 Checkbox 组件

在鸿蒙系统中，RadioButton 组件用于实现多选一的操作，在使用时需要搭配 RadioContainer 实现单选效果。RadioContainer 是 RadioButton 的容器，在其包裹下的 RadioButton 保证只有一个备选项。RadioButton 组件自有 XML 属性如下。

- marked：当前状态（选中或未选中）。
- text_color_on：处于选中状态的文本颜色。
- text_color_off：处于未选中状态的文本颜色。
- check_element：状态标志样式。

Checkbox 组件可以实现选中和取消选中的功能，与 RadioButton 相比，Checkbox 组件可以实现多选功能。Checkbox 组件的自有 XML 属性如下。

- marked：当前状态（选中或未选中）。
- text_color_on：处于选中状态的文本颜色。
- text_color_off：处于未选中状态的文本颜色。
- check_element：状态标志样式。

实例 4-7，演示了使用 Checkbox 组件的过程。

实例 4-7：使用 Checkbox 组件（源码路径 :codes\4\TabList）

编写文件 src/main/resources/base/layout/main_ability_slice.xml，定义了一个 HarmonyOS 的界面布局，包含一些文本和复选框，用于展示问题和选择答案。

```
<?xml version="1.0" encoding="utf-8"?>
<DirectionalLayout
    xmlns:ohos="http://schemas.huawei.com/res/ohos"
    ohos:height="match_parent"
    ohos:width="match_parent"
    ohos:orientation="vertical"
    ohos:left_padding="40vp"
    ohos:top_padding="40vp">
```

```xml
<DirectionalLayout
    ohos:height="match_content"
    ohos:width="match_parent"
    ohos:orientation="horizontal">

    <Text
        ohos:height="match_content"
        ohos:width="match_content"
        ohos:text_size="18fp"
        ohos:text="Which of the following are fruits?"/>

    <Text
        ohos:id="$+id:text_answer"
        ohos:height="match_content"
        ohos:width="match_content"
        ohos:left_margin="20vp"
        ohos:text_size="20fp"
        ohos:text_color="#FF3333"
        ohos:text="[]" />
</DirectionalLayout>

<Checkbox
    ohos:id="$+id:check_box_1"
    ohos:top_margin="40vp"
    ohos:height="match_content"
    ohos:width="match_content"
    ohos:text="A Apples"
    ohos:text_size="20fp"
    ohos:text_color_on="#FF3333"
    ohos:text_color_off="#000000"/>

<Checkbox
    ohos:id="$+id:check_box_2"
    ohos:top_margin="40vp"
    ohos:height="match_content"
    ohos:width="match_content"
    ohos:text="B Bananas"
    ohos:text_size="20fp"
    ohos:text_color_on="#FF3333"
    ohos:text_color_off="#000000"/>

<Checkbox
    ohos:id="$+id:check_box_3"
    ohos:top_margin="40vp"
    ohos:height="match_content"
    ohos:width="match_content"
    ohos:text="C Strawberries"
    ohos:text_size="20fp"
    ohos:text_color_on="#FF3333"
    ohos:text_color_off="#000000" />

<Checkbox
```

```
            ohos:id="$+id:check_box_4"
            ohos:top_margin="40vp"
            ohos:height="match_content"
            ohos:width="match_content"
            ohos:text="D Potatoes"
            ohos:text_size="20fp"
            ohos:text_color_on="#FF3333"
            ohos:text_color_off="black" />
</DirectionalLayout>
```

对上述代码的具体说明如下。

- 在第一个 DirectionalLayout 中，有两个 Text 组件，一个用于显示问题，另一个用于显示答案。问题文本使用默认样式，而答案文本的 ID 设置为 text_answer，初始内容为"[]"。
- 紧随其后的是四个 Checkbox 组件，分别代表不同的选项。每个 Checkbox 都有一个唯一的 ID（例如，check_box_1），设置了显示文本、字体大小、选中和未选中时的文本颜色等属性。

例如，在图 4-19 所示的执行效果中选中了 A 和 B 两个选项。

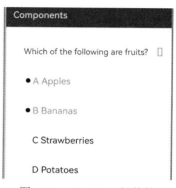

图 4-19 Checkbox 组件的执行效果

4.3.8 ProgressBar、RoundProgressBar 和 Slider 组件

进度条（ProgressBar）是一种用户界面元素，用于显示某个任务或操作的进度。通常，进度条是一个长条，显示了一个任务的完成度或剩余工作量的百分比。它是用户界面中常见的一种反馈机制，用于向用户展示长时间任务的执行情况。

1. ProgressBar 组件

在鸿蒙系统中，进度条组件 ProgressBar 用于显示内容或操作的进度。ProgressBar 组件的自有 XML 属性如下。

- divider_lines_enabled：是否显示分割线。
- divider_lines_number：分割线数量。
- infinite：是否使用不确定模式。
- infinite_element：不确定模式图样。
- max：最大值。
- max_height：最大高度。
- max_width：最大宽度。
- min：最小值。
- orientation：排列方向。
- progress：当前进度。
- background_instruct_element：背景。
- progress_width：进度条宽度。

- progress_color：进度条颜色。
- progress_element：进度条背景。
- progress_hint_text：进度提示文本。
- progress_hint_text_alignment：进度提示文本对齐方式。
- progress_hint_text_color：进度提示文本颜色。
- vice_progress：副本进度。
- vice_progress_element：副本进度条背景。
- step：进度的步长。
- progress_hint_text_size：进度提示文本大小。

2. RoundProgressBar 组件

RoundProgressBar 组件继承自 ProgressBar，拥有 ProgressBar 的属性，在设置同样属性时用法和 ProgressBar 一致，用于显示环形进度。RoundProgressBar 组件的自有 XML 属性如下。

- start_angle：圆形进度条的起始角度。
- max_angle：圆形进度条的最大角度。

3. Slider 组件

Slider 的功能和进度条组件 ProgressBar 的功能十分相似，区别是 ProgressBar 不能拖动，只有显示功能。Slider 组件的常用属性如下。

- 设置拖动条方向：ohos:orientation="horizontal"，表示水平方向。
- 设置最小值：ohos:min="0"。
- 设置最大值：ohos:max="100"。
- 设置当前值：ohos:progress="66"。
- 设置背景颜色：ohos:background_element="#000000"，表示黑色。
- 设置进度条颜色：ohos:progress_color="#00FF00"，表示绿色。

实例 4-8，演示了使用 Slider 和 ProgressBar 组件的过程。

实例 4-8：使用 Slider 和 ProgressBar 组件（源码路径 :codes\4\ProgressBar）

（1）编写文件 src/main/resources/base/layout/main_ability_slice.xml，使用 Slider、Text、Button 和 ProgressBar 等组件创建了一个进度条界面，允许用户设置进度条的相关参数并启动进度。

（2）编写文件 src/main/java/ohos/samples/progress/utils/LogUtil.java，实现一个简单的日志工具类 LogUtil，用于在 HarmonyOS 应用程序中记录调试日志。

```
public class LogUtil {
    private static final String TAG_LOG = "Sample_Progress";
    private static final int DOMAIN_ID = 0xD000F00;
    private static final HiLogLabel LABEL_LOG = new HiLogLabel(3, DOMAIN_ID, LogUtil.TAG_LOG);
    private static final String LOG_FORMAT = "%{public}s: %{public}s";
    private LogUtil() {
        /* Do nothing */
    }
```

```java
    /**
     * Print debug log
     *
     * @param tag log tag
     * @param msg log message
     */
    public static void debug(String tag, String msg) {
        HiLog.debug(LABEL_LOG, LOG_FORMAT, tag, msg);
    }
}
```

对上述代码的具体说明如下。
- HiLogLabel：HiLog 的标签类，它用于标识日志记录的来源。在这里，使用 HiLogLabel 创建了一个标签 LABEL_LOG，包括标签级别（3），域 ID（DOMAIN_ID），以及标签名称（TAG_LOG）。
- HiLog：HarmonyOS 提供的日志输出工具。在 debug 方法中，通过 HiLog.debug 记录日志。传递的参数包括日志标签 LABEL_LOG、日志格式字符串 LOG_FORMAT 和日志消息。
- LOG_FORMAT：一个日志格式化字符串，其中 %{public}s 表示字符串参数。
- debug 方法：用于输出调试日志，接受两个参数，分别是日志标签 tag 和日志消息 msg。该方法使用 HiLog.debug 记录日志，使用预定义的格式 LOG_FORMAT。
- 私有构造方法：LogUtil 类包含一个私有的构造方法，确保该类不能被实例化。这是常见的实现方式，使工具类只能通过静态方法访问，而不能被实例化。

（3）编写文件 src/main/java/ohos/samples/progress/slice/MainAbilitySlice.java，定义 HarmonyOS 应用程序的切片类 MainAbilitySlice，负责展示和控制与进度条相关的 UI。具体来说，类 MainAbilitySlice 通过监听滑块组件的变化，实现了一个简单的进度条功能，并通过定时任务更新进度，同时处理了滑块的触摸事件。

```java
    public class MainAbilitySlice extends AbilitySlice implements Slider.ValueChangedListener {
        private static final String TAG = MainAbilitySlice.class.getSimpleName();
        private static final int EVENT_ID = 0x12;
        private static final int DELAY_TIME = 1000;
        private static final int PERIOD = 1000;
        private Slider currentSlider;
        private Slider maxSlider;
        private Slider speedSlider;
        private Text currentText;
        private Text maxText;
        private Text speedText;
        private ProgressBar progressBar;
        private TimerTask timerTask;
        private Timer timer;
        private int frequencyValue;

        private final EventHandler handler = new EventHandler(EventRunner.current()) {
            @Override
```

```java
            protected void processEvent(InnerEvent event) {
                if (event.eventId == EVENT_ID) {
                    progressBar.setProgressValue(progressBar.getProgress() + frequencyValue);
                    if (progressBar.getProgress() >= maxSlider.getProgress()) {
                        finishTask();
                        new ToastDialog(MainAbilitySlice.this).setText("Progress Finish").show();
                    }
                }
            }
        };

        @Override
        public void onStart(Intent intent) {
            super.onStart(intent);
            super.setUIContent(ResourceTable.Layout_main_ability_slice);
            initComponents();
        }

        private void initComponents() {
            currentSlider = (Slider) findComponentById(ResourceTable.Id_current_value_slider);
            currentText = (Text) findComponentById(ResourceTable.Id_current_value_text);
            maxSlider = (Slider) findComponentById(ResourceTable.Id_max_value_slider);
            maxText = (Text) findComponentById(ResourceTable.Id_max_value_text);
            speedSlider = (Slider) findComponentById(ResourceTable.Id_speed_value_slider);
            speedText = (Text) findComponentById(ResourceTable.Id_speed_value_text);
            Component startProgressButton = findComponentById(ResourceTable.Id_start_progress_button);
            progressBar = (ProgressBar) findComponentById(ResourceTable.Id_progressbar);
            currentSlider.setValueChangedListener(this);
            maxSlider.setValueChangedListener(this);
            speedSlider.setValueChangedListener(this);
            progressBar.setProgressColor(Color.RED);
            startProgressButton.setClickedListener(component -> startProgress());
        }

        private void startProgress() {
            if (currentSlider.getProgress() > maxSlider.getProgress()) {
                new ToastDialog(this).setText("Error:Max < current").show();
                return;
            }
            frequencyValue = (maxSlider.getProgress() - currentSlider.getProgress())
                    / (speedSlider.getProgress());
            progressBar.setProgressValue(currentSlider.getProgress());
```

```java
        finishTask();
        startTask();
    }

    private void startTask() {
        timerTask = new TimerTask() {
            @Override
            public void run() {
                handler.sendEvent(EVENT_ID);
            }
        };
        timer = new Timer();
        timer.schedule(timerTask, DELAY_TIME, PERIOD);
    }

    private void finishTask() {
        if (timer != null && timerTask != null) {
            timer.cancel();
            timer = null;
            timerTask = null;
        }
    }

    @Override
    public void onProgressUpdated(Slider slider, int position, boolean isUpdated) {
        switch (slider.getId()) {
            case ResourceTable.Id_current_value_slider: {
                progressBar.setProgressValue(position);
                currentText.setText(String.valueOf(currentSlider.getProgress()));
                break;
            }
            case ResourceTable.Id_max_value_slider: {
                maxText.setText(String.valueOf(maxSlider.getProgress()));
                progressBar.setMaxValue(position);
                break;
            }
            case ResourceTable.Id_speed_value_slider: {
                speedText.setText(String.valueOf(speedSlider.getProgress()));
                break;
            }
            default:
                break;
        }
    }

    @Override
    public void onTouchStart(Slider slider) {
        LogUtil.debug(TAG, "Slider Touch Start");
    }

    @Override
    public void onTouchEnd(Slider slider) {
```

```
            LogUtil.debug(TAG, "Slider Touch End");
        }
   }
}
```

上述代码的实现流程如下。
- 首先，类 MainAbilitySlice 是 HarmonyOS 应用程序的切片类，负责展示和控制与进度条相关的 UI。
- 其次，在 onStart 方法中，设置了初始的 UI 内容，并调用了 initComponents 方法进行组件的初始化。
- 再次，在 initComponents 方法中对应用程序中的各个组件进行初始化操作。如 currentSlider、maxSlider 和 speedSlider 分别对应当前值、最大值和速度的滑块组件，而 progressBar 则是一个进度条组件。通过 setValueChangedListener 方法为滑块组件设置了值变化的监听器。
- 复次，startProgress 方法用于开始进度。在该方法中，首先检查当前值是否大于最大值，如果是，则显示错误提示。然后计算进度的增加频率，通过设置 progressBar 的 ProgressValue 属性，启动任务并开始定时更新进度。
- 最后，在 onProgressUpdated 方法中，根据滑块组件的变化更新了相应的 UI 元素，如 progressBar、currentText、maxText 和 speedText。

执行效果，如图 4-20 所示。

图 4-20　进度条组件的执行效果

4.3.9　ToastDialog、PopupDialog 和 CommonDialog 组件

对话框（Dialog）是用户界面设计中常见的一种元素，用于在应用程序和用户之间进行交互。它通常以弹出窗口的形式出现，覆盖在应用程序的主界面上。对话框可以包含文本、按钮、输入框、选择框等交互元素，用于向用户显示信息、接收用户输入或请求用户进行某种操作。下文将详细介绍鸿蒙系统中常用的对话框组件。

1. ToastDialog

ToastDialog 组件的功能是在窗口上方弹出的对话框，是通知操作的简单反馈。

ToastDialog 会在一段时间后消失，在此期间，用户还可以操作当前窗口的其他组件。ToastDialog 组件中的常用方法如下。

- setAlignment(int gravity)：设置对话框的对齐属性。
- setComponent(Component component)：自定义内容区域。
- setOffset(int offsetX, int offsetY)：设置对话框偏移量。
- setSize(int width, int height)：设置对话框尺寸
- setText(String textContent)：设置对话框显示内容。
- show()：显示对话框。

2. PopupDialog

PopupDialog 是指在当前界面之上弹出的气泡对话框，可以参照相对组件或者屏幕显示。显示时会获取焦点，中断用户操作，被覆盖的其他组件无法交互。气泡对话框内容一般简单明了，并提示用户需要确认的一些信息。组件 PopupDialog 中的构造方法如下。

- PopupDialog(Context context, Component contentComponent)：创建一个气泡对话框实例，并传入需要相对显示的组件。
- PopupDialog(Context context, Component contentComponent, int width, int height)：创建一个气泡对话框实例，初始化气泡对话框尺寸并传入需要相对显示的组件。

组件 PopupDialog 中的常用方法如下。

- setArrowOffset(int offset)：设置当前气泡对话框箭头的偏移量。
- setArrowSize(int width, int height)：设置当前气泡对话框箭头的尺寸。
- setBackColor(Color color)：设置当前气泡对话框的背景颜色。
- setCustomComponent(Component customComponent)：自定义内容区域。
- setHasArrow(boolean status)：设置是否显示气泡对话框的箭头。
- setMode(int mode)：设置气泡对话框的对齐模式。
- setText(String text)：设置气泡对话框的内容。
- showOnCertainPosition(int alignment, int x, int y)：设置气泡对话框相对屏幕显示的位置。
- alignment 为相对屏幕对齐模式，x 和 y 为偏移量。
- show()：显示气泡对话框。

3. CommonDialog

组件 CommonDialog 是一种在弹出框消失之前，用户无法操作其他界面内容的对话框。通常用来展示用户当前需要关注的信息或操作。对话框的内容通常是不同组件进行组合布局，如文本、列表、输入框、网格、图标或图片，常用于选择或确认信息。组件 CommonDialog 只有一个构造方法 CommonDialog(Context context)，功能是创建一个对话框实例。

组件 CommonDialog 中的常用方法如下。

- setButton(int buttonNum, String text, IDialog.ClickedListener listener)：设置按钮区的按钮，可设置按钮的位置、文本及相关单击事件。
- setContentCustomComponent(Component component)：自定义内容区域。
- setContentImage(int resId)：设置要在内容区域显示的图标。

- setContentText(String text):设置要在内容区域中显示的文本。
- setDestroyedListener(CommonDialog.DestroyedListener destroyedListener):设置对话框销毁监听器。
- setImageButton(int buttonNum, int resId, IDialog.ClickedListener listener):设置对话框的图像按钮。
- setMovable(boolean movable):设置对话框是否可以拖动。
- setTitleCustomComponent(DirectionalLayout component):自定义标题区域。
- setTitleIcon(int resId, int iconId):设置标题区域图标。
- setTitleSubText(String text):在标题区域设置补充文本信息。
- setTitleText(String text):在标题区域中设置标题文本。
- setAlignment(int alignment):设置对话框的对齐模式。
- setAutoClosable(boolean closable):设置是否启用触摸对话框外区域关闭对话框的功能。
- setCornerRadius(float radius):设置对话框圆角的半径。
- setDuration(int ms):设置对话框自动关闭前的持续时间。
- setOffset(int offsetX, int offsetY):设置对话框的偏移量。
- setSize(int width, int height):设置对话框的大小。
- setTransparent(boolean isEnable):设置是否为对话框启用透明背景。
- show():显示对话框。

实例4-9,演示了使用CommonDialog组件的过程。

实例4-9:使用CommonDialog组件(源码路径:codes\4\Dialog)

(1)编写文件src/main/resources/base/layout/custom_dialog_content.xml,定义了一个简单的自定义对话框,其中包含一个标题、一组文本字段和一个确认按钮,这样的对话框可用于输入一系列数字或信息。

(2)编写文件src/main/java/ohos/samples/dialog/slice/MainAbilitySlice.java,通过HarmonyOS提供的对话框组件实现了通用对话框、列表对话框、多选列表对话框和自定义对话框的展示,并根据用户的操作更新界面上的文本信息。

```java
public class MainAbilitySlice extends AbilitySlice {
    /**
     * DIALOG_BOX_CORNER_RADIUS
     */
    public static final float DIALOG_BOX_CORNER_RADIUS = 36.0f;
    /**
     * DIALOG_BOX_WIDTH
     */
    public static final int DIALOG_BOX_WIDTH = 984;
    private Text resultText;

    @Override
    public void onStart(Intent intent) {
        super.onStart(intent);
        super.setUIContent(ResourceTable.Layout_main_ability_slice);
```

```java
            initComponents();
    }

    private void initComponents() {
        Component commonDialogButton = findComponentById(ResourceTable.Id_common_dialog);
        Component listDialogButton = findComponentById(ResourceTable.Id_list_dialog);
        Component multiSelectDialogButton = findComponentById(ResourceTable.Id_multiselect_dialog);
        Component customDialogButton = findComponentById(ResourceTable.Id_custom_dialog);
        resultText = (Text) findComponentById(ResourceTable.Id_result_text);

        commonDialogButton.setClickedListener(component -> showCommonDialog());
        listDialogButton.setClickedListener(component -> showListDialog());
        multiSelectDialogButton.setClickedListener(component -> showMultiSelectDialog());
        customDialogButton.setClickedListener(component -> showCustomDialog());
    }

    private void showCommonDialog() {
        CommonDialog commonDialog = new CommonDialog(this);
        commonDialog.setTitleText("This Is Common Dialog");
        commonDialog.setContentText("Hello common dialog");
        commonDialog.setCornerRadius(DIALOG_BOX_CORNER_RADIUS);
        commonDialog.setAlignment(TextAlignment.CENTER);
        commonDialog.setSize(DIALOG_BOX_WIDTH, MATCH_CONTENT);
        commonDialog.setAutoClosable(true);
        commonDialog.setButton(IDialog.BUTTON1, "Yes", (iDialog, var) -> {
            resultText.setText("You Clicked Yes Button");
            iDialog.destroy();
        });
        commonDialog.setButton(IDialog.BUTTON2, "No", (iDialog, var) -> {
            resultText.setText("You Clicked No Button");
            iDialog.destroy();
        });
        commonDialog.show();
    }

    private void showListDialog() {
        String[] items = new String[]{"item 1", "item 2", "item 3"};
        ListDialog listDialog = new ListDialog(this);
        listDialog.setAlignment(TextAlignment.CENTER);
        listDialog.setSize(DIALOG_BOX_WIDTH, MATCH_CONTENT);
        listDialog.setTitleText("This Is List Dialog");
        listDialog.setAutoClosable(true);
        listDialog.setItems(items);
        listDialog.setOnSingleSelectListener((iDialog, index) -> {
            resultText.setText(items[index]);
            iDialog.destroy();
        });
        listDialog.show();
```

```java
        }
        private void showMultiSelectDialog() {
            String[] itemsString = new String[]{"item 1", "item 2", "item 3", "item 4"};
            boolean[] areSelected = new boolean[]{false, false, false, false};
            List<String> selectedItems = new ArrayList<>();
            ListDialog listDialog = new ListDialog(this);
            listDialog.setTitleText("This Is MultiSelect Dialog");
            listDialog.setAlignment(TextAlignment.CENTER);
            listDialog.setSize(DIALOG_BOX_WIDTH, MATCH_CONTENT);
            listDialog.setAutoClosable(true);
            listDialog.setMultiSelectItems(itemsString, areSelected);
            listDialog.setOnMultiSelectListener((iDialog, index, isSelected) ->
                    multiSelect(itemsString[index], selectedItems,
            listDialog.getItemComponent(index)));
            listDialog.setDialogListener(() -> {
                resultText.setText("");
                for (String selectedItem : selectedItems) {
                    resultText.append(selectedItem);
                }
                return false;
            });
            listDialog.show();
        }

        private void multiSelect(String string, List<String> selectedItems,
Component itemComponent) {
            if (selectedItems.contains(string)) {
                selectedItems.remove(string);
                itemComponent.setBackground(ElementScatter.getInstance(this)
                        .parse(ResourceTable.Graphic_multi_unselected_background));
            } else {
                selectedItems.add(string);
                itemComponent.setBackground(ElementScatter.getInstance(this)
                        .parse(ResourceTable.Graphic_multi_selected_background));
            }
        }

        private void showCustomDialog() {
            CustomDialog customDialog = new CustomDialog(this);
            customDialog.setTitle("This Is Custom Dialog");
            customDialog.setAutoClosable(true);
            customDialog.setOnConfirmListener(string -> {
                resultText.setText(string);
                customDialog.destroy();
            });
            customDialog.show();
        }
    }
```

对上述代码的具体说明如下。

- initComponents 方法：初始化界面组件，通过 findComponentById 获取按钮和文本组件的引用，用于后续设置单击监听器和显示对话框时更新文本。
- showCommonDialog 方法：创建并显示一个通用对话框（CommonDialog），设置对话框的标题、内容、圆角半径、对齐方式等属性。添加两个按钮（Yes 和 No），并为它们设置单击监听器。在显示对话框后，根据用户单击的按钮更新 resultText 中的文本。
- showListDialog 方法：创建并显示一个列表对话框（ListDialog），设置对话框的标题、对齐方式等属性。添加单选列表项，根据用户的选择更新 resultText 中的文本。
- showMultiSelectDialog 方法：创建并显示一个多选列表对话框（ListDialog），设置对话框的标题、对齐方式等属性。添加多选列表项，根据用户的选择更新 resultText 中的文本。使用自定义的 multiSelect 方法处理多选项的选择状态。
- multiSelect 方法：处理多选列表项的选择状态，根据用户选择更新背景，通过回调函数更新 selectedItems 列表。
- showCustomDialog 方法：创建并显示一个自定义对话框（CustomDialog），设置对话框的标题和单击确认按钮的监听器。在用户单击确认按钮后，更新 resultText 中的文本。

执行效果，如图 4-21 所示。

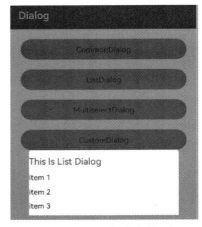

图 4-21　对话框执行效果

第 5 章
Ark UI 开发

ArkTS 是 HarmonyOS 的主力应用开发语言。为了便于熟悉 Web 前端的开发者快速上手，HarmonyOS 的 UI 开发框架中提供了"兼容 JS 的类 Web 开发范式"。这种范式通过模板、样式、逻辑的三段式方法来构建相应的应用 UI 界面。HarmonyOS 的运行时环境对这种开发模式进行了优化，确保了应用程序能够提供流畅且高效的用户体验。华为官方建议使用 ArkTS 来开发 HarmonyOS 应用程序。

5.1 方舟开发框架概述

方舟开发框架（ArkUI），是一套构建 HarmonyOS 应用界面的 UI 开发框架，它提供了极简的 UI 语法，包括 UI 组件、动画机制、事件交互等在内的 UI 开发基础设施，以满足应用开发者的可视化界面开发需求。

5.1.1 框架说明

在 ArkUI 框架中，组件是界面搭建与显示的最小单位。开发者通过多种组件的组合，构建出满足自身应用诉求的完整界面。page 页面是 ArkUI 框架最小的调度分割单位。开发者可以将应用设计为多个功能页面，每个页面进行单独的文件管理，并通过页面路由 API 完成页面间的调度管理，以实现应用内功能的解耦。

ArkUI 框架的主要特征如下。

- UI 组件：方舟开发框架内置了丰富的多态组件，包括文本、图片、按钮等基础组件，可包含一个或多个子组件的容器组件，满足开发者自定义绘图需求的绘制组件，以及提供视频播放能力的媒体组件等。其中"多态"是指组件针对不同类型设备进行了设计，提供了在不同平台上的样式适配能力。
- 布局：UI 界面设计离不开布局的参与。方舟开发框架提供了多种布局方式，不仅保留了经典的弹性布局能力，也提供了列表、宫格、栅格布局和适应多分辨率场景开发的原子布局能力。
- 动画：方舟开发框架对于 UI 界面的美化，除了组件内置动画效果外，也提供了属性动画、转场动画和自定义动画能力。
- 绘制：方舟开发框架提供了多种绘制能力，以满足开发者的自定义绘图需求，支持绘制形状、颜色填充、绘制文本、变形与裁剪、嵌入图片等。
- 交互事件：方舟开发框架提供了多种交互能力，以满足应用在不同平台通过不同输入设备进行 UI 交互响应的需求，默认适配了触摸手势、遥控器按键输入、键鼠输入，同时提供了相应的事件回调以便开发者添加交互逻辑。
- 平台 API 通道：方舟开发框架提供了 API 扩展机制，可通过该机制对平台能力进行封装，提供风格统一的 JS 接口。
- 两种开发范式：方舟开发框架针对不同的应用场景以及不同技术背景的开发者提供了两种开发范式，分别是基于 ArkTS 的声明式开发范式（声明式开发范式）和兼容 JS 的类 Web 开发范式（类 Web 开发范式）。

ArkUI 框架的结构，如图 5-1 所示。

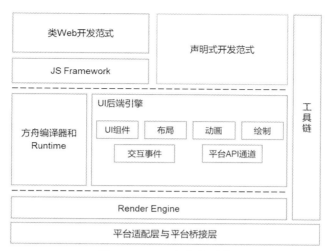

图 5-1　ArkUI 框架的结构

从图 5-1 可以看出，类 Web 开发范式与声明式开发范式的 UI 后端引擎和语言运行时是共用的，其中，UI 后端引擎实现了方舟开发框架的六种基本能力。声明式开发范式无须 JS Framework 进行页面 DOM 管理，渲染更新链路更为精简，占用内存更少，因此，推荐开发者选用声明式开发范式来搭建应用 UI 界面。

5.1.2　基本语法

ArkTS 通过装饰器 @Component 和 @Entry 装饰 struct 关键字声明的数据结构，构成一个自定义组件。自定义组件中提供了一个 build 函数，开发者需在该函数内以链式调用的方式进行基本的 UI 描述，UI 描述的方法请参考 UI 描述规范。

1. 基本概念

- struct：自定义组件可以基于 struct 实现，它们不能有继承关系。对于 struct 的实例化，可以省略 new 关键字。
- 装饰器：装饰器能够给被装饰的对象赋予特定的能力。它们不仅可以装饰类或结构体，还可以装饰类的属性。多个装饰器可以叠加到目标元素上，可以定义在同一行中或者分开多行，推荐分开多行定义。

```
@Entry
@Component
struct MyComponent {
}
```

- build 函数：自定义组件必须定义 build 函数，并且禁止自定义构造函数。build 函数满足 Builder 构造器接口的定义，用于定义组件的声明式 UI 描述。

```
interface Builder {
    build: () => void
}
```

- @Component：这个装饰器用于装饰 struct，使其具有基于组件的能力。装饰后的 struct 需要实现 build 方法来创建 UI。

- @Entry：这个装饰器用于装饰 struct，被装饰的组件作为页面的入口。当页面加载时，该组件将被渲染显示。
- @Preview：这个装饰器用于装饰 struct，被 @Preview 装饰的自定义组件可以在 DevEco Studio 的预览器上进行实时预览。加载页面时，将创建并显示 @Preview 装饰的自定义组件。

注意：在单个源文件中，最多可以使用 10 个 @Preview 装饰的自定义组件。更多说明请参考 ArkTS 组件预览效果。

- 链式调用：通过以 "." 连接的链式调用方式配置 UI 组件的属性方法、事件方法等。

2. UI 描述规范

1）无参数构造配置

如果组件的接口定义中不包含必选构造参数，组件后面的 "()" 中不需要配置任何内容。例如，Divider 组件不包含构造参数：

```
Column() {
    Text('item 1')
    Divider()
    Text('item 2')
}
```

2）必选参数构造配置

如果组件的接口定义中包含必选构造参数，则在组件后面的 "()" 中必须配置相应参数，参数可以使用常量进行赋值。例如，Image 组件的必选参数 src：

```
Image('https://xyz/test.jpg')
```

Text 组件的必选参数 content：

```
Text('test')
```

变量或表达式也可以用于参数赋值，其中表达式返回的结果类型必须满足参数类型要求。变量的定义详见页面级变量的状态管理与应用级变量的状态管理。例如，设置变量或表达式来构造 Image 和 Text 组件的参数：

```
Image(this.imagePath)
Image('https://' + this.imageUrl)
Text(`count: ${this.count}`)
```

3）属性配置

使用属性方法配置组件的属性，属性方法紧随组件，并用 "." 运算符连接。

- 配置 Text 组件的字体大小属性：

```
Text('test')
    .fontSize(12)
```

- 使用 "." 运算符进行链式调用并同时配置组件的多个属性：

```
Image('test.jpg')
    .alt('error.jpg')
    .width(100)
```

```
.height(100)
```

- 除了直接传递常量参数外，还可以传递变量或表达式：

```
Text('hello')
    .fontSize(this.size)
Image('test.jpg')
    .width(this.count % 2 === 0 ? 100 : 200)
    .height(this.offset + 100)
```

- 对于系统内置组件，框架还为其属性预定义了一些枚举类型供开发人员调用，枚举类型可以作为参数传递，且必须满足参数类型要求。例如，可以按以下方式配置 Text 组件的颜色和字体属性：

```
Text('hello')
    .fontSize(20)
    .fontColor(Color.Red)
    .fontWeight(FontWeight.Bold)
```

4）事件配置

通过事件方法可以配置组件支持的事件，事件方法紧随组件，并用 "." 运算符连接。
使用 lambda 表达式配置组件的事件方法：

```
Button('add counter')
    .onClick(() => {
        this.counter += 2
    })
```

使用匿名函数表达式配置组件的事件方法，要求使用 bind，以确保函数体中的 this 引用包含的组件：

```
Button('add counter')
    .onClick(function () {
        this.counter += 2
    }.bind(this))
```

使用组件的成员函数配置组件的事件方法：

```
myClickHandler(): void {
  this.counter += 2
}
...
Button('add counter')
  .onClick(this.myClickHandler)
```

5）子组件配置

对于支持子组件配置的组件，例如容器组件，在 "{ ... }" 里为组件添加子组件的 UI 描述。Column、Row、Stack、Grid、List 等组件都是容器组件。例如，以下是简单的 Column 示例：

```
Column() {
```

```
    Text('Hello')
        .fontSize(100)
    Divider()
    Text(this.myText)
        .fontSize(100)
        .fontColor(Color.Red)
}
```

容器组件之间也可以互相嵌套，实现相对复杂的多级嵌套效果：

```
Column() {
  Row() {
    Image('test1.jpg')
      .width(100)
      .height(100)
    Button('click +1')
      .onClick(() => {
        console.info('+1 clicked!')
      })
  }

  Divider()
  Row() {
    Image('test2.jpg')
      .width(100)
      .height(100)
    Button('click +2')
      .onClick(() => {
        console.info('+2 clicked!')
      })
  }

  Divider()
  Row() {
    Image('test3.jpg')
      .width(100)
      .height(100)
    Button('click +3')
      .onClick(() => {
        console.info('+3 clicked!')
      })
  }
}
```

5.1.3 创建自定义组件

在 ArkUI 中，UI 显示的内容均为组件，由框架直接提供的称为系统组件，由开发者定义的称为自定义组件。在进行 UI 界面开发时，通常不是简单地将系统组件进行组合使用，而是需要考虑代码的可复用性、业务逻辑与 UI 分离，以及后续版本演进等因素。因此，将 UI 和部分业务逻辑封装成自定义组件是不可或缺的能力。

下面的代码展示了自定义组件的基本用法，我们创建了一个自定义组件 HelloComponent。

```
@Component
struct HelloComponent {
  @State message: string = 'Hello, World!';

  build() {
    // HelloComponent 自定义组件组合系统组件 Row 和 Text
    Row() {
      Text(this.message)
        .onClick(() => {
          // 状态变量message的改变驱动UI刷新，UI从'Hello, World!'刷新为'Hello, ArkUI!'
          this.message = 'Hello, ArkUI!';
        })
    }
  }
}
```

接下来，可以在其他自定义组件的 build() 函数中多次创建组件 HelloComponent，实现自定义组件的重用。

```
@Entry
@Component
struct ParentComponent {
  build() {
    Column() {
      Text('ArkUI message')
      HelloComponent({ message: 'Hello, World!' });
      Divider()
      HelloComponent({ message: '你好!' });
    }
  }
}
```

接下来，开始介绍上面两段代码的含义。

（1）struct：自定义组件基于 struct 实现，struct + 自定义组件名 + {...} 的组合构成自定义组件，不能有继承关系。对于 struct 的实例化，可以省略 new。

（2）@Component：这个装饰器仅能装饰 struct 关键字声明的数据结构。struct 被 @Component 装饰后具备组件化的能力，需要实现 build() 方法描述 UI，一个 struct 只能被一个 @Component 装饰。

（3）build() 函数：用于定义自定义组件的声明式 UI 描述，自定义组件必须定义 build() 函数。

（4）@Entry：被 @Entry 装饰的自定义组件将作为 UI 页面的入口。在单个 UI 页面中，最多可以使用 @Entry 装饰一个自定义组件。@Entry 可以接受一个可选的 LocalStorage 参数。

所有在 build() 函数中声明的语言，我们统称为 UI 描述。UI 描述需要遵循以下规则。

- @Entry 装饰的自定义组件，其 build() 函数下的根节点是唯一且必要的，必须为容器组件。其中 ForEach 禁止作为根节点。

- @Component 装饰的自定义组件，其 build() 函数下的根节点也是唯一且必要的，但可以是非容器组件。同样，ForEach 禁止作为根节点。

```
@Entry
@Component
struct MyComponent {
  build() {
    // 根节点唯一且必要，必须为容器组件
    Row() {
      ChildComponent()
    }
  }
}

@Component
struct ChildComponent {
  build() {
    // 根节点唯一且必要，可为非容器组件
    Image('test.jpg')
  }
}
```

- 不允许在 build() 函数中声明本地变量，例如：

```
build() {
  // 反例：不允许声明本地变量
  let a: number = 1;
}
```

- 不允许在 UI 描述里直接使用 console.info，但允许在方法或者函数里使用，例如：

```
build() {
  // 反例：不允许在 UI 描述里直接使用 console.info
  console.info('print debug log');
}
```

- 不允许在 build() 函数中创建本地作用域，例如：

```
build() {
  // 反例：不允许创建本地作用域
  {
    ...
  }
}
```

- 不能调用没有用 @Builder 装饰的方法。系统组件的参数可以是 TypeScript 方法的返回值。例如：

```
@Component
struct ParentComponent {
  doSomeCalculations() {
  }
```

```
  calcTextValue(): string {
    return 'Hello World';
  }

  @Builder doSomeRender() {
    Text(`Hello World`)
  }

  build() {
    Column() {
      // 反例：不能调用没有使用 @Builder 装饰的方法
      this.doSomeCalculations();
      // 正例：可以调用使用 @Builder 装饰的方法
      this.doSomeRender();
      // 正例：参数可以为调用 TypeScript 方法的返回值
      Text(this.calcTextValue())
    }
  }
}
```

- 不允许使用 switch 语法。如果需要使用条件判断，请使用 if，例如：

```
build() {
  Column() {
    // 反例：不允许使用 switch 语法
    switch (expression) {
      case 1:
        Text('...')
        break;
      case 2:
        Image('...')
        break;
      default:
        Text('...')
        break;
    }
  }
}
```

- 不允许在 UI 描述中使用表达式，例如：

```
build() {
  Column() {
    // 反例：不允许使用表达式
    (this.aVar > 10) ? Text('...') : Image('...')
  }
}
```

5.2 UI 布局

布局指的是使用特定的组件或属性来管理用户页面中 UI 组件的大小和位置。在实际开发过程中，需要确定页面的布局结构，分析页面中的元素构成，然后选用合适的布局容器组件或属性来控制页面中各个元素的位置和大小约束。

5.2.1 布局结构

布局结构通常是分层的，它代表了用户界面中的整体架构。一个常见的页面布局结构，如图 5-2 所示。

为实现上述效果，开发者需要在页面中声明对应的元素。其中，Page 表示页面的根节点，Column/Row 等元素为系统组件。针对不同的页面结构，ArkUI 提供了不同的布局组件来帮助开发者实现相应的布局效果，例如，Row 用于实现线性布局。通过使用与布局相关的容器组件，可以形成相应的布局效果。例如，List 组件可以构成线性布局。

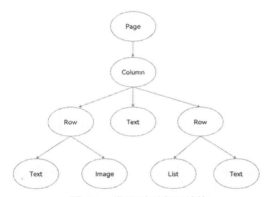

图 5-2 常见页面布局结构

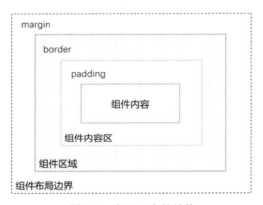

图 5-3 布局元素的结构

布局元素的结构，如图 5-3 所示，各个元素的具体说明如下。

- 组件区域（蓝色方块）：组件区域表明组件的大小，width、height 属性设置该区域的大小。
- 组件内容区（黄色方块）：组件内容区大小为组件区域大小减去组件的 border 值，组件内容区大小将作为组件内容（或子组件）进行大小测算时的布局测算限制。
- 组件内容区（绿色方块）：组件内容本身占用的大小，例如，文本内容占用的大小。组件内容和组件内容区不一定匹配，例如，设置了固定的 width 和 height，此时组件内容的大小就是设置的 width 和 height 减去 padding 和 border 值，但文本内容则是通过文本布局引擎测算后得到的大小，可能出现文本真实大小小于设置的组件内容区大小。当组件内容和组件内容区大小不一致时，align 属性生效，定义组件内容在组件内容区的对齐方式，如居中对齐。
- 组件布局边界（虚线部分）：组件通过 margin 属性设置外边距时，组件布局边界就是组件区域加上组件的 margin 值。

5.2.2 线性布局

线性布局（LinearLayout）是开发中最常用的布局，通过线性容器 Column 和 Row 构建。线性布局是其他布局的基础，其子元素在线性方向上（水平方向或垂直方向）依次排列。线性布局的排列方向由所选容器组件决定，Column 容器内子元素按照垂直方向排列（见图 5-4），Row 容器内子元素按照水平方向排列（见图 5-5）。根据不同的排列方向，开发者可以选择使用 Column 或 Row 容器创建线性布局。

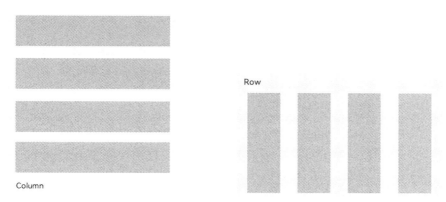

图 5-4　Column 容器内子元素按照垂直排列方向　　图 5-5　Row 容器内子元素按照水平方向排列

和 UI 布局相关的概念如下。
- 布局容器：具有布局能力的容器组件，可以承载其他元素作为其子元素，布局容器会对其子元素进行尺寸计算和布局排列。
- 布局子元素：布局容器内部的元素。
- 主轴：线性布局容器在布局方向上的轴线，子元素默认沿主轴排列。Row 容器主轴为水平方向，Column 容器主轴为垂直方向。
- 交叉轴：垂直于主轴方向的轴线。Row 容器交叉轴为垂直方向，Column 容器交叉轴为水平方向。
- 间距：布局子元素的间距。

在 LinearLayout 线性布局容器内，可以使用 space 属性设置排列方向上子元素的间距，使各子元素在排列方向上有等间距效果，如图 5-6 所示。

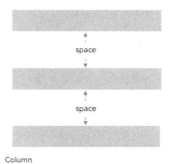

图 5-6　Column 容器内排列方向上子元素的间距

以下演示代码，使用 LinearLayout 线性布局创建一个包含文本和多个行的列布局，并设置相应的样式和属性。

```
Column({ space: 20 }) {
  Text('space: 20').fontSize(15).fontColor(Color.Gray).width('90%')
  Row().width('90%').height(50).backgroundColor(0xF5DEB3)
  Row().width('90%').height(50).backgroundColor(0xD2B48C)
  Row().width('90%').height(50).backgroundColor(0xF5DEB3)
}.width('100%')
```

对上述代码的具体说明如下。
- Column({ space: 20 })：创建一个垂直列布局，其中的子元素之间的间距为 20。
- Text('space: 20').fontSize(15).fontColor(Color.Gray).width('90%')：在列布局中添加一个文本元素，设置文本内容为 'space: 20'，字体大小为 15，字体颜色为灰色，宽度为父元素的 90%。
- Row().width('90%').height(50).backgroundColor(0xF5DEB3)：在列布局中添加一个行布局，设置宽度为父元素的 90%，高度为 50，背景颜色为 0xF5DEB3。
- Row().width('90%').height(50).backgroundColor(0xD2B48C)：在列布局中再添加一个行布局，设置宽度为父元素的 90%，高度为 50，背景颜色为 0xD2B48C。
- Row().width('90%').height(50).backgroundColor(0xF5DEB3)：在列布局中再添加一个行布局，设置宽度为父元素的 90%，高度为 50，背景颜色为 0xF5DEB3。
- width('100%')：设置整个列布局的宽度为父元素的 100%。

上述代码的执行效果，如图 5-7 所示。

图 5-7 线性布局效果

1. Column 容器内子元素在垂直方向上的排列

在 Column 容器内，可以设置子元素在水平方向上的排列方式，具体说明如下。
- HorizontalAlign.Start：子元素在水平方向左对齐。
- HorizontalAlign.Center：子元素在水平方向居中对齐。
- HorizontalAlign.End：子元素在水平方向右对齐。

2. Row 容器内子元素在水平方向上的排列

在 Row 容器内，也可以设置子元素在垂直方向上的排列方式，具体说明如下。
- VerticalAlign.Top：子元素在垂直方向顶部对齐。
- VerticalAlign.Center：子元素在垂直方向居中对齐。
- VerticalAlign.Bottom：子元素在垂直方向底部对齐。

3. Row 容器内子元素在水平方向上的排列

在布局容器内，也可以通过属性 justifyContent 设置子元素在容器主轴上的排列方式。可以从主轴起始位置开始排布，也可以从主轴结束位置开始排布，或者均匀分割主轴的空间。具体说明如下。

- justifyContent(FlexAlign.Center)：元素在垂直方向中心对齐，第一个元素与行首的距离与最后一个元素与行尾距离相同。
- justifyContent(FlexAlign.End)：元素在垂直方向尾部对齐，最后一个元素与行尾对齐，其他元素与后一个对齐。
- justifyContent(FlexAlign.Spacebetween)：垂直方向均匀分配元素，相邻元素之间距离相同。第一个元素与行首对齐，最后一个元素与行尾对齐。
- justifyContent(FlexAlign.SpaceAround)：垂直方向均匀分配元素，相邻元素之间距离相同。第一个元素到行首的距离和最后一个元素到行尾的距离是相邻元素之间距离的一半。
- justifyContent(FlexAlign.SpaceEvenly)：垂直方向均匀分配元素，相邻元素之间的距离、第一个元素与行首的间距、最后一个元素到行尾的间距都完全一样。

实例 5-1，演示了使用 justifyContent(FlexAlign.Center) 属性设置元素在水平方向中心对齐，第一个元素与行首的距离与最后一个元素与行尾距离相同。

实例 5-1：设置 LinearLayout 内的元素在水平方向居中对齐（源码路径 :codes\5\LinearLayout）

编写文件 src/main/ets/MainAbility/pages/index.ets，功能是使用 LinearLayout 创建一个行布局，其中包含三个列布局，每个列布局都有不同的宽度、高度和背景颜色。整个行布局的样式也被设置为特定的宽度、高度、背景颜色和主轴对齐方式。

扫码看视频

```
Row({}) {
  Column() {
  }.width('20%').height(30).backgroundColor(0xF5DEB3)

  Column() {
  }.width('20%').height(30).backgroundColor(0xD2B48C)

  Column() {
  }.width('20%').height(30).backgroundColor(0xF5DEB3)
}.width('100%').height(200).backgroundColor('rgb(242,242,242)').
justifyContent(FlexAlign.Center)
```

在上述代码的 Column() {} 中，设置在行布局中添加如下三个列布局。

- .width('20%').height(30).backgroundColor(0xF5DEB3)：第一个列布局设置宽度为父元素的 20%，高度为 30，背景颜色为 0xF5DEB3。
- .width('20%').height(30).backgroundColor(0xD2B48C)：第二个列布局设置宽度为父元素的 20%，高度为 30，背景颜色为 0xD2B48C。
- .width('20%').height(30).backgroundColor(0xF5DEB3)：第三个列布局设置宽度为父元素的 20%，高度为 30，背景颜色为 0xF5DEB3。

执行效果如图 5-8 所示。

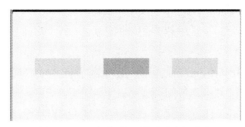

图 5-8　元素在水平方向居中对齐

5.2.3　层叠布局

层叠布局（StackLayout）用于在屏幕上预留一块区域，以显示组件中的元素，并提供元素可以重叠的布局方式。层叠布局通过 Stack 容器组件实现元素的固定定位与层叠效果，容器中的子元素（子组件）依次入栈，后一个子元素覆盖前一个子元素，子元素可以叠加，也可以设置位置。

层叠布局具有强大的页面层叠和位置定位能力，适用于广告、卡片层叠效果等场景。例如，在图 5-9 所示的结构中，Stack 作为容器，容器内的子元素（子组件）的排列顺序为 Item1->Item2->Item3。

Stack 组件是一个容器组件，可以在其中包含各种子组件，子组件默认进行居中堆叠。子元素被约束在 Stack 内，进行自己的样式定义和排列。运行以下代码，执行效果如图 5-10 所示。

```
Column(){
  Stack({ }) {
    Column(){}.width('90%').height('100%').backgroundColor ('#ff58b87c')
    Text('text').width('60%').height('60%').backgroundColor ('#ffc3f6aa')
      Button('button').width('30%').height('30%').backgroundColor('#ff8ff3eb'). fontColor('#000')
  }.width('100%').height(150).margin({ top: 50 })
}
```

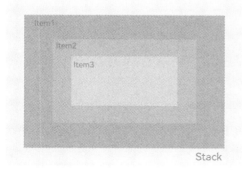

图 5-9　层叠布局

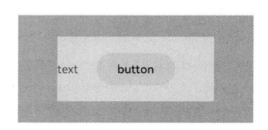

图 5-10　子元素进行自己的样式定义及排列

1. 对齐方式

Stack 组件通过参数 alignContent 实现位置的相对移动，支持 9 种对齐方式，如图 5-11 所示。

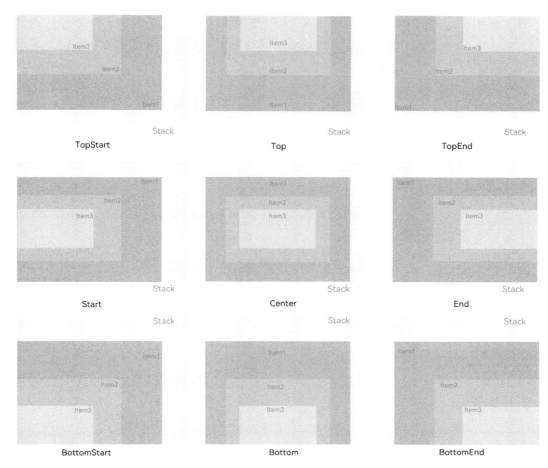

图 5-11 Stack 容器支持的 9 种对齐方式

2. Z 序控制

Stack 容器中的兄弟组件的显示层级关系,可以通过 Z 序控制的 zIndex 属性来改变。zIndex 值越大,显示层级越高,即 zIndex 值大的组件会覆盖在 zIndex 值小的组件上方。在层叠布局中,如果后面子元素的尺寸大于前面子元素的尺寸,则前面子元素会被完全隐藏。运行以下代码,执行效果如图 5-12 所示。

```
Stack({ alignContent: Alignment.BottomStart }) {
  Column() {
    Text('Stack 子元素 1').textAlign(TextAlign.End).fontSize(20)
  }.width(100).height(100).backgroundColor(0xffd306)

  Column() {
    Text('Stack 子元素 2').fontSize(20)
  }.width(150).height(150).backgroundColor(Color.Pink)

  Column() {
    Text('Stack 子元素 3').fontSize(20)
  }.width(200).height(200).backgroundColor(Color.Grey)
}.margin({ top: 100 }).width(350).height(350).backgroundColor(0xe0e0e0)
```

在上图中,最后的子元素 3 的尺寸大于前面的所有子元素,所以前面两个元素完全被隐藏。如果改变子元素 1 和子元素 2 的 zIndex 属性,可以将元素展示出来。例如下面的代码,执行效果如图 5-13 所示。

图 5-12　Z 序控制组件的显示层级关系

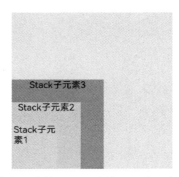

图 5-13　改变元素的 zIndex 属性可以将元素展示出来

```
@Entry
@Component
struct StackSample {
  private arr: string[] = ['APP1', 'APP2', 'APP3', 'APP4', 'APP5', 'APP6', 'APP7', 'APP8'];

  build() {
    Stack({ alignContent: Alignment.Bottom }) {
      Flex({ wrap: FlexWrap.Wrap }) {
        ForEach(this.arr, (item) => {
          Text(item)
            .width(100)
            .height(100)
            .fontSize(16)
            .margin(10)
            .textAlign(TextAlign.Center)
            .borderRadius(10)
            .backgroundColor(0xFFFFFF)
        }, item => item)
      }.width('100%').height('100%')

      Flex({ justifyContent: FlexAlign.SpaceAround, alignItems: ItemAlign.Center }) {
        Text(' 联系人 ').fontSize(16)
        Text(' 设置 ').fontSize(16)
        Text(' 短信 ').fontSize(16)
      }
      .width('50%')
      .height(50)
      .backgroundColor('#16302e2e')
      .margin({ bottom: 15 })
      .borderRadius(15)
    }.width('100%').height('100%').backgroundColor('#CFD0CF')
  }
}
```

实例 5-2，使用 Stack 布局创建了一个界面，其中包含一个水平堆叠布局，堆叠布局中包含一个水平弹性布局和一个垂直弹性布局。水平弹性布局中包含三个文本元素，而垂直弹性布局中包含了一个文本列表。

实例 5-2：使用 Stack 布局创建一个手机按键页面（源码路径 :codes\5\Stack）

编写文件 src/main/ets/MainAbility/pages/index.ets，功能是创建一个包含文本元素和按钮的界面。界面中的元素以垂直堆叠的方式排列，其中部分元素在水平方向上以弹性布局方式呈现。整个界面的样式通过设置宽度、高度、背景颜色和其他样式属性来定义。

扫码看视频

```
@Entry
@Component
struct StackSample {
   private arr: string[] = ['APP1', 'APP2', 'APP3', 'APP4', 'APP5', 'APP6', 'APP7', 'APP8'];

   build() {
     Stack({ alignContent: Alignment.Bottom }) {
       Flex({ wrap: FlexWrap.Wrap }) {
         ForEach(this.arr, (item) => {
           Text(item)
             .width(100)
             .height(100)
             .fontSize(16)
             .margin(10)
             .textAlign(TextAlign.Center)
             .borderRadius(10)
             .backgroundColor(0xFFFFFF)
         }, item => item)
       }.width('100%').height('100%')

       Flex({ justifyContent: FlexAlign.SpaceAround, alignItems: ItemAlign.Center }) {
         Text(' 联系人 ').fontSize(16)
         Text(' 设置 ').fontSize(16)
         Text(' 短信 ').fontSize(16)
       }
       .width('50%')
       .height(50)
       .backgroundColor('#16302e2e')
       .margin({ bottom: 15 })
       .borderRadius(15)
     }.width('100%').height('100%').backgroundColor('#CFD0CF')
   }
}
```

上述代码的实现流程如下。

- 首先，创建了一个垂直堆叠布局（Stack），并设置其内容的垂直对齐方式为底部（alignContent: Alignment.Bottom）。
- 其次，在堆叠布局中创建了一个水平弹性布局（Flex），允许其中的内容进行换行

（wrap: FlexWrap.Wrap）。
- 再次，通过 ForEach 遍历了一个字符串数组，为数组中的每个元素创建一个 Text 组件，显示元素的内容，并设置一些样式，如宽度、高度、字体大小、边距等。
- 复次，在垂直堆叠布局中创建了一个水平弹性布局，其中包含三个 Text 组件，显示'联系人'、'设置'和'短信'，并设置了相应的样式属性，如宽度、高度、背景颜色、下边距和圆角。
- 最后，为整个垂直堆叠布局设置宽度、高度和背景颜色。

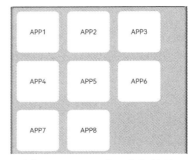

图 5-14　Stack 布局执行效果

执行效果，如图 5-14 所示。

5.2.4　弹性布局

弹性布局（Flex）提供了一种高效的方式，用于对容器中的子元素进行排列、对齐和分配剩余空间。Flex 容器默认存在主轴与交叉轴，子元素默认沿主轴排列。子元素在主轴方向的尺寸称为主轴尺寸，而在交叉轴方向的尺寸称为交叉轴尺寸。弹性布局在开发场景中的应用非常广泛，例如，页面头部导航栏的均匀分布、页面框架的搭建、多行数据的排列等。图 5-15 所示为主轴为水平方向的 Flex 容器布局效果。

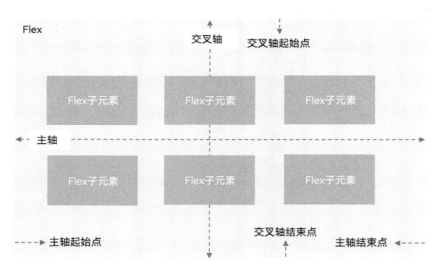

图 5-15　主轴为水平方向的 Flex 容器布局

图 5-15 涉及以下两个概念。
- 主轴：Flex 组件布局方向的轴线，子元素默认沿着主轴排列。主轴开始的位置称为主轴起始点，结束位置称为主轴结束点。
- 交叉轴：垂直于主轴方向的轴线。交叉轴开始的位置称为交叉轴起始点，结束位置称为交叉轴结束点。

1. 布局方向

在弹性布局中，容器的子元素可以按照任意方向排列。通过设置参数 direction，可以决定主轴的方向，从而控制子组件的排列方向。弹性布局（Flex）的布局方向说明，如图 5-16 所示。

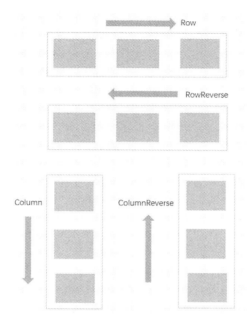

图 5-16　弹性布局方向说明

在弹性布局（Flex）中，各个布局属性的具体说明如下。
- FlexDirection.Row（默认值）：主轴为水平方向，子组件从起始端沿着水平方向开始排布。
- FlexDirection.RowReverse：主轴为水平方向，子组件从终点端沿着 FlexDirection.Row 相反的方向开始排布。
- FlexDirection.Column：主轴为垂直方向，子组件从起始端沿着垂直方向开始排布。
- FlexDirection.ColumnReverse：主轴为垂直方向，子组件从终点端沿着 FlexDirection.Column 相反的方向开始排布。

2. 布局换行

弹性布局分为单行布局和多行布局。在默认情况下，Flex 容器中的子元素排列在一条线上，即主轴上。当子元素的尺寸大于容器主轴尺寸时，可以使用 wrap 属性来控制是单行布局还是多行布局。在多行布局时，子元素会根据交叉轴方向来确定新行的堆叠方向。各个换行属性的具体说明如下。
- FlexWrap. NoWrap（默认值）：不换行。如果子组件的宽度总和大于父元素的宽度，则子组件会被压缩宽度。
- FlexWrap. Wrap：换行，每一行子组件按照主轴方向排列。
- FlexWrap. WrapReverse：换行，每一行子组件按照主轴反方向排列。

3. 主轴对齐方式

在弹性布局（Flex）中，可以通过参数 justifyContent 设置在主轴方向的对齐方式，如图 5-17 所示。

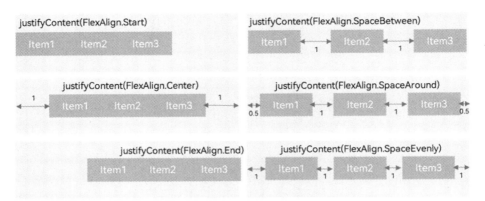

图 5-17 通过参数 justifyContent 设置在主轴方向的对齐方式

各种对齐方式的具体说明如下。
- FlexAlign.Start（默认值）：子组件在主轴方向起始端对齐，第一个子组件与父元素边沿对齐，其他元素与前一个元素对齐。
- FlexAlign.Center：子组件在主轴方向居中对齐。
- FlexAlign.End：子组件在主轴方向终点端对齐，最后一个子组件与父元素边沿对齐，其他元素与后一个元素对齐。
- FlexAlign.SpaceBetween：Flex 主轴方向均匀分配弹性元素，相邻子组件之间距离相同。第一个子组件和最后一个子组件与父元素边沿对齐。
- FlexAlign.SpaceAround：Flex 主轴方向均匀分配弹性元素，相邻子组件之间距离相同。第一个子组件到主轴起始端的距离和最后一个子组件到主轴终点端的距离是相邻元素之间距离的一半。
- FlexAlign.SpaceEvenly：Flex 主轴方向元素等间距布局，相邻子组件之间的间距、第一个子组件与主轴起始端的间距、最后一个子组件到主轴终点端的间距均相等。

实例 5-3，演示了使用弹性布局（Flex）的过程。

实例 5-3：使用弹性布局（Flex）排列元素（源码路径 :codes\5\Flex）

编写文件 src/main/ets/MainAbility/pages/index.ets，使用弹性布局（Flex）创建一个包含多个文本元素的界面，这些元素按照水平弹性布局的方式排列在一起，同时界面的布局结构由多个嵌套的列布局组成。

扫码看视频

```
@Entry
@Component
struct FlexExample {
  build() {
    Column() {
      Column({ space: 5 }) {
        Flex({ direction: FlexDirection.Row, wrap: FlexWrap.NoWrap,
justifyContent: FlexAlign.SpaceBetween, alignItems: ItemAlign.Center }) {
          Text('1').width('30%').height(50).backgroundColor(0xF5DEB3)
          Text('2').width('30%').height(50).backgroundColor(0xD2B48C)
          Text('3').width('30%').height(50).backgroundColor(0xF5DEB3)
        }
        .height(70)
```

```
          .width('90%')
          .backgroundColor(0xAFEEEE)
      }.width('100%').margin({ top: 5 })
    }.width('100%')
  }
}
```

通过上述代码，在内部列布局中创建一个水平弹性布局，设置主轴方向为水平、不允许换行、主轴上的对齐方式为均匀分布、交叉轴上的对齐方式为居中。执行效果，如图 5-18 所示。

图 5-18　弹性布局（Flex）的执行效果

5.2.5　相对布局

相对布局（RelativeContainer）用于创建采用相对布局的容器，支持容器内部的子元素设置相对位置关系。子元素可以指定兄弟元素作为锚点，也可以指定父容器作为锚点，基于锚点进行相对位置布局。图 5-19 所示为一个 RelativeContainer 的概念图，图中的虚线表示位置的依赖关系。相对布局涉及了两个概念。

- 锚点：通过锚点设置当前元素基于哪个元素确定位置。
- 对齐方式：通过对齐方式，设置当前元素是基于锚点的上中下对齐，还是基于锚点的左中右对齐。

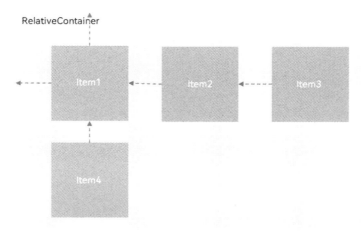

图 5-19　相对布局概念示意

注意：子元素的关系并不完全是图 5-19 所示的依赖关系。例如，Item4 可以以 Item2 为依赖锚点，也可以以 RelativeContainer 父容器为依赖锚点。

1）锚点设置

锚点设置是指设置子元素相对于父元素或兄弟元素的位置依赖关系。在水平方向上，可以设置 left、middle、right 的锚点。在竖直方向上，可以设置 top、center、bottom 的锚点。为了明确定义锚点，必须为 RelativeContainer 及其子元素设置 ID，用于指定锚点信息。ID 默认为"__container__"，其余子元素的 ID 通过 id 属性设置。未设置 ID 的子元素在 RelativeContainer 中不会显示。

- RelativeContainer 父组件为锚点，__container__ 代表父容器的 id。例如，下面的代码创建了一个相对布局容器，其中包含两个水平行布局，它们分别位于相对布局容器的左上角和右上角。效果如图 5-20 所示。

```
RelativeContainer() {
  Row()
    // 添加其他属性
    .alignRules({
      top: { anchor: '__container__', align: VerticalAlign.Top },
      left: { anchor: '__container__', align: HorizontalAlign.Start }
    })
    .id("row1")

  Row()
    ...
    .alignRules({
      top: { anchor: '__container__', align: VerticalAlign.Top },
      right: { anchor: '__container__', align: HorizontalAlign.End }
    })
    .id("row2")
}
...
```

- 以子元素为锚点：例如下面的代码，首先，创建一个相对布局容器，其中包含多个水平行布局。其次，通过对其中一个行布局的相对定位规则，将另一个组件定位于该行布局的底部位置。最后，设置了容器的宽度、高度、左边距和边框样式。效果如图 5-21 所示。

```
RelativeContainer() {
  ...
  top: { anchor: 'row1', align: VerticalAlign.Bottom },
  ...
}
.width(300).height(300)
.margin({ left: 20 })
.border({ width: 2, color: '#6699FF' })
```

图 5-20　水平行布局分别位于相对容器的左上角和右上角

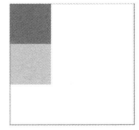

图 5-21　将另一个组件定位于该行布局的底部位置

2）设置相对于锚点的对齐位置

在设置了锚点之后，可以通过 align 设置相对于锚点的对齐位置。在水平方向上，对齐位置可以设置为 HorizontalAlign.Start、HorizontalAlign.Center、HorizontalAlign.End。如图 5-22 所示。

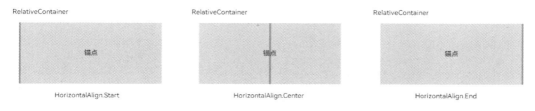

图 5-22　水平方向上的对齐

在竖直方向上，对齐位置可以设置为 VerticalAlign.Top、VerticalAlign.Center、VerticalAlign.Bottom。如图 5-23 所示。

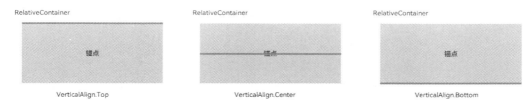

图 5-23　竖直方向上的对齐

实例 5-4，演示了使用相对布局（RelativeContainer）的过程。

实例 5-4：使用相对布局（RelativeContainer）构建一个图案（源码路径 :codes\5\RelativeContainer）

编写文件 src/main/resources/base/layout/main_ability_slice.xml，使用 RelativeContainer 相对布局创建了一个相对布局容器，其中包含多个水平行布局，每个行布局中包含不同样式和位置的组件。

扫码看视频

```
@Entry
@Component
struct Index {
  build() {
    Row() {
      RelativeContainer() {
        Row()
          .width(100)
          .height(100)
          .backgroundColor('#FF3333')
          .alignRules({
            top: { anchor: '__container__', align: VerticalAlign.Top },
// 以父容器为锚点，竖直方向顶头对齐
            middle: { anchor: '__container__', align: HorizontalAlign.Center }   // 以父容器为锚点，水平方向居中对齐
          })
```

```
          .id('row1')    // 设置锚点为row1

      Row() {
        Image($r('app.media.icon'))
      }
      .height(100).width(100)
      .alignRules({
          top: { anchor: 'row1', align: VerticalAlign.Bottom },    // 以
row1 组件为锚点，竖直方向低端对齐
          left: { anchor: 'row1', align: HorizontalAlign.Start }    // 以
row1 组件为锚点，水平方向开头对齐
      })
      .id('row2')    // 设置锚点为row2

      Row()
        .width(100)
        .height(100)
        .backgroundColor('#FFCC00')
        .alignRules({
          top: { anchor: 'row2', align: VerticalAlign.Top }
        })
        .id('row3')    // 设置锚点为row3

      Row()
        .width(100)
        .height(100)
        .backgroundColor('#FF9966')
        .alignRules({
          top: { anchor: 'row2', align: VerticalAlign.Top },
          left: { anchor: 'row2', align: HorizontalAlign.End },
        })
        .id('row4')    // 设置锚点为row4

      Row()
        .width(100)
        .height(100)
        .backgroundColor('#FF66FF')
        .alignRules({
          top: { anchor: 'row2', align: VerticalAlign.Bottom },
          middle: { anchor: 'row2', align: HorizontalAlign.Center }
        })
        .id('row5')    // 设置锚点为row5
    }
    .width(300).height(300)
    .border({ width: 2, color: '#6699FF' })
  }
  .height('100%').margin({ left: 30 })
 }
}
```

在上述代码中，通过 Row() 创建了一个垂直行布局，具体说明如下。

- RelativeContainer()：创建了一个相对布局容器，其中包含多个水平行布局。

- 第一个 Row()：创建了一个水平行布局，设置了宽度、高度、背景颜色和相对布局规则，使得该行布局顶头对齐、水平居中，并设置了唯一标识符为 'row1'。
- 第二个 Row()：创建了一个包含 Image 组件的行布局，设置了高度、宽度、相对布局规则，使其底端对齐于 'row1' 的顶部、水平开头对齐，并设置了唯一标识符为 'row2'。
- 第三个至第五个 Row()：分别创建了三个水平行布局，设置了宽度、高度、背景颜色和相对布局规则，以 'row2' 为锚点，实现了不同的位置关系，并分别设置了唯一标识符为 'row3'、'row4' 和 'row5'。
- RelativeContainer 的样式设置：设置了相对布局容器的宽度、高度，并添加了边框样式。

外层 Row() 设置了外层垂直行布局的高度为 '100%'，并在左侧添加了一个左边距为 30，这个外层布局容纳了之前的相对布局容器。

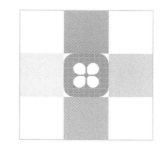

图 5-24 相对布局的执行效果

执行效果如图 5-24 所示。

5.2.6 栅格布局

栅格布局（GridRow/GridCol）是一种通用的辅助定位工具，对移动设备的界面设计具有很好的参考作用。

1. 栅格系统断点

栅格系统以设备的水平宽度（屏幕密度像素值，单位为 vp）作为断点依据，定义设备的宽度类型，形成了一套断点规则。开发者可根据需求在不同的断点区间实现不同的页面布局效果。栅格系统默认断点将设备宽度分为 xs、sm、md、lg 四类，尺寸范围如表 5-1 所示。

表 5-1 栅格系统断点的尺寸范围

断点名称	取值范围（vp）	设备描述
xs	[0, 320)	最小宽度类型设备
sm	[320, 520)	小宽度类型设备
md	[520, 840)	中等宽度类型设备
lg	[840, +∞)	大宽度类型设备

在 GridRow 栅格组件中，允许开发者使用 breakpoints 自定义修改断点的取值范围，最多支持 6 个断点。除了表 5-1 中默认的 4 个断点外，还可以启用 xl、xxl 两个断点，支持 6 种不同尺寸（xs、sm、md、lg、xl、xxl）设备的布局设置。

- xl：特大宽度类型设备。
- xxl：超大宽度类型设备。

（1）针对断点位置，开发者根据实际使用场景，通过一个单调递增数组设置。由于 breakpoints 最多支持 6 个断点，单调递增数组长度最大为 5。例如，下面的代码表示启用 xs、sm、md 共 3 个断点，小于 100vp 为 xs，100vp—200vp 为 sm，大于 200vp 为 md。

```
breakpoints: {value: ['100vp', '200vp']}
```

例如，下面的代码表示启用 xs、sm、md、lg、xl 共 5 个断点，小于 320vp 为 xs，320vp—520vp 为 sm，520vp—840vp 为 md，840vp—1080vp 为 lg，大于 1080vp 为 xl。

```
breakpoints: {value: ['320vp', '520vp', '840vp', '1080vp']}
```

（2）栅格系统通过监听窗口或容器的尺寸变化进行断点，通过 reference 设置断点切换参考物。考虑到应用可能以非全屏窗口的形式显示，以应用窗口宽度为参照物更为通用。例如，使用栅格的默认列数 12 列，通过断点设置将应用宽度分成六个区间，在各区间中，每个栅格子元素占用的列数均不同。例如，在实例 5-5 中，实现了一个响应式颜色网格效果，能够适配不同屏幕宽度的布局设计。使用栅格的默认列数 12 列，通过断点设置将应用宽度分成 6 个区间，在各区间中，每个栅格子元素占用的列数均不同。

实例 5-5：响应式颜色网格效果（源码路径 :codes\5\GridRow）

编写文件 src/main/ets/MainAbility/pages/index.ets，使用栅格布局创建了一个网格布局，每个网格元素包含一个垂直布局，其中显示了一个文本元素，代表了颜色在数组中的索引。不同的颜色和布局列数会根据屏幕宽度的不同而自适应调整。

```
@State bgColors: Color[] = [Color.Red, Color.Orange, Color.Yellow,
Color.Green, Color.Pink, Color.Grey, Color.Blue, Color.Brown];
...
GridRow({
  breakpoints: {
    value: ['200vp', '300vp', '400vp', '500vp', '600vp'],
    reference: BreakpointsReference.WindowSize
  }
}) {
  ForEach(this.bgColors, (color, index) => {
    GridCol({
      span: {
        xs: 2, // 在最小宽度类型设备上，栅格子组件占据的栅格容器 2 列。
        sm: 3, // 在小宽度类型设备上，栅格子组件占据的栅格容器 3 列。
        md: 4, // 在中等宽度类型设备上，栅格子组件占据的栅格容器 4 列。
        lg: 6, // 在大宽度类型设备上，栅格子组件占据的栅格容器 6 列。
        xl: 8, // 在特大宽度类型设备上，栅格子组件占据的栅格容器 8 列。
        xxl: 12 // 在超大宽度类型设备上，栅格子组件占据的栅格容器 12 列。
      }
    }) {
      Row() {
        Text(`${index}`)
      }.width("100%").height('50vp')
    }.backgroundColor(color)
  })
}
```

在上述代码中，首先，定义了一个状态变量 bgColors，其中包含了多种颜色。其次，在 GridRow 组件内部具体内容如下。

- 创建了一个网格行布局，使用 GridRow 组件，并通过断点配置指定了不同屏幕宽度下栅格列的数量。
- 遍历了 bgColors 数组，为每一种颜色创建一个网格列布局，使用 GridCol 组件，并根据不同屏幕宽度设置了占据的栅格容器列数。
- 在每个列布局中创建了一个垂直行布局，其中包含了一个文本元素，显示了当前颜色在 bgColors 数组中的索引。
- 设置了垂直行布局的宽度为 100%，高度为界面视图的 50%。
- 为每个列布局设置了背景颜色，颜色来自遍历到的 bgColors 数组中的元素。

最后，整个实例的目的是创建一个响应式的网格布局，其中的每个网格元素都包含了一个显示索引的文本元素，并且根据不同屏幕宽度调整了显示列数和颜色。这种设计使布局在不同设备上都能适应不同的屏幕尺寸和显示效果。执行效果如图 5-25 所示。

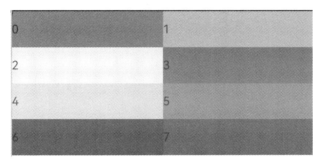

图 5-25　响应式颜色网格效果

2. 布局的总列数

在 GridRow 中通过 columns 设置栅格布局的总列数。
- columns 默认值为 12，即在未设置 columns 时，任何断点下，栅格布局被分成 12 列。例如，下面的代码创建一个响应式的网格布局，其中的每个网格元素都包含了一个显示索引的文本元素，并且根据不同屏幕宽度调整了显示列数和颜色。执行效果，如图 5-26 所示。

```
@State bgColors: Color[] = [Color.Red, Color.Orange, Color.Yellow,
Color.Green, Color.Pink, Color.Grey, Color.Blue, Color.Brown,Color.Red,
Color.Orange, Color.Yellow, Color.Green];
  ...
  GridRow() {
    ForEach(this.bgColors, (item, index) => {
      GridCol() {
        Row() {
          Text(`${index + 1}`)
        }.width('100%').height('50')
      }.backgroundColor(item)
    })
  }
```

图 5-26　设置栅格布局的列数

- 当 columns 为自定义值，栅格布局在任何尺寸设备下都被分为 columns 列。例如，下面的代码分别设置了栅格布局列数为 4 和 8，子元素默认占一列，效果如图 5-27 所示。

```
@State bgColors: Color[] = [Color.Red, Color.Orange, Color.Yellow,
Color.Green, Color.Pink, Color.Grey, Color.Blue, Color.Brown];
@State currentBp: string = 'unknown';
...
Row() {
  GridRow({ columns: 4 }) {
    ForEach(this.bgColors, (item, index) => {
      GridCol() {
        Row() {
          Text(`${index + 1}`)
        }.width('100%').height('50')
      }.backgroundColor(item)
    })
  }
  .width('100%').height('100%')
  .onBreakpointChange((breakpoint) => {
    this.currentBp = breakpoint
  })
}
.height(160)
.border({ color: Color.Blue, width: 2 })
.width('90%')

Row() {
  GridRow({ columns: 8 }) {
    ForEach(this.bgColors, (item, index) => {
      GridCol() {
        Row() {
          Text(`${index + 1}`)
        }.width('100%').height('50')
      }.backgroundColor(item)
    })
  }
  .width('100%').height('100%')
  .onBreakpointChange((breakpoint) => {
    this.currentBp = breakpoint
  })
}
.height(160)
.border({ color: Color.Blue, width: 2 })
.width('90%')
```

- 当 columns 类型为 GridRowColumnOption 时，支持 6 种不同尺寸（xs、sm、md、lg、xl、xxl）设备的总列数设置，各个尺寸下数值可不同。例如，下面的代码创建了一个响应式的网格布局，其中的每个网格元素都包含了一个显示索引的文本元素，并且根据不同颜色自适应调整了显示效果。在小型和中型屏幕上，列数分别为 4 和 8。效果如图 5-28 所示。

```
  @State bgColors: Color[] = [Color.Red, Color.Orange, Color.Yellow,
Color.Green, Color.Pink, Color.Grey, Color.Blue, Color.Brown]
  GridRow({ columns: { sm: 4, md: 8 }, breakpoints: { value: ['200vp',
'300vp', '400vp', '500vp', '600vp'] } }) {
    ForEach(this.bgColors, (item, index) => {
      GridCol() {
        Row() {
          Text(`${index + 1}`)
        }.width('100%').height('50')
      }.backgroundColor(item)
    })
  }
```

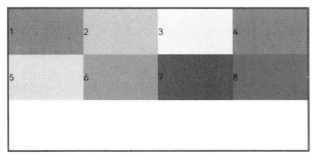

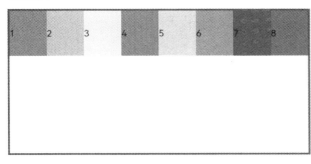

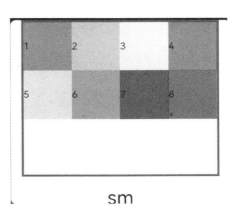

图 5-27　columns 为自定义值　　　　图 5-28　columns 类型为 GridRowColumnOption
　　　　　　　　　　　　　　　　　　　　　　时支持不同尺寸设备的总列数设置

注意：如果只设置 sm, md 的栅格总列数，则较小的尺寸使用默认 columns 值 12，较大的尺寸使用前一个尺寸的 columns。这里只设置 sm:4, md:8，则较小尺寸的 xs:12，较大尺寸参照 md 的设置，lg:8, xl:8, xxl:8。

3. 栅格组件的嵌套使用

我们也可以嵌套使用栅格组件，以完成一些复杂的布局。例如，在实例 5-6 中，栅格将整个空间分为 12 份。第一层 GridRow 嵌套 GridCol，分为中间大区域以及 "footer" 区域。第二层 GridRow 嵌套 GridCol，分为 "left" 和 "right" 区域。子组件空间按照上一层父组件的空间划分，粉色的区域是屏幕空间的 12 列，绿色和蓝色的区域是父组件 GridCol 的 12 列，依次进行空间的划分。

实例 5-6：实现一个通用网页布局模板（源码路径 :codes\5\web）

编写文件 src/main/ets/MainAbility/pages/index.ets，使用栅格布局实现了一个通用 Web 页面的效果。在屏幕中创建了一个网格布局，每个网格元素包含一个垂直行布局，其中显示了一个文本元素，代表了颜色在数组中的索引。不同的颜色和布局列数会根据屏幕宽度的不同而自适应调整。

```
@Entry
@Component
struct GridRowExample {
  build() {
    GridRow() {
      GridCol({ span: { sm: 12 } }) {
        GridRow() {
          GridCol({ span: { sm: 2 } }) {
            Row() {
              Text('left').fontSize(24)
            }
            .justifyContent(FlexAlign.Center)
            .height('90%')
          }.backgroundColor('#ff41dbaa')

          GridCol({ span: { sm: 10 } }) {
            Row() {
              Text('right').fontSize(24)
            }
            .justifyContent(FlexAlign.Center)
            .height('90%')
          }.backgroundColor('#ff4168db')
        }
        .backgroundColor('#19000000')
        .height('100%')
      }

      GridCol({ span: { sm: 12 } }) {
        Row() {
          Text('footer').width('100%').textAlign(TextAlign.Center)
        }.width('100%').height('10%').backgroundColor(Color.Pink)
      }
    }.width('100%').height(300)
  }
}
```

在上述代码中创建了一个内部网格行布局 GridRow()，具体说明如下。

- 在内部网格行布局中，使用两个网格列布局 GridCol({ span: { sm: 2 } }) 和 GridCol({ span: { sm: 10 } })：
- 第一个内部网格列布局中包含一个垂直行布局 Row()，其中有一个文本元素 Text("left")，并设置了居中对齐、高度为 "90%"。这个列布局的背景颜色为 "#ff41dbaa"。
- 第二个内部网格列布局中包含一个垂直行布局 Row()，其中有一个文本元素

Text("right"),并设置了居中对齐、高度为"90%"。这个列布局的背景颜色为"#ff4168db"。
- 内部网格行布局的背景颜色为"#19000000",高度为"100%"。
- 创建了另一个网格列布局 GridCol({ span: { sm: 12 } }),其中包含一个垂直行布局 Row(),其中有一个文本元素 Text("footer"),并设置了宽度为"100%"、高度为"10%",以及背景颜色为 Color.Pink。

执行效果,如图 5-29 所示。

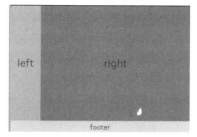

图 5-29 网页布局模板效果

5.2.7 列表布局

列表布局(List)是一种复杂的容器,当列表项达到一定数量,内容超过屏幕大小时,可以自动提供滚动功能。它适合用于呈现同类数据类型或数据类型集,如图片和文本。在列表中显示数据集合是许多应用程序中的常见要求(如通讯录、音乐列表、购物清单等)。使用列表可以轻松高效地显示结构化、可滚动的信息。通过在 List 组件中按垂直或水平方向线性排列子组件 ListItemGroup 或 ListItem,为列表中的行或列提供单个视图,或使用 ForEach 迭代一组行或列,或混合任意数量的单个视图和 ForEach 结构,构建一个列表。List 组件支持使用条件渲染、循环渲染、懒加载等渲染控制方式生成子组件。

列表作为一种容器,会自动按其滚动方向排列子组件,向列表中添加组件或从列表中移除组件会重新排列子组件,如图 5-30 所示,在垂直列表中,List 按垂直方向自动排列 ListItemGroup 或 ListItem。

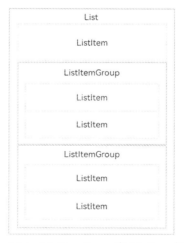

图 5-30 List、ListItemGroup 和 ListItem 组件关系

ListItemGroup 用于列表数据的分组展示,其子组件也是 ListItem。ListItem 表示单个列表项,可以包含单个子组件。

List 除了提供垂直和水平布局能力、超出屏幕时可以滚动的自适应延伸能力之外,还提供了自适应交叉轴方向上排列个数的布局能力。利用垂直布局能力可以构建单列或者多列垂直滚动列表,如图 5-31 所示。

图 5-31 垂直滚动列表（左：单列；右：多列）

利用水平布局能力可以是构建单行或多行水平滚动列表，如图 5-32 所示。

图 5-32 水平滚动列表（左：单行；右：多行）

实例 5-7 中，使用列表布局（List）创建了一个联系人列表。

实例 5-7：带有素材图片的联系列表（源码路径 :codes\5\List）

编写文件 src/main/ets/MainAbility/pages/index.ets，使用列表布局（List）创建一个简单的列表界面，其中包含了两个列表项，每个列表项包含一个头像和一个文本元素。

扫码看视频

```
@Entry
@Component
struct Index {
  @State message: string = 'Hello World'

  build() {
    List() {
      ListItem() {
        Row() {
          Image($r('app.media.iconE'))
            .width(40)
            .height(40)
            .margin(10)

          Text(' 小明 ')
            .fontSize(20)
        }
      }

      ListItem() {
        Row() {
          Image($r('app.media.iconF'))
            .width(40)
            .height(40)
            .margin(10)
```

```
            Text(' 小红 ')
                .fontSize(20)
            }
        }
    }
    .height('100%')
    }
}
```

在上述代码中创建了一个列表布局 List()，其中包含两个列表项 ListItem()。每个列表项包含一个水平行布局 Row()，具体说明如下。

- 在每个水平行布局中，包含一个图像元素 Image($r('app.media.iconE'))，设置了宽度、高度和外边距。
- 紧随其后的是一个文本元素 Text(' 小明 ')，设置了字体大小。
- 类似地，第二个列表项中的水平行布局包含了另一张图像元素和一个文本元素，代表了另一个人物 ' 小红 '。

执行效果如图 5-33 所示。

图 5-33　联系人列表效果

5.2.8　网格布局

网格（Grid/GridItem）布局是由"行"和"列"分割的单元格组成，通过指定"项目"所在的单元格作出各种各样的布局。网格布局具有较强的页面均分能力和子组件占比控制能力，是一种重要的自适应布局，其使用场景有九宫格图片展示、日历、计算器等。

在 ArkUI 中提供了 Grid 容器组件和子组件 GridItem，用于构建网格布局。Grid 用于设置网格布局相关参数，GridItem 定义子组件相关特征。Grid 组件支持使用条件渲染、循环渲染、懒加载等渲染控制方式生成子组件。

1. 布局与约束

Grid 组件为网格容器，其中容器内每一个条目对应一个 GridItem 组件，如图 5-34 所示。

网格布局是一种二维布局，Grid 组件支持自定义行列数和每行每列尺寸占比、设置子组件横跨几行或者几列，同时提供了垂直和水平布局能力。当网格容器组件尺寸发生变化时，所有子组件及间距会等比例调整，从而实现网格布局的自适应能力。根据 Grid 的这些布局能力，可以构建出不同样式的网格布局，如图 5-35 所示。

图 5-34　Grid 容器与 GridItem 组件关系

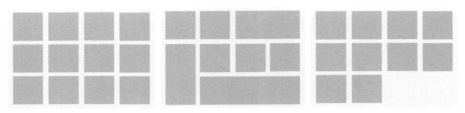

图 5-35　不同样式的网格布局

如果 Grid 组件设置了宽高属性,则其尺寸为设置值。如果没有设置宽高属性,Grid 组件的尺寸默认适应其父组件的尺寸。Grid 组件根据行、列数量与占比属性的设置,可以分为三种布局情况。

- 行、列数量与占比同时设置:Grid 只展示固定行列数的元素,其余元素不展示,且 Grid 不可滚动。(推荐使用该种布局方式)
- 只设置行、列数量与占比中的一个:元素按照设置的方向进行排布,超出的元素可通过滚动的方式展示。
- 行、列数量与占比都不设置:元素在布局方向上排布,其行列数由布局方向、单个网格的宽高等多个属性共同决定。超出行列容纳范围的元素不展示,且 Grid 不可滚动。

2. 设置行、列数量与占比

通过设置行、列数量与尺寸占比可以确定网格布局的整体排列方式。Grid 组件提供了 rowsTemplate 和 columnsTemplate 属性用于设置网格布局行列数量与尺寸占比。rowsTemplate 和 columnsTemplate 的属性值是一个由多个空格和 ' 数字 +fr' 间隔拼接的字符串,'fr' 的个数即网格布局的行或列数,'fr' 前面的数值大小,用于计算该行或列在网格布局宽度上的占比,最终决定该行或列的宽度。如图 5-36 所示,构建的是一个三行三列的网格布局,其在垂直方向上分为三等份,每行占一份;在水平方向上分为四等份,第一列占一份,第二列占两份,第三列占一份。

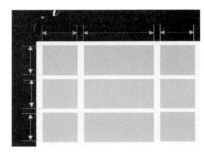

图 5-36　行、列数量占比示例

例如,在下面的代码中,只要将 rowsTemplate 的值设置为 '1fr 1fr 1fr',同时将 columnsTemplate 的值设置为 '1fr 2fr 1fr',即实现上述网格布局。

```
Grid() {
  ...
}
.rowsTemplate('1fr 1fr 1fr')
.columnsTemplate('1fr 2fr 1fr')
```

3. 设置子组件所占行列数

除了大小相同的等比例网格布局,由不同大小的网格组成不均匀分布的网格布局场景在实际应用中十分常见,如图 5-37 所示。在 Grid 组件中,通过设置 GridItem 的 rowStart、rowEnd、columnStart 和 columnEnd 可以实现如图 5-37 所示的单个网格横跨多行或多列的场景。

实例 5-8，使用网格（Grid/GridItem）布局创建了一个包含多个网格项的网格布局，其中每个网格项包含了一个文本元素，展示了不同的数字和样式。

实例 5-8：创建包含多个网格项的网格布局（源码路径 :codes\5\gun）

编写文件 src/main/ets/MainAbility/pages/index.ets，首先，声明了一个状态变量 numbers，包含了一个长度为 16 的字符串数组，每个元素为对应的索引值。其次，创建了一个垂直列布局 Column()。最后在列布局内部，使用网格布局 Grid() 设置页面元素的内容。

图 5-37　不均匀网格布局

扫码看视频

```
@Entry
@Component
struct GridItemExample {
  @State numbers: string[] = Array.apply(null, { length: 16 }).map(function (item, i) {
    return i.toString()
  })

  build() {
    Column() {
      Grid() {
        GridItem() {
          Text('4')
            .fontSize(16)
            .backgroundColor(0xFAEEE0)
            .width('100%')
            .height('100%')
            .textAlign(TextAlign.Center)
        }.rowStart(1).rowEnd(2).columnStart(1).columnEnd(2)  // 同时设置合理的
行列号

        ForEach(this.numbers, (item) => {
          GridItem() {
            Text(item)
              .fontSize(16)
              .backgroundColor(0xF9CF93)
              .width('100%')
              .height('100%')
              .textAlign(TextAlign.Center)
          }
        }, item => item)

        GridItem() {
          Text('5')
            .fontSize(16)
            .backgroundColor(0xDBD0C0)
            .width('100%')
            .height('100%')
            .textAlign(TextAlign.Center)
```

```
        }.columnStart(1).columnEnd(4)  // 只设置列号,不会从第 1 列开始布局
      }
      .columnsTemplate('1fr 1fr 1fr 1fr 1fr')
      .rowsTemplate('1fr 1fr 1fr 1fr 1fr')
      .width('90%').height(300)
    }.width('100%').margin({ top: 5 })
  }
}
```

对上述代码的具体说明如下。

- 第一个网格项 GridItem() 包含了一个文本元素 "4",设置了字体大小、背景颜色、宽度、高度和文本对齐方式,并通过 rowStart、rowEnd、columnStart、columnEnd 设置了合理的行列号。
- 使用 ForEach 遍历了 numbers 数组,为每个数字创建一个网格项 GridItem(),每个网格项包含了一个显示数字的文本元素,并设置了相应的样式。
- 最后一个网格项 GridItem() 包含了一个文本元素 "5",设置了字体大小、背景颜色、宽度、高度和文本对齐方式,并通过 columnStart、columnEnd 设置了列号,从第 1 列开始布局。
- 对整个网格布局使用了 columnsTemplate 和 rowsTemplate 分别设置了列和行的布局模板,然后设置了布局的宽度和高度。
- 最外层的 Column() 设置了自身的宽度和外边距。

执行效果,如图 5-38 所示。

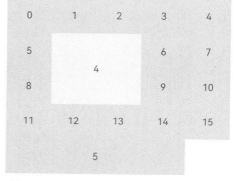

图 5-38　包含多个网格项的网格布局效果

5.3 基本组件

在 ArkTS 中,UI 组件是构建用户界面的基本构建块,它们可以用于实现各种交互和展示效果。每个组件都具有特定的功能和样式,可以在应用程序中组合使用,形成复杂的用户界面。

5.3.1 按钮组件

按钮(Button)组件通常用于响应用户的点击操作,其类型包括胶囊按钮、圆形按钮和普通按钮。当 Button 作为容器使用时,可以通过添加子组件实现包含文字、图片等元素的按钮。

1. 创建按钮

Button 可以通过调用接口来创建,接口调用有以下两种形式。

1)创建不包含子组件的按钮

```
Button(label?: string, options?: { type?: ButtonType, stateEffect?: boolean })
```

该接口用于创建不包含子组件的按钮,其中 label 用于设置按钮文字,type 用于设置按

钮类型，stateEffect 属性设置 Button 是否开启点击效果。例如，以下代码创建了一个不包含子组件的按钮，效果如图 5-39 所示。

```
Button('Ok', { type: ButtonType.Normal, stateEffect: true })
    .borderRadius(8)
    .backgroundColor(0x317aff)
    .width(90)
    .height(40)
```

图 5-39　不包含子组件的按钮

2）创建包含子组件的按钮

```
Button(options?: {type?: ButtonType, stateEffect?: boolean})
```

该接口用于创建包含子组件的按钮，只支持包含一个子组件，子组件可以是基础组件或容器组件。例如，以下代码创建了一个包含子组件的按钮，效果如图 5-40 所示。

```
Button({ type: ButtonType.Normal, stateEffect: true }) {
    Row() {
        Image($r('app.media.loading')).width(20).height(40).margin({ left: 12 })
        Text('loading').fontSize(12).fontColor(0xffffff).margin({ left: 5, right: 12 })
    }.alignItems(VerticalAlign.Center)
}.borderRadius(8).backgroundColor(0x317aff).width(90).height(40)
```

图 5-40　包含子组件的按钮

2. 设置按钮类型

Button 有三种可选类型，分别为 Capsule（胶囊按钮）、Circle（圆形按钮）和 Normal（普通按钮），它们通过 type 进行设置。

1）胶囊按钮（默认类型）

胶囊按钮的圆角自动设置为高度的一半，不支持通过 borderRadius 属性重新设置圆角。以下代码执行效果，如图 5-41 所示。

```
Button('Disable', { type: ButtonType.Capsule, stateEffect: false })
    .backgroundColor(0x317aff)
    .width(90)
    .height(40)
```

2）圆形按钮

圆形按钮不支持通过 borderRadius 属性重新设置圆角。以下代码执行效果，如图 5-42 所示。

```
Button('Circle', { type: ButtonType.Circle, stateEffect: false })
  .backgroundColor(0x317aff)
  .width(90)
  .height(90)
```

3）普通按钮

普通按钮默认圆角为 0，支持通过 borderRadius 属性重新设置圆角。以下代码执行效果，如图 5-43 所示。

```
Button('Ok', { type: ButtonType.Normal, stateEffect: true })
  .borderRadius(8)
  .backgroundColor(0x317aff)
  .width(90)
  .height(40)
```

图 5-41　胶囊按钮　　　图 5-42　圆形按钮　　　图 5-43　普通按钮

实例 5-9，功能是使用 Button 创建多个不同样式的按钮。

实例 5-9：使用 Button 创建多个不同样式的按钮（源码路径 :codes\5\Button）

编写文件 src/main/ets/MainAbility/pages/index.ets，通过 Flex 弹性布局组织了不同类型的按钮，并对每个按钮进行了样式和状态的设置。这个组件展示了如何创建不同类型的按钮，包括普通按钮、胶囊按钮和圆形按钮，以及这些按钮在不同状态下的展示。

扫码看视频

```
// xxx.ets
@Entry
@Component
struct ButtonExample {
  build() {
    Flex({ direction: FlexDirection.Column, alignItems: ItemAlign.Start,
justifyContent: FlexAlign.SpaceBetween }) {
      Text('Normal button').fontSize(9).fontColor(0xCCCCCC)
      Flex({ alignItems: ItemAlign.Center, justifyContent: FlexAlign.
SpaceBetween }) {
        Button('OK', { type: ButtonType.Normal, stateEffect: true })
          .borderRadius(8)
          .backgroundColor(0x317aff)
          .width(90)
          .onClick(() => {
            console.log('ButtonType.Normal')
          })
        Button({ type: ButtonType.Normal, stateEffect: true }) {
          Row() {
            LoadingProgress().width(20).height(20).margin({ left: 12 }).
color(0xFFFFFF)
            Text('loading').fontSize(12).fontColor(0xffffff).margin({
left: 5, right: 12 })
```

```
            }.alignItems(VerticalAlign.Center)
          }.borderRadius(8).backgroundColor(0x317aff).width(90).height(40)

          Button('Disable', { type: ButtonType.Normal, stateEffect: false
}).opacity(0.4)
            .borderRadius(8).backgroundColor(0x317aff).width(90)
        }

        Text('Capsule button').fontSize(9).fontColor(0xCCCCCC)
        Flex({ alignItems: ItemAlign.Center, justifyContent: FlexAlign.
SpaceBetween }) {
          Button('OK', { type: ButtonType.Capsule, stateEffect: true }).
backgroundColor(0x317aff).width(90)
          Button({ type: ButtonType.Capsule, stateEffect: true }) {
            Row() {
              LoadingProgress().width(20).height(20).margin({ left: 12
}).color(0xFFFFFF)
              Text('loading').fontSize(12).fontColor(0xffffff).margin({ left:
5, right: 12 })
            }.alignItems(VerticalAlign.Center).width(90).height(40)
          }.backgroundColor(0x317aff)

          Button('Disable', { type: ButtonType.Capsule, stateEffect: false
}).opacity(0.4)
            .backgroundColor(0x317aff).width(90)
        }

        Text('Circle button').fontSize(9).fontColor(0xCCCCCC)
        Flex({ alignItems: ItemAlign.Center, wrap: FlexWrap.Wrap }) {
          Button({ type: ButtonType.Circle, stateEffect: true }) {
            LoadingProgress().width(20).height(20).color(0xFFFFFF)
          }.width(55).height(55).backgroundColor(0x317aff)

          Button({ type: ButtonType.Circle, stateEffect: true }) {
            LoadingProgress().width(20).height(20).color(0xFFFFFF)
          }.width(55).height(55).margin({ left: 20 }).backgroundColor(0xF55A42)
        }
      }.height(400).padding({ left: 35, right: 35, top: 35 })
    }
  }
```

上述代码创建了 3 种类型的按钮。

（1）普通按钮（Normal button）：包含一个文本为"Normal button"的标题，包含三个不同状态的普通按钮。

- 第一个按钮标签为"OK"，点击时触发 onClick 事件，控制台输出"ButtonType.Normal"。
- 第二个按钮包含加载进度条和文本"loading"，表示加载状态。
- 第三个按钮标签为"Disable"，被禁用并设置不透明度为 0.4。

（2）胶囊按钮（Capsule button）：包含一个文本为"Capsule button"的标题，包含三个不同状态的胶囊按钮。

- 第一个按钮标签为"OK"。
- 第二个按钮包含加载进度条和文本"loading"。
- 第三个按钮标签为"Disable",被禁用并设置不透明度为 0.4。

(3)圆形按钮(Circle button):包含一个文本为 "Circle button" 的标题,包含两个圆形按钮,每个按钮包含一个加载进度条。

执行效果如图 5-44 所示。

图 5-44 多个不同样式的按钮效果

5.3.2 单选框组件

单选框(Radio)组件,通常用于提供用户交互选择项。在同一组的 Radio 中,只能有一个被选中。在 ArkUI 中,可以通过调用接口来创建 Radio。调用接口的语法格式如下:

```
Radio(options: {value: string, group: string})
```

该接口用于创建一个单选框。其中,value 是单选框的名称,group 是单选框所属的群组名称。checked 属性可以设置单选框的状态,状态有 false 和 true 两种,设置为 true 时表示单选框被选中。Radio 仅支持选中和未选中两种样式,不支持自定义颜色和形状。以下代码的执行效果如图 5-45 所示。

图 5-45 单选框未选中和被选中

```
Radio({ value: 'Radio1', group: 'radioGroup' })
  .checked(false)
Radio({ value: 'Radio2', group: 'radioGroup' })
  .checked(true)
```

除了支持通用事件外,Radio 通常用于选中后触发某些操作,可以绑定 onChange 事件来响应选中操作后的自定义行为。例如，以下代码创建了两个单选框（Radio）组件，它们都属于同一个单选框组（radioGroup）。每个单选框都有一个值（value），表示选中时的标识，并且它们都共享同一组名（group）。每个单选框都设置了一个 onChange 事件监听器，当单选框的选中状态发生变化时，这个监听器会被触发。在监听器中，通过 isChecked 参数判断当前单选框是否被选中。

```
Radio({ value: 'Radio1', group: 'radioGroup' })
  .onChange((isChecked: boolean) => {
    if(isChecked) {
      // 需要执行的操作
    }
  })
Radio({ value: 'Radio2', group: 'radioGroup' })
  .onChange((isChecked: boolean) => {
    if(isChecked) {
      // 需要执行的操作
    }
  })
```

在上述代码中，当某个单选框被选中时（isChecked 为 true），注释中的 "// 需要执行的操作" 部分表示在单选框被选中时需要执行的自定义操作。这可以是任何与单选框状态相关的操作，如更新界面、触发其他事件等。实例 5-10 演示了使用 Radio 单选框组件的过程，创建了一个包含三个单选框（Radio）的 UI 界面。

实例 5-10：创建一个包含三个单选框的 UI 界面（源码路径 :codes\5\Radio）

编写文件 src/main/ets/MainAbility/pages/index.ets，通过 Flex 弹性布局展示了三个带有文本标签的单选框。每个单选框都附带有一个状态改变的事件监听器，用于在选中状态发生变化时输出相应的信息到控制台。

扫码看视频

```
@Entry
@Component
struct RadioExample {
  build() {
    Flex({ direction: FlexDirection.Row, justifyContent: FlexAlign.Center, alignItems: ItemAlign.Center }) {
      Column() {
        Text('Radio1')
        Radio({ value: 'Radio1', group: 'radioGroup' }).checked(true)
          .height(50)
          .width(50)
          .onChange((isChecked: boolean) => {
            console.log('Radio1 status is ' + isChecked)
          })
      }
      Column() {
        Text('Radio2')
        Radio({ value: 'Radio2', group: 'radioGroup' }).checked(false)
          .height(50)
          .width(50)
          .onChange((isChecked: boolean) => {
```

```
            console.log('Radio2 status is ' + isChecked)
          })
      }
      Column() {
        Text('Radio3')
        Radio({ value: 'Radio3', group: 'radioGroup' }).checked(false)
          .height(50)
          .width(50)
          .onChange((isChecked: boolean) => {
            console.log('Radio3 status is ' + isChecked)
          })
      }
    }.padding({ top: 30 })
  }
}
```

在上述代码中创建了一个 Flex 弹性布局，其中包含三个列布局（Column），每个列布局包含文本（Text）和一个单选框（Radio）。其中第一个列布局包含"Radio1"文本和一个默认选中的单选框，第二个列布局包含"Radio2"文本和一个默认未选中的单选框，第三个列布局包含"Radio3"文本和一个默认未选中的单选框。设置了每个单选框的高度（height）、宽度（width）和状态改变时的事件监听器（onChange）。在事件监听器中，使用 isChecked 参数判断单选框的选中状态，并在控制台输出相应的状态信息。执行效果如图 5-46 所示。

图 5-46　三个单选框

5.3.3　进度条组件

在 ArkUI 中，进度条（Progress）组件用于显示某次目标操作的当前进度。

1. 创建进度条

Progress 通过调用接口来创建，调用接口形式如下：

```
Progress(options: {value: number, total?: number, type?: ProgressType})
```

通过上述格式调用接口，可以创建 type 样式的进度条，其中 value 用于设置初始进度值，total 用于设置进度总长度，type 决定 Progress 样式。以下代码创建了一个进度总长为 100、初始进度值为 24 的线性进度条，效果如图 5-47 所示。

```
Progress({ value: 24, total: 100, type: ProgressType.Linear })
```

图 5-47　创建线性进度条

2. 设置进度条样式

Progress 有 5 种可选类型，在创建时通过设置 ProgressType 枚举类型给 type 可选项指定 Progress 类型。其分别为：ProgressType.Linear（线性样式）、ProgressType.Ring（环形无刻度样式）、ProgressType.ScaleRing（环形有刻度样式）、ProgressType.Eclipse（圆形样式）和 ProgressType.Capsule（胶囊样式）。

- Linear：线性样式，从 API version9 开始，高度大于宽度时自适应垂直显示。
- Ring：环形无刻度样式，环形圆环逐渐显示至完全填充效果。
- Eclipse8：圆形样式，显示类似月圆月缺的进度展示效果，从月牙逐渐变化至满月。
- ScaleRing：环形有刻度样式，显示类似时钟刻度形式的进度展示效果。从 API version9 开始，刻度外圈出现重叠的时候自动转换为环形无刻度进度条。
- Capsule：胶囊样式，头尾两端圆弧处的进度展示效果与 Eclipse 相同；中段处的进度展示效果与 Linear 相同。高度大于宽度时自适应垂直显示。

实例 5-11 使用 Progress 组件创建了多种样式类型的进度条。

实例 5-11：创建了多种样式类型的进度条（源码路径 :codes\5\Progress）

编写文件 src/main/ets/MainAbility/pages/index.ets，创建 4 种类型的进度条，每个进度条都通过 Progress 组件创建，并设置了不同的属性，如总进度、颜色、当前值、宽度、高度等。

扫码看视频

```
@Entry
@Component
struct ProgressExample {
  build() {
    Column({ space: 15 }) {
      Text('Linear Progress').fontSize(9).fontColor(0xCCCCCC).width('90%')
      Progress({ value: 10, type: ProgressType.Linear }).width(200)
      Progress({ value: 20, total: 150, type: ProgressType.Linear }).color(Color.Grey).value(50).width(200)


      Text('Eclipse Progress').fontSize(9).fontColor(0xCCCCCC).width('90%')
      Row({ space: 40 }) {
        Progress({ value: 10, type: ProgressType.Eclipse }).width(100)
        Progress({ value: 20, total: 150, type: ProgressType.Eclipse }).color(Color.Grey).value(50).width(100)
      }

      Text('ScaleRing Progress').fontSize(9).fontColor(0xCCCCCC).width('90%')
      Row({ space: 40 }) {
        Progress({ value: 10, type: ProgressType.ScaleRing }).width(100)
        Progress({ value: 20, total: 150, type: ProgressType.ScaleRing })
          .color(Color.Grey).value(50).width(100)
          .style({ strokeWidth: 15, scaleCount: 15, scaleWidth: 5 })
      }

      // scaleCount 和 scaleWidth 效果对比
      Row({ space: 40 }) {
        Progress({ value: 20, total: 150, type: ProgressType.ScaleRing })
          .color(Color.Grey).value(50).width(100)
          .style({ strokeWidth: 20, scaleCount: 20, scaleWidth: 5 })
        Progress({ value: 20, total: 150, type: ProgressType.ScaleRing })
          .color(Color.Grey).value(50).width(100)
```

```
      .style({ strokeWidth: 20, scaleCount: 30, scaleWidth: 3 })
    }
    Text('Ring Progress').fontSize(9).fontColor(0xCCCCCC).width('90%')
    Row({ space: 40 }) {
      Progress({ value: 10, type: ProgressType.Ring }).width(100)
      Progress({ value: 20, total: 150, type: ProgressType.Ring })
        .color(Color.Grey).value(50).width(100)
        .style({ strokeWidth: 20, scaleCount: 30, scaleWidth: 20 })
    }

     Text('Capsule Progress').fontSize(9).fontColor(0xCCCCCC).width('90%')
    Row({ space: 40 }) {
      Progress({ value: 10, type: ProgressType.Capsule }).width(100).height(50)
      Progress({ value: 20, total: 150, type: ProgressType.Capsule })
        .color(Color.Grey)
        .value(50)
        .width(100)
        .height(50)
    }
  }.width('100%').margin({ top: 30 })
  }
}
```

本实例的目的是演示不同类型的进度条以及它们的样式和设置方法，在上述代码中创建了4种类型的进度条，具体说明如下。

- 线性进度条（Linear Progress）：包含两个线性进度条，一个采用默认样式，另一个则设置了总进度、颜色和当前值。
- 圆形进度条（Eclipse Progress）：包含两个圆形进度条，一个采用默认样式，另一个则设置了总进度、颜色和当前值。
- 刻度环形进度条（ScaleRing Progress）：包含两个刻度环形进度条，一个采用默认样式，另一个则设置了总进度、颜色、当前值以及刻度的数量和宽度。
- 环形进度条（Ring Progress）：包含一个环形进度条，采用默认样式，并设置了总进度、颜色、当前值以及刻度的数量和宽度。
- 胶囊形进度条（Capsule Progress）：包含两个胶囊形进度条，一个采用默认样式，另一个则设置了总进度、颜色和当前值。

执行效果如图 5-48 所示。

图 5-48　多种进度条效果

5.3.4　切换按钮组件

在 ArkUI 中，切换按钮（Toggle）组件提供了状态按钮样式、勾选框样式及开关样式，一般用于两种状态之间的切换。Toggle 通过调用接口来创建，调用接口形式如下：

```
Toggle(options: { type: ToggleType, isOn?: boolean })
```

该接口用于创建切换按钮，其中 ToggleType 为枚举类型，包括 Button、Checkbox 和 Switch，isOn 为切换按钮的状态。ToggleType 枚举值的具体说明如下。

- Checkbox：提供单选框样式，通用属性 margin 的默认值为：

```
{
top: '14px',
right: '14px',
bottom: '14px',
left: '14px'
}
```

- Button：提供状态按钮样式，如果子组件包含文本设置，则相应的文本内容将显示在按钮内部。
- Switch：提供开关样式，通用属性 margin 的默认值为：

```
{
top: '6px',
right: '14px',
bottom: '6px',
left: '14px
}
```

实例 5-12 功能是使用 Toggle 切换按钮组件创建一个蓝牙开关按钮。

实例 5-12：创建蓝牙开关按钮（源码路径 :codes\5\Toggle）

编写文件 src/main/ets/MainAbility/pages/index.ets，使用 Toggle 组件实现一个蓝牙开关按钮的功能，在状态变化时，通过 promptAction.showToast 显示相应的提示信息。

扫码看视频

```
import promptAction from '@ohos.promptAction';
@Entry
@Component
struct ToggleExample {
  build() {
    Column() {
      Row() {
        Text("Bluetooth Mode")
          .height(50)
          .fontSize(16)
      }
      Row() {
        Text("Bluetooth")
          .height(50)
          .padding({left: 10})
          .fontSize(16)
          .textAlign(TextAlign.Start)
          .backgroundColor(0xFFFFFF)
        Toggle({ type: ToggleType.Switch })
          .margin({left: 200, right: 10})
          .onChange((isOn: boolean) => {
            if(isOn) {
              promptAction.showToast({ message: 'Bluetooth is on.' })
```

```
                } else {
                    promptAction.showToast({ message: 'Bluetooth is off.' })
                }
            })
        }
        .backgroundColor(0xFFFFFF)
    }
    .padding(10)
    .backgroundColor(0xDCDCDC)
    .width('100%')
    .height('100%')
  }
}
```

在上述代码中,第二个 Row 包含 "Bluetooth" 文本和一个切换按钮(Toggle),用于模拟蓝牙的开关状态。切换按钮的类型为 ToggleType.Switch。当切换按钮的状态发生变化时,通过 onChange 事件监听器触发相应的操作。如果切换按钮被打开,则显示蓝牙打开的提示,如果切换按钮被关闭,则显示蓝牙关闭的提示。执行效果,如图 5-49 所示。

5.3.5 文本显示组件

在 ArkUI 中,文本显示(Text)组件通常用于展示用户的视图,如显示文章的文字。除了通用属性外,还支持以下属性。

- textAlign:设置文本段落在水平方向的对齐方式,默认值为 TextAlign.Start。可通过 align 属性控制文本段落在垂直方向上的位置,但在此组件中,align 属性不可用于控制文本段落在水平方向上的位置,即 align 属性中的 Alignment.TopStart、Alignment.Top、Alignment.TopEnd 效果相同,用于控制内容在顶部。

图 5-49 切换按钮开关显示蓝牙开关效果

- textOverflow:设置文本超长时的显示方式。文本截断是按字截断的,例如,英文以单词为最小单位进行截断。若需要以字母为单位进行截断,可在字母间添加零宽空格:\u200B。
- maxLines:设置文本的最大行数。
- decoration:设置文本装饰线样式及颜色。
- baselineOffset:设置文本基线的偏移量,默认值为 0。
- letterSpacing:设置文本字符间距。
- minFontSize:设置文本最小显示字号。
- maxFontSize:设置文本最大显示字号。
- textCase:设置文本大小写。

- copyOption：组件支持设置文本是否可复制粘贴，当设置 copyOptions 为 CopyOptions.InApp 或 CopyOptions.LocalDevice 时，长按文本会弹出文本选择菜单，可选中文本并进行复制、全选操作。

实例 5-13 功能是使用 Text 组件显示不同样式文本的过程，通过 textAlign，textOverflow，maxLines，lineHeight 设置文本的显示样式。

实例 5-13：使用 Text 组件显示不同样式的文本（源码路径 :codes\5\Text）

编写文件 src/main/ets/MainAbility/pages/index.ets，通过使用 Text 组件演示文本在不同样式设置下的显示效果，包括对齐方式、超长显示方式、行高以及其他样式设置。

```
@Entry
@Component
struct TextExample2 {
  build() {
    Flex({ direction: FlexDirection.Column, alignItems: ItemAlign.Start,
justifyContent: FlexAlign.SpaceBetween }) {
        Text('decoration').fontSize(9).fontColor(0xCCCCCC)
      Text('This is the text content with the decoration set to LineThrough
and the color set to Red.')
        .decoration({
          type: TextDecorationType.LineThrough,
          color: Color.Red
        })
        .fontSize(12)
        .border({ width: 1 })
        .padding(10)
        .width('100%')


      Text('This is the text content with the decoration set to Overline
and the color set to Red.')
        .decoration({
          type: TextDecorationType.Overline,
          color: Color.Red
        })
        .fontSize(12)
        .border({ width: 1 })
        .padding(10)
        .width('100%')



      Text('This is the text content with the decoration set to
Underline and the color set to Red.')
        .decoration({
          type: TextDecorationType.Underline,
          color: Color.Red
        })
        .fontSize(12)
```

```
  .border({ width: 1 })
  .padding(10)
  .width('100%')

// 文本基线偏移
Text('baselineOffset').fontSize(9).fontColor(0xCCCCCC)
Text('This is the text content with baselineOffset 0.')
  .baselineOffset(0)
  .fontSize(12)
  .border({ width: 1 })
  .padding(10)
  .width('100%')
Text('This is the text content with baselineOffset 30.')
  .baselineOffset(30)
  .fontSize(12)
  .border({ width: 1 })
  .padding(10)
  .width('100%')
Text('This is the text content with baselineOffset -20.')
  .baselineOffset(-20)
  .fontSize(12)
  .border({ width: 1 })
  .padding(10)
  .width('100%')

// 文本字符间距
Text('letterSpacing').fontSize(9).fontColor(0xCCCCCC)
Text('This is the text content with letterSpacing 0.')
  .letterSpacing(0)
  .fontSize(12)
  .border({ width: 1 })
  .padding(10)
  .width('100%')
Text('This is the text content with letterSpacing 3.')
  .letterSpacing(3)
  .fontSize(12)
  .border({ width: 1 })
  .padding(10)
  .width('100%')
Text('This is the text content with letterSpacing -1.')
  .letterSpacing(-1)
  .fontSize(12)
  .border({ width: 1 })
  .padding(10)
  .width('100%')

Text('textCase').fontSize(9).fontColor(0xCCCCCC)
Text('This is the text content with textCase set to Normal.')
  .textCase(TextCase.Normal)
  .fontSize(12)
  .border({ width: 1 })
  .padding(10)
```

```
            .width('100%')
        // 文本全小写展示
        Text('This is the text content with textCase set to LowerCase.')
            .textCase(TextCase.LowerCase)
            .fontSize(12)
            .border({ width: 1 })
            .padding(10)
            .width('100%')
        // 文本全大写展示
        Text('This is the text content with textCase set to UpperCase.')
            .textCase(TextCase.UpperCase)
            .fontSize(12).border({ width: 1 }).padding(10)

    }.height(700).width(350).padding({ left: 35, right: 35, top: 35 })
  }
}
```

对上述代码的具体说明如下。

- 文本水平方向对齐方式设置：分别展示了文本水平方向对齐方式为居中（TextAlign.Center）、起始（TextAlign.Start）、末尾（TextAlign.End）的单行文本和多行文本。
- 文本超长时显示方式设置：展示了当文本超过一行时，设置了 TextOverflow.Clip 进行截断展示，以及设置了 TextOverflow.Ellipsis 显示省略号。
- 行高设置（lineHeight）：展示了设置行高的文本。
- 其他设置：包含了一些其他的文本样式设置，如字体大小、字体颜色、边框、内边距等。

执行效果如图 5-50 所示。

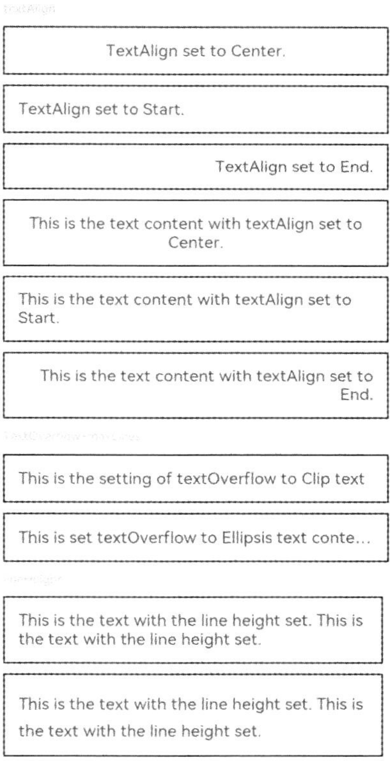

5.3.6 文本输入框

在 ArkUI 中，文本输入框（TextInput/TextArea）组件通常用于响应用户的输入操作，如评论区的输入、聊天框的输入、表格的输入等，也可以结合其他组件构建功能页面，例如登录注册页面。TextInput 为单行输入框，TextArea 为多行输入框，其通过以下接口来创建。

图 5-50 不同样式的文本效果

```
    TextArea(value?:{placeholder?: ResourceStr, text?: ResourceStr,
controller?: TextAreaController})
    TextInput(value?:{placeholder?: ResourceStr, text?: ResourceStr,
controller?: TextInputController})
```

各个参数的具体说明如下。

- placeholder：设置无输入时的提示文本，输入内容后，提示文本不显示。
- text：设置输入框当前的文本内容。当使用 stateStyles 等刷新属性时，建议通过

onChange 事件将状态变量与文本实时绑定，避免组件刷新时 TextArea 中的文本内容异常。
- controller：设置 TextArea 控制器。

TextInput 组件支持以下属性。
- type：设置输入框类型。
- placeholderColor：设置 placeholder 文本颜色。
- placeholderFont：设置 placeholder 文本样式。
- enterKeyType：设置输入法回车键类型。
- caretColor：设置输入框光标颜色。
- maxLength：设置文本的最大输入字符数。
- inputFilter：正则表达式，匹配表达式的输入允许显示，不匹配的输入将被过滤。目前仅支持单个字符匹配，不支持字符串匹配。其中 value 用于设置正则表达式，error 表示当正则匹配失败时返回被过滤的内容。
- copyOption：设置输入的文本是否可复制。当设置 CopyOptions.None 时，当前 TextInput 中的文字无法被复制或剪切，仅支持粘贴。
- showPasswordicon：密码输入模式时，输入框末尾的图标是否显示。
- textAlign：设置输入文本在输入框中的对齐方式。

实例 5-14 实现了一个简单的会员登录文本框界面，包含用户名和密码的文本输入框及登录按钮。用户可以在文本输入框中输入用户名和密码，并通过按钮触发登录操作。

实例 5-14：会员登录文本框界面（源码路径 :codes\5\TextInput）

编写文件 src/main/ets/MainAbility/pages/index.ets，创建一个包含两个文本输入框（TextInput）和一个按钮（Button）的登录界面组件。

扫码看视频

```
@Entry
@Component
struct TextInputSample {
  build() {
    Column() {
      TextInput({ placeholder: 'input your username' }).margin({ top: 20 })
        .onSubmit((EnterKeyType)=>{
          console.info(EnterKeyType+' 输入法回车键的类型值')
        })
      TextInput({ placeholder: 'input your password' }).type(InputType.Password).margin({ top: 20 })
        .onSubmit((EnterKeyType)=>{
          console.info(EnterKeyType+' 输入法回车键的类型值')
        })
      Button('Sign in').width(150).margin({ top: 20 })
    }.padding(20)
  }
}
```

对上述代码的具体说明如下。
- 第一个文本输入框：设置了占位符（placeholder）为 "input your username"，通过 margin 方法设置了上边距。通过 onSubmit 事件监听器，监听输入法 Enter 键的操作。

- 第二个文本输入框：设置了占位符（placeholder）为 "input your password"，通过 type 方法将输入框的类型设置为密码输入框。通过 margin 方法设置了上边距。通过 onSubmit 事件监听器，监听输入法 Enter 键的操作。
- 按钮：创建了一个 "Sign in" 的按钮，宽度设置为 150，上边距设置为 20。

执行效果如图 5-51 所示。

图 5-51 会员登录文本框界面

TextArea 组件支持以下属性。
- placeholderColor：设置 placeholder 文本颜色。
- placeholderFont：设置 placeholder 文本样式，包括字体大小、字体粗细、字体族、字体风格。目前仅支持默认字体族。
- textAlign：设置文本在输入框中的水平对齐方式，默认值：TextAlign.Start。
- caretColor：设置输入框光标颜色。
- inputFilter：通过正则表达式设置输入过滤器。匹配表达式的输入允许显示，不匹配的输入将被过滤。仅支持单个字符匹配，不支持字符串匹配。
- copyOption：设置输入的文本是否可复制，当设置为 CopyOptions.None 时，当前 TextArea 中的文字无法被复制或剪切，仅支持粘贴。

实例 5-15 功能是使用 TextArea 组件实现一个留言板发布系统。

实例 5-15：留言板发布系统（源码路径 :codes\5\TextArea）

编写文件 src/main/ets/MainAbility/pages/index.ets，创建一个包含文本区域（TextArea）的发布留言表单界面，并通过按钮发布留言。所输入的文本内容也会实时显示在页面上。

```
@Entry
@Component
struct TextAreaExample {
  @State text: string = ''
  controller: TextAreaController = new TextAreaController()

  build() {
    Column() {
      TextArea({
        placeholder: '请输入你的留言内容',
        controller: this.controller
      })
        .placeholderFont({ size: 16, weight: 400 })
        .width(336)
        .height(100)
        .margin(20)
        .fontSize(16)
        .fontColor('#182431')
```

```
        .backgroundColor('#FFFFFF')
        .onChange((value: string) => {
          this.text = value
        })
      Text(this.text)
      Button(' 发布留言 ')
        .backgroundColor('#007DFF')
        .margin(15)
        .onClick(() => {
          // 设置光标位置到第一个字符后
          this.controller.caretPosition(1)
        })
    }.width('100%').height('100%').backgroundColor('#F1F3F5')
  }
}
```

执行效果如图 5-52 所示。

5.3.7 视频播放组件

视频播放（Video）组件用于播放视频文件并控制其播放状态，常用于短视频应用和应用内部视频的列表页面。当视频完整出现时会自动播放，用户点击视频区域则会暂停播放，同时显示播放进度条，通过拖动播放进度条指定视频播放到具体位置。

图 5-52　留言板发布系统

Video 通过调用接口来创建，调用接口的语法格式如下：

```
Video(value: {src?: string | Resource, currentProgressRate?: number | string | PlaybackSpeed, previewUri?: string | PixelMap | Resource, controller?: VideoController})
```

该接口用于创建视频播放组件。其中，src 指定视频播放源的路径，currentProgressRate 用于设置视频播放倍速，previewUri 指定视频未播放时的预览图片路径，controller 设置视频控制器，用于自定义控制视频。

1．加载本地视频

● 加载普通本地视频：在加载本地视频时，首先在本地 rawfile 目录指定对应的文件，如图 5-53 所示。

图 5-53　本地视频位置

接下来，使用资源访问符 $rawfile() 引用视频资源。例如，在下面的代码中，通过设置不同的属性，可以播放指定源的视频，并显示预览图。播放器的行为可以通过传递的 VideoController 控制器进行控制。

```
@Component
export struct VideoPlayer{
    private controller:VideoController;
    private previewUris: Resource = $r ('app.media.preview');
    private innerResource: Resource = $rawfile('videoTest.mp4');
    build(){
      Column() {
        Video({
          src: this.innerResource,
          previewUri: this.previewUris,
          controller: this.controller
        })
      }
    }
  }
```

- Data Ability 提供的视频路径带有 dataability:// 前缀,在使用时需要确保对应视频资源存在。例如,在下面的视频播放器代码中,通过设置不同的属性,可以播放指定数据能力 URI 的视频,并显示预览图。播放器的行为可以通过传递的 VideoController 控制器进行控制。

```
@Component
export struct VideoPlayer{
    private controller:VideoController;
    private previewUris: Resource = $r ('app.media.preview');
    private videosrc: string= 'dataability://device_id/com.domainname.dataability.videodata/video/10'
    build(){
      Column() {
        Video({
          src: this.videosrc,
          previewUri: this.previewUris,
          controller: this.controller
        })
      }
    }
  }
```

2. 加载沙箱路径视频

在 ArkUI 中,支持 file:///data/storage 路径前缀的字符串,用于读取应用沙箱路径内的资源。需要保证应用沙箱目录路径下的文件存在并且有可读权限。例如,下面的代码实现了一个本地视频播放器,通过设置不同的属性,可以播放指定本地路径的视频文件,并通过传递的 VideoController 控制器进行操作。

```
@Component
export struct VideoPlayer {
    private controller: VideoController;
    private videosrc: string = 'file:///data/storage/el2/base/haps/entry/files/show.mp4'
```

```
build() {
  Column() {
    Video({
      src: this.videosrc,
      controller: this.controller
    })
  }
}
}
```

3. 加载网络视频

在 ArkUI 中加载网络视频时，需要申请权限 ohos.permission.INTERNET，具体申请方式请参考权限申请声明。此时，Video 的 src 属性为网络视频的链接。例如，在下面的视频播放器代码中，通过设置不同的属性，可以播放指定 URL 的远程视频文件，并通过传递的 VideoController 控制器进行操作。请注意，属性 videosrc 中的 URL 是示例网址，需要替换为实际视频加载网址。

```
@Component
export struct VideoPlayer{
    private controller:VideoController;
    private previewUris: Resource = $r ('app.media.preview');
     private videosrc: string= 'https://www.example.com/example.mp4'
// 使用时请替换为实际视频加载网址
    build(){
      Column() {
        Video({
          src: this.videosrc,
          previewUri: this.previewUris,
          controller: this.controller
        })
      }
    }
}
```

4. VideoController 控制器

在 ArkUI 中，VideoController 控制器主要用于控制视频的状态，包括播放、暂停、停止及设置进度等操作。一个 VideoController 对象可以控制一个或多个视频（video），使用时需要先导入对象：

```
controller: VideoController = new VideoController()
```

VideoController 包含如下内置方法。

- start()：开始播放。
- pause()：暂停播放，显示当前帧，再次播放时从当前位置继续播放。
- stop()：停止播放，显示当前帧，再次播放时从头开始播放。
- setCurrentTime()：指定视频播放的进度位置。

实例 5-16 功能是使用视频播放组件 Video 播放指定的视频。

实例 5-16：播放指定的视频（源码路径 :codes\5\Video）

编写文件 src/main/ets/MainAbility/pages/index.ets，实现一个基本的视频播放器，支持加载、播放、暂停、停止视频，切换不同的视频源和预览图，调整播放速度，并提供控制按钮进行这些操作。

扫码看视频

```
@Entry
@Component
struct VideoCreateComponent {
  @State videoSrc: Resource = $rawfile('snow.mp4')
  @State previewUri: Resource = $r('app.media.poster1')
  @State curRate: PlaybackSpeed = PlaybackSpeed.Speed_Forward_1_00_X
  @State isAutoPlay: boolean = false
  @State showControls: boolean = true
  controller: VideoController = new VideoController()

  build() {
    Column() {
      Video({
        src: this.videoSrc,
        previewUri: this.previewUri,
        currentProgressRate: this.curRate,
        controller: this.controller
      }).width('100%').height(600)
        .autoPlay(this.isAutoPlay)
        .controls(this.showControls)
        .onStart(() => {
          console.info('onStart')
        })
        .onPause(() => {
          console.info('onPause')
        })
        .onFinish(() => {
          console.info('onFinish')
        })
        .onError(() => {
          console.info('onError')
        })
        .onPrepared((e) => {
          console.info('onPrepared is ' + e.duration)
        })
        .onSeeking((e) => {
          console.info('onSeeking is ' + e.time)
        })
        .onSeeked((e) => {
          console.info('onSeeked is ' + e.time)
        })
        .onUpdate((e) => {
          console.info('onUpdate is ' + e.time)
        })

      Row() {
```

```
        Button('src').onClick(() => {
          this.videoSrc = $rawfile('snow1.mp4') // 切换视频源
        }).margin(5)
        Button('previewUri').onClick(() => {
          this.previewUri = $r('app.media.poster2') // 切换视频预览海报
        }).margin(5)
        Button('controls').onClick(() => {
          this.showControls = !this.showControls // 切换是否显示视频控制栏
        }).margin(5)
      }

      Row() {
        Button('start').onClick(() => {
          this.controller.start() // 开始播放
        }).margin(5)
        Button('pause').onClick(() => {
          this.controller.pause() // 暂停播放
        }).margin(5)
        Button('stop').onClick(() => {
          this.controller.stop() // 结束播放
        }).margin(5)
        Button('setTime').onClick(() => {
          this.controller.setCurrentTime(10, SeekMode.Accurate) // 精准跳
转到视频的 10s 位置
        }).margin(5)
      }

      Row() {
        Button('rate 0.75').onClick(() => {
          this.curRate = PlaybackSpeed.Speed_Forward_0_75_X // 0.75 倍速播放
        }).margin(5)
        Button('rate 1').onClick(() => {
          this.curRate = PlaybackSpeed.Speed_Forward_1_00_X // 原倍速播放
        }).margin(5)
        Button('rate 2').onClick(() => {
          this.curRate = PlaybackSpeed.Speed_Forward_2_00_X // 2 倍速播放
        }).margin(5)
      }
    }
  }
}
```

在上述代码中包含多个状态属性。

- videoSrc：用于指定视频文件的源，使用 $rawfile 函数加载 snow.mp4 视频文件。
- previewUri：用于指定视频的预览图，使用 $r 函数加载 app.media.poster1 图片资源。
- curRate：用于指定视频的播放速度，默认为正常速度。
- isAutoPlay：用于指定是否自动播放，默认为 false。
- showControls：用于指定是否显示视频控制栏，默认为 true。
- controller：VideoController 类型的实例，用于控制视频的播放、暂停、停止等操作。

在上述代码的 build 方法中包含一个 Video 组件，用于播放视频，具体说明如下。
- src：使用 videoSrc 指定视频文件源。
- previewUri：使用 previewUri 指定视频的预览图。
- currentProgressRate：使用 curRate 指定当前播放速度。
- controller：使用 controller 实例进行视频控制。
- 设置视频的宽度、高度、自动播放和控制栏显示等属性。
- 监听视频的各种事件，如开始、暂停、结束、错误等，并在控制台输出相关信息。

在上述代码中也包含多个 Button 组件，用于切换视频源、切换预览图、切换控制栏的显示状态，以及执行视频播放的相关操作，具体说明如下。

- 通过点击按钮触发相应的操作，如切换视频源、切换预览图、切换是否显示视频控制栏等。
- 提供控制视频播放的操作，如开始播放、暂停播放、停止播放、精准跳转等。
- 提供切换播放速度的操作，支持 0.75x、1x、2x 三种速度。

执行效果如图 5-54 所示。

图 5-54 视频播放器

5.3.8 气泡提示

在 ArkUI 中，气泡提示（Popup）属性可绑定在组件上显示气泡弹窗提示，设置弹窗内容、交互逻辑和显示状态。主要用于屏幕录制、信息弹出提醒等显示状态。气泡分为两种类型，一种是系统提供的气泡 PopupOptions，一种是开发者自定义的气泡 CustomPopupOptions。其中 PopupOptions 为系统提供的气泡，通过配置 primaryButton、secondaryButton 来设置带按钮的气泡。CustomPopupOptions 通过配置 builder 参数来设置自定义的气泡。

1. 文本提示气泡

文本提示气泡常用于只展示带有文本信息的提示，不带有任何交互的场景。Popup 属性需绑定组件，当 bindPopup 属性中参数 show 为 true 时会弹出气泡提示。请看下面的代码，在 Button 组件上绑定 Popup 属性，每次点击 Button 按钮，handlePopup 会切换布尔值，当其为 true 时，触发 bindPopup 弹出气泡。执行效果，如图 5-55 所示。

图 5-55 文本提示气泡效果

```
@Entry
@Component
struct PopupExample {
  @State handlePopup: boolean = false

  build() {
    Column() {
```

```
      Button('PopupOptions')
        .onClick(() => {
          this.handlePopup = !this.handlePopup
        })
        .bindPopup(this.handlePopup, {
          message: 'This is a popup with PopupOptions',
        })
    }.width('100%').padding({ top: 5 })
  }
}
```

2. 带按钮的提示气泡

通过使用属性 primaryButton、secondaryButton，可为气泡最多设置两个 Button 按钮，通过此按钮进行简单的交互；开发者可以通过配置 action 参数来设置想要触发的操作。请看下面的代码，实现了带按钮的提示气泡效果，执行效果，如图 5-56 所示。

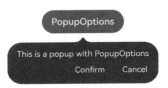

图 5-56 带按钮的提示气泡效果

```
@Entry
@Component
struct PopupExample22 {
  @State handlePopup: boolean = false
  build() {
    Column() {
      Button('PopupOptions').margin({top:200})
        .onClick(() => {
          this.handlePopup = !this.handlePopup
        })
        .bindPopup(this.handlePopup, {
          message: 'This is a popup with PopupOptions',
          primaryButton:{
            value:'Confirm',
            action: () => {
              this.handlePopup = !this.handlePopup
              console.info('confirm Button click')
            }
          },
          secondaryButton: {
            value: 'Cancel',
            action: () => {
              this.handlePopup = !this.handlePopup
            }
          },
        })
    }.width('100%').padding({ top: 5 })
  }
}
```

3. 自定义气泡

开发者可以使用构建器 CustomPopupOptions 创建自定义气泡，@Builder 中可以放置自定义的内容。除此之外，还可以通过 popupColor 等参数控制气泡样式。例如，下面的代码创建了一个按钮，当按钮被点击时显示一个自定义气泡效果，该气泡包含一个图像和文本。通过调整构造器、弹出位置和气泡背景色等参数，可以实现不同样式和内容的自定义气泡。执行效果如图 5-57 所示。

图 5-57　自定义气泡效果

```
@Entry
@Component
struct Index {
  @State customPopup: boolean = false
  // popup 构造器定义弹框内容
  @Builder popupBuilder() {
    Row({ space: 2 }) {
      Image($r("app.media.icon")).width(24).height(24).margin({ left: 5 })
      Text('This is Custom Popup').fontSize(15)
    }.width(200).height(50).padding(5)
  }
  build() {
    Column() {
      Button('CustomPopupOptions')
        .position({x:100,y:200})
        .onClick(() => {
          this.customPopup = !this.customPopup
        })
        .bindPopup(this.customPopup, {
          builder: this.popupBuilder, // 气泡的内容
          placement:Placement.Bottom, // 气泡的弹出位置
          popupColor:Color.Pink // 气泡的背景色
        })
    }
    .height('100%')
  }
}
```

5.3.9　菜单

在 ArkUI 中，菜单（Menu）是菜单接口，一般用于鼠标右键弹窗、点击弹窗等操作。Menu 中的常用属性如下。

- bindMenu：给组件绑定菜单，点击后弹出菜单。弹出菜单项支持文本和自定义两种功能。
- bindContextMenu：给组件绑定菜单，触发方式为长按或者右键点击，弹出菜单项需要自定义。

实例 5-17 演示了使用鸿蒙应用框架创建菜单（Menu）的过程，模拟实现一个快捷键菜单。

实例 5-17：模拟实现一个快捷键菜单（源码路径 :codes\5\Menu）

扫码看视频

编写文件 src/main/ets/MainAbility/pages/index.ets，功能是创建一个简单的垂直菜单，每个菜单选项都包含一个图像和文本，点击菜单项会触发相应的事件。可以通过修改

listData 数组以及菜单项的内容和样式，来适应我们的具体需求。

```
@Entry
@Component
struct MenuExample {
  @State listData: number[] = [0, 0, 0]

  @Builder MenuBuilder() {
      Flex({ direction: FlexDirection.Column, justifyContent: FlexAlign.Center,
alignItems: ItemAlign.Center }) {
          ForEach(this.listData, (item, index) => {
            Column() {
              Row() {
                Image($r("app.media.icon")).width(20).height(20).margin({ right: 5 })
                Text(`选项${index + 1}`).fontSize(20)
              }
              .width('100%')
              .height(30)
              .justifyContent(FlexAlign.Center)
              .align(Alignment.Center)
              .onClick(() => {
                console.info(`菜单${index + 1} 被点击了！`)
              })

              if (index != this.listData.length - 1) {
                Divider().height(10).width('80%').color('#ccc')
              }
            }.padding(5).height(40)
          })
       }.width(100)
  }

  build() {
    Column() {
      Text('点击菜单选项')
        .fontSize(20)
        .margin({ top: 20 })
        .bindMenu(this.MenuBuilder)
    }
    .height('100%')
    .width('100%')
    .backgroundColor('#f0f0f0')
  }
}
```

执行效果，如图 5-58 所示。

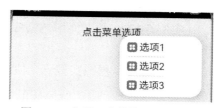

图 5-58　实现一个快捷键菜单效果

第 6 章
图形、图像开发

在 HarmonyOS 移动设备中,图形图像开发是一个非常重要的领域。图形图像开发涵盖了显示图形、绘制几何图形和画布等方面的知识,对用户界面 (UI) 的呈现和用户体验至关重要。本章将详细讲解在 HarmonyOS 系统中开发图形图像应用程序的知识。

6.1 显示图片

在 HarmonyOS 系统中，开发者经常需要在应用程序中显示图片，例如，按钮中的图标、网络图片、本地图片等。大多数情况下，需要使用 Image 组件来实现显示图片的功能，Image 组件支持多种图片格式，包括 png、jpg、bmp、svg 和 gif，具体用法请参考 Image 组件文档。

注意：使用网络图片时，需要申请权限 ohos.permission.INTERNET。

6.1.1 Image 组件介绍

在 HarmonyOS 系统中，Image 组件用于显示指定的图片，其语法格式如下。

```
Image(src: string | PixelMap | Resource)
```

参数 src 表示图片的数据源，支持本地图片和网络图片。可以使用以下三种格式的数据源。

（1）string 格式：用于加载网络图片和本地图片，常用于加载网络图片。当使用相对路径引用本地图片时，如 Image ("common/test.jpg")，不支持跨包/跨模块调用该 Image 组件，建议使用 Resource 格式来管理需全局使用的图片资源。

- 支持 Base64 字符串格式：data:image/[png|jpeg|bmp|webp];base64,[base64 data]，其中 [base64 data] 为 Base64 字符串数据。
- 支持 file:// 路径前缀的字符串：用于读取本应用安装目录下 files 文件夹下的图片资源，需要保证目录下文件有可读权限。

（2）PixelMap 格式：为像素图，常用于图片编辑的场景。

（3）Resource 格式：可以跨包/跨模块访问资源文件，是访问本地图片的推荐方式。

通过图片数据源参数 src 获取图片，然后进行渲染展示。当 Image 组件加载图片失败或图片尺寸为 0 时，图片组件大小自动为 0，不跟随父组件的布局约束。从 API version 9 开始，该接口支持在 ArkTS 卡片中使用。

1. 属性

Image 组件除了支持 HarmonyOS 系统的通用属性外，还支持如下几项专有属性。

（1）alt：加载时显示的占位图，支持本地图片（png、jpg、bmp、svg 和 gif 类型），不支持网络图片。默认值为 null。从 API version 9 开始，该接口支持在 ArkTS 卡片中使用。

（2）objectFit：设置图片的填充效果。默认值为 ImageFit.Cover。从 API version 9 开始，该接口支持在 ArkTS 卡片中使用。

（3）objectRepeat：设置图片的重复样式。从中心点向两边重复，剩余空间不足放下一张图片时会被截断。默认值为 ImageRepeat.NoRepeat。从 API version 9 开始，该接口支持在 ArkTS 卡片中使用。注意，svg 类型图源不支持该属性。

（4）interpolation：设置图片的插值效果，即减轻低清晰度图片在放大显示时出现的锯齿问题。默认值为 ImageInterpolation.None。从 API version 9 开始，该接口支持在 ArkTS 卡片中使用。注意，svg 类型图源不支持该属性，PixelMap 资源不支持该属性。

（5）renderMode：设置图片的渲染模式为原色或黑白。默认值为 ImageRenderMode.Original。从 API version 9 开始，该接口支持在 ArkTS 卡片中使用。注意，svg 类型图源不支持该属性。

（6）sourceSize：设置图片解码尺寸，降低图片的分辨率，常用于图片显示尺寸比组件尺寸更小的场景。和 ImageFit.None 配合使用时，可在组件内显示小图，单位为 px。从 API version 9 开始，该接口支持在 ArkTS 卡片中使用。注意，仅在目标尺寸小于图源尺寸时生效，svg 类型图源不支持该属性，PixelMap 资源不支持该属性。

（7）matchTextDirection：设置图片是否跟随系统语言方向，在 RTL 语言环境下显示镜像翻转显示效果。默认值为 false。从 API version 9 开始，该接口支持在 ArkTS 卡片中使用。

（8）fitOriginalSize：当图片组件尺寸未设置时，显示尺寸是否跟随图源尺寸。注意，当组件不设置宽和高或仅设置宽或高时，该属性不生效。默认值为 false。从 API version 9 开始，该接口支持在 ArkTS 卡片中使用。

（9）fillColor：设置填充颜色，设置后填充颜色会覆盖在图片上。从 API version 9 开始，该接口支持在 ArkTS 卡片中使用。注意，仅对 svg 图源生效，设置后会替换 svg 图片的填充颜色。

（10）autoResize：设置图片解码过程中是否对图源自动缩放。设置为 true 时，组件会根据显示区域的尺寸决定用于绘制的图源尺寸，有利于减少内存占用。如原图大小为 1920×1080px，而显示区域大小为 200×200px，则图片会自动解码到 200×200px 的尺寸，大幅节省图片占用的内存。默认值为 true。从 API version 9 开始，该接口支持在 ArkTS 卡片中使用。

（11）syncLoad：设置是否同步加载图片，默认为异步加载。同步加载时阻塞 UI 线程，不会显示占位图。默认值为 false。从 API version 9 开始，该接口支持在 ArkTS 卡片中使用。注意，建议加载尺寸较小的本地图片时将 syncLoad 设为 true，因为耗时较短，在主线程上执行即可。

（12）copyOption：设置图片是否可复制。当 copyOption 设置为非 CopyOptions.None 时，支持使用长按（在触摸屏设备上）、鼠标右击（在桌面设备上）以及通过快捷键"CTRL+C"来执行复制操作。默认值为 CopyOptions.None。从 API version 9 开始，该接口支持在 ArkTS 卡片中使用。注意，svg 图片不支持复制。

（13）colorFilter：给图像设置颜色滤镜效果，入参为一个 4×5 的 RGBA 转换矩阵。矩阵第一行表示 R（红色）的向量值，第二行表示 G（绿色）的向量值，第三行表示 B（蓝色）的向量值，第四行表示 A（透明度）的向量值，4 行分别代表不同的 RGBA 的向量值。RGBA 值分别为 0 和 1 之间的浮点数字，当矩阵对角线值为 1 时，保持图片原有色彩。计算规则如下。

如果输入的滤镜矩阵为：

```
[ r_1, r_2, r_3, r_4, r_5,
  g_1, g_2, g_3, g_4, g_5,
```

```
        b_1, b_2, b_3, b_4, b_5,
        a_1, a_2, a_3, a_4, a_5 ]
```

像素点为 [R, G, B, A]，
则过滤后的颜色为 [R', G', B', A']，计算公式为：

```
R' = r_1*R + r_2*G + r_3*B + r_4*A + r_5
G' = g_1*R + g_2*G + g_3*B + g_4*A + g_5
B' = b_1*R + b_2*G + b_3*B + b_4*A + b_5
A' = a_1*R + a_2*G + a_3*B + a_4*A + a_5
```

从 API version 9 开始，该接口支持在 ArkTS 卡片中使用。

（14）draggable：设置组件默认拖曳效果，设置为 true 时，组件可拖曳。注意，不能和拖曳事件同时使用。默认值为 false。

2. 图像插值接口

从 API version 9 开始，可以在 ArkTS 卡片中使用图像插值（ImageInterpolation）接口。图像插值主要用于调整图像的大小或者在图像处理中进行平滑处理。ImageInterpolation 的取值级别说明如下。

- 无（None）：不使用图像插值。这意味着在图像处理过程中，不进行额外的插值操作，直接使用原始图像的像素值。这样做可能会导致在图像调整大小或缩放时出现锯齿状边缘。
- 高（High）：高级别的图像插值。在这个级别上，插值质量最高，通常会使用更复杂的算法来确保图像调整大小或者其他处理操作时能够保持较高的质量。然而，这可能会影响到图片渲染的速度。
- 中（Medium）：中级别的图像插值。在质量和性能之间寻找一种平衡，适用于一般的图像处理需求。
- 低（Low）：低级别的图像插值。在这个级别上，可能采用更简单的插值方法，以提高处理速度，但可能会在图像质量上产生一些损失。

选择适当的插值级别通常取决于具体的应用场景和性能要求。高级别的插值可以提供更好的图像质量，但可能会牺牲一些性能。在特定情况下，根据应用的需求和目标，开发者可以选择适当的插值级别。

3. 渲染模式接口

从 API 9 开始，可以在 ArkTS 卡片中使用渲染模式（ImageRenderMode）接口。该接口定义了两种渲染模式：原色渲染模式（Original）和 黑白渲染模式（Template）。

6.1.2 Image 组件实战：手机相册系统

在实例中，基于 ArkTS 技术实现了一个手机相册系统。本项目是一个基于鸿蒙操作系统的电子相册应用程序，旨在为用户提供直观、流畅的图片浏览和管理体验。应用具备多项功能，包括主界面展示、图片列表、详细图片展示等，通过巧妙的页面设计和手势操作，使用户能够轻松地浏览、分享和管理个人相册。

本项目的主要功能如下。

- 主界面展示：通过精心设计的主界面，呈现清晰的分类标签和图片预览，让用户能够迅速定位感兴趣的图片集。
- 图片列表和详细展示：支持水平滚动的图片列表，用户可随心所欲地浏览多组图片。详细展示图片页面则提供了手势缩放、拖曳等功能，使用户能够更近距离地查看和管理图片。
- 智能导航和交互体验：利用鸿蒙操作系统的强大功能，项目充分利用导航模块和手势控制，提供直观的交互体验。用户可通过滑动、点击和手势缩放等方式实现快速导航和详细浏览。
- 应用性能和适配：项目充分考虑不同设备类型，通过适配和性能优化，确保在各种设备上都能够流畅运行，为用户提供一致的使用体验。

总之，本手机相册系统旨在为用户提供一种轻松、高效的图片浏览和管理方式，使用户能够更好地组织、回顾和分享自己的珍贵瞬间。

实例 6-1：手机相册系统（源码路径 :codes\6\Album）

（1）编写文件 src/main/module.json5，这是一个典型的鸿蒙项目 module.json5 文件，它设置了应用程序的模块信息。这个文件主要定义了应用程序的入口模块，设置入口文件的路径是"./ets/entryability/EntryAbility.ts"。

扫码看视频

```
{
  "module": {
    "name": "entry",
    "type": "entry",
    "description": "$string:module_desc",
    "mainElement": "EntryAbility",
    "deviceTypes": [
      "phone"
    ],
    "deliveryWithInstall": true,
    "installationFree": false,
    "pages": "$profile:main_pages",
    "abilities": [
      {
        "name": "EntryAbility",
        "srcEntry": "./ets/entryability/EntryAbility.ts",
        "description": "$string:EntryAbility_desc",
        "icon": "$media:icon",
        "label": "$string:EntryAbility_label",
        "startWindowIcon": "$media:icon",
        "startWindowBackground": "$color:start_window_background",
        "exported": true,
        "skills": [
          {
            "entities": [
              "entity.system.home"
            ],
            "actions": [
              "action.system.home"
            ]
          }
        ]
      }
```

```
                ]
            }
        ]
    }
}
```

（2）在 src/main/resources/base/media 目录中准备需要的素材照片，这些照片将作为在手机相册中显示的图片，如图 6-1 所示。

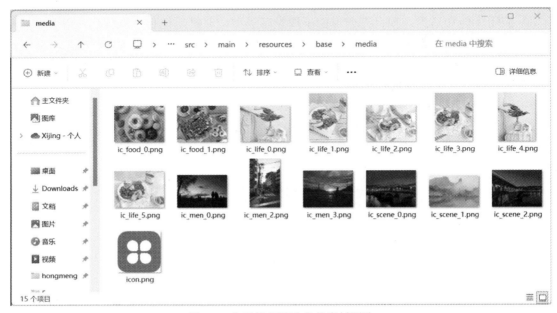

图 6-1　为手机相册准备的素材图片

（3）编写文件 src/main/ets/common/constants/Constants.ets，这是一个鸿蒙项目中的 TypeScript 类，定义项目中需要的一些常量和数组。这些常量和数值用于定义页面布局、动画、图片资源等，通过在整个应用程序中引用这些常量，可以更容易地进行调整和维护。

（4）编写文件 src/main/ets/entryability/EntryAbility.ts，定义了一个鸿蒙应用程序 EntryAbility 类，这个类继承自 UIAbility 类。EntryAbility 类主要定义了在应用程序的生命周期中不同阶段需要执行的操作，包括创建、销毁、窗口阶段的处理以及前台和后台的切换。hilog 被用来记录日志，方便开发者调试和了解应用程序的运行状态。

```
import UIAbility from '@ohos.app.ability.UIAbility';
import hilog from '@ohos.hilog';
import Window from '@ohos.window';

export default class EntryAbility extends UIAbility {
    onCreate(want, launchParam) {
        hilog.isLoggable(0x0000, 'testTag', hilog.LogLevel.INFO);
        hilog.info(0x0000, 'testTag', '%{public}s', 'Ability onCreate');
        hilog.info(0x0000, 'testTag', '%{public}s', 'want 参数: ' + JSON.stringify(want) ?? '');
        hilog.info(0x0000, 'testTag', '%{public}s', 'launchParam: ' + JSON.stringify(launchParam) ?? '');
```

```
    }
    onDestroy() {
        hilog.isLoggable(0x0000, 'testTag', hilog.LogLevel.INFO);
        hilog.info(0x0000, 'testTag', '%{public}s', 'Ability onDestroy');
    }

    onWindowStageCreate(windowStage: Window.WindowStage) {
        // 主窗口创建，为该能力设置主页
        hilog.isLoggable(0x0000, 'testTag', hilog.LogLevel.INFO);
        hilog.info(0x0000, 'testTag', '%{public}s', 'Ability onWindowStageCreate');

        windowStage.loadContent('pages/SuoPage', (err, data) => {
            if (err.code) {
                hilog.isLoggable(0x0000, 'testTag', hilog.LogLevel.ERROR);
                hilog.error(0x0000, 'testTag', '加载内容失败。原因：%{public}s', JSON.stringify(err) ?? '');
                return;
            }
            hilog.isLoggable(0x0000, 'testTag', hilog.LogLevel.INFO);
            hilog.info(0x0000, 'testTag', '成功加载内容。数据：%{public}s', JSON.stringify(data) ?? '');
        });
    }

    onWindowStageDestroy() {
        // 主窗口销毁，释放与 UI 相关的资源
        hilog.isLoggable(0x0000, 'testTag', hilog.LogLevel.INFO);
        hilog.info(0x0000, 'testTag', '%{public}s', 'Ability onWindowStageDestroy');
    }

    onForeground() {
        // 能力进入前台
        hilog.isLoggable(0x0000, 'testTag', hilog.LogLevel.INFO);
        hilog.info(0x0000, 'testTag', '%{public}s', 'Ability onForeground');
    }

    onBackground() {
        // 能力进入后台
        hilog.isLoggable(0x0000, 'testTag', hilog.LogLevel.INFO);
        hilog.info(0x0000, 'testTag', '%{public}s', 'Ability onBackground');
    }
}
```

对上述代码的具体说明如下。

- EntryAbility 类：继承自 UIAbility 类，表示应用程序的入口能力。
- onCreate 方法：当 EntryAbility 被创建时调用此方法，使用 hilog 记录，包括 want 和 launchParam 的信息。
- onDestroy 方法：当 EntryAbility 销毁时调用此方法，使用 hilog 记录信息。
- onWindowStageCreate 方法：当窗口阶段创建时调用此方法，加载名为 'pages/

SuoPage' 的内容，并在回调中处理加载结果。
- onWindowStageDestroy 方法：当窗口阶段销毁时调用此方法。
- onForeground 方法：当 EntryAbility 进入前台时调用此方法，使用 hilog 记录信息。
- onBackground 方法：当 EntryAbility 进入后台时调用此方法，使用 hilog 记录信息。

（5）编写文件 src/main/ets/pages/SuoPage.ets，定义了一个鸿蒙应用程序的入口页面 SuoPage，包含轮播图、网格布局和标题栏。通过导入模块和使用注解，实现了页面的布局和用户交互逻辑。页面中的图片资源和跳转链接通过常量和路由进行管理。

```
import router from '@ohos.router';
import Constants from '../common/constants/Constants';
import PhotoItem from '../view/PhotoItem';

@Entry
@Component
struct SuoPage {
  swiperController: SwiperController = new SwiperController();
  scroller: Scroller = new Scroller();
  @State currentIndex: number = 0;
  @State angle: number = 0;

  build() {
    Column() {
      Row() {
        Text($r('app.string.EntryAbility_label'))
          .fontSize($r('app.float.title_font_size'))
          .fontWeight(Constants.TITLE_FONT_WEIGHT)
      }
      .height($r('app.float.navi_bar_height'))
      .alignItems(VerticalAlign.Center)
      .justifyContent(FlexAlign.Start)
      .margin({ top: $r('app.float.grid_padding') })
      .padding({ left: $r('app.float.title_padding') })
      .width(Constants.FULL_PERCENT)

      Swiper(this.swiperController) {
        ForEach(Constants.BANNER_IMG_LIST, (item: Resource) => {
          Row() {
            Image(item)
              .width(Constants.FULL_PERCENT)
              .height(Constants.FULL_PERCENT)
          }
          .width(Constants.FULL_PERCENT)
          .aspectRatio(Constants.BANNER_ASPECT_RATIO)
        }, (item: Resource, index?: number) => JSON.stringify(item) + index)
      }
      .autoPlay(true)
      .loop(true)
      .margin($r('app.float.grid_padding'))
      .borderRadius($r('app.float.img_border_radius'))
      .clip(true)
```

```
      .duration(Constants.BANNER_ANIMATE_DURATION)
      .indicator(false)

    Grid() {
      ForEach(Constants.IMG_ARR, (photoArr: Array<Resource>) => {
        GridItem() {
          PhotoItem({ photoArr })
        }
        .width(Constants.FULL_PERCENT)
        .aspectRatio(Constants.STACK_IMG_RATIO)
        .onClick(() => {
          router.pushUrl({
            url: Constants.URL_LIST_PAGE,
            params: { photoArr: photoArr }
          });
        })
      }, (item: Array<Resource>, index?: number) => JSON.stringify(item) + index)
    }
    .columnsTemplate(Constants.INDEX_COLUMNS_TEMPLATE)
    .columnsGap($r('app.float.grid_padding'))
    .rowsGap($r('app.float.grid_padding'))
    .padding({ left: $r('app.float.grid_padding'), right: $r('app.float.grid_padding') })
    .width(Constants.FULL_PERCENT)
    .layoutWeight(1)
  }
  .width(Constants.FULL_PERCENT)
  .height(Constants.FULL_PERCENT)
 }
}
```

上述代码的实现流程如下。

- 首先，代码导入了鸿蒙框架的模块，包括路由（router）、常量定义（Constants）和自定义的照片项组件（PhotoItem）。
- 其次，定义了一个名为 SuoPage 的结构体，使用 @Entry 和 @Component 注解，表明它是一个入口页面且是一个组件。结构体内部包含状态如当前索引（currentIndex）和角度（angle），以及控制器实例，如轮播控制器（SwiperController）和滚动控制器（Scroller）。
- 再次，结构体 SuoPage 定义了一个 build 方法，用于构建页面布局，创建了一个包含标题、轮播图和网格布局的垂直列（Column）。
- 复次，通过 Row 和 Text 创建了页面标题栏，并设置了字体大小、字体粗细等样式属性。使用 Swiper 创建了轮播图，加载了 Constants.BANNER_IMG_LIST 中的图片资源，并设置了自动播放、循环等属性。
- 最后，通过 Grid 创建了一个网格布局，遍历 Constants.

图 6-2　入口页面

IMG_ARR 中的图片数组，并使用 GridItem 和自定义的 PhotoItem 组件展示图片。每个网格项设置了单击事件，点击时通过路由跳转到指定的列表页面。

执行效果，如图 6-2 所示。

（6）编写文件 src/main/ets/pages/ListPage.ets，定义了一个名为 ListPage 的鸿蒙页面，作为应用程序的一个入口页面。页面包含导航栏、网格布局和图片列表。

总之，整个代码实现了一个简单的图片列表页面，用户可以点击图片查看详细列表信息。执行效果如图 6-3 所示。

图 6-3　图片列表页面

（7）编写文件 src/main/ets/pages/XlistPage.ets，定义了一个名为 XListPage 的鸿蒙页面，包含两个水平滚动的图片列表和底部导航栏。页面通过导入显示、路由模块和常量定义，实现了图片列表的联动显示和手势缩放功能。页面展示了小图和大图的交互效果，用户可以水平滚动查看图片的详细信息。

```
import display from '@ohos.display';
import router from '@ohos.router';
import Constants from '../common/constants/Constants';

enum scrollTypeEnum {
  STOP = 'onScrollStop',
  SCROLL = 'onScroll'
};

@Entry
@Component
struct XListPage {
  private smallScroller: Scroller = new Scroller();
  private bigScroller: Scroller = new Scroller();
  @State deviceWidth: number = Constants.DEFAULT_WIDTH;
  @State smallImgWidth: number = (this.deviceWidth - Constants.LIST_ITEM_SPACE
* (Constants.SHOW_COUNT - 1)) /
    Constants.SHOW_COUNT;
  @State imageWidth: number = this.deviceWidth + this.smallImgWidth;
```

```
    private photoArr: Array<ResourceStr> = (router.getParams() as
Record<string, Array<ResourceStr>>)[`${Constants.PARAM_PHOTO_ARR_KEY}`];
    private smallPhotoArr: Array<ResourceStr> = new Array<ResourceStr>().
concat(Constants.CACHE_IMG_LIST,
      (router.getParams() as Record<string, Array<ResourceStr>>)[`${Constants.
PARAM_PHOTO_ARR_KEY}`],
      Constants.CACHE_IMG_LIST)
    @StorageLink('selectedIndex') selectedIndex: number = 0;

    @Builder
    SmallImgItemBuilder(img: Resource, index?: number) {
      if (index && index > (Constants.CACHE_IMG_SIZE - 1) && index < (this.
smallPhotoArr.length - Constants.CACHE_IMG_SIZE)) {
        Image(img)
          .onClick(() => this.smallImgClickAction(index))
      }
    }

    aboutToAppear() {
      let displayClass: display.Display = display.getDefaultDisplaySync();
      let width = displayClass?.width / displayClass.densityPixels ?? Constants.
DEFAULT_WIDTH;
      this.deviceWidth = width;
      this.smallImgWidth = (width - Constants.LIST_ITEM_SPACE * (Constants.
SHOW_COUNT - 1)) / Constants.SHOW_COUNT;
      this.imageWidth = this.deviceWidth + this.smallImgWidth;
    }

    onPageShow() {
      this.smallScroller.scrollToIndex(this.selectedIndex);
      this.bigScroller.scrollToIndex(this.selectedIndex);
    }

    goXPage(): void {
      router.pushUrl({
        url: Constants.URL_DETAIL_PAGE,
        params: { photoArr: this.photoArr }
      });
    }

    smallImgClickAction(index: number): void {
      this.selectedIndex = index - Constants.CACHE_IMG_SIZE;
      this.smallScroller.scrollToIndex(this.selectedIndex);
      this.bigScroller.scrollToIndex(this.selectedIndex);
    }

    smallScrollAction(type: scrollTypeEnum): void {
      this.selectedIndex = Math.round(((this.smallScroller.currentOffset().
xOffset as number) +
          this.smallImgWidth / Constants.DOUBLE_NUMBER) / (this.smallImgWidth
+ Constants.LIST_ITEM_SPACE));
      if (type === scrollTypeEnum.SCROLL) {
```

```
            this.bigScroller.scrollTo({ xOffset: this.selectedIndex * this.
imageWidth, yOffset: 0 });
      } else {
            this.smallScroller.scrollTo({ xOffset: this.selectedIndex * this.
smallImgWidth, yOffset: 0 });
      }
    }

    bigScrollAction(type: scrollTypeEnum): void {
      let smallWidth = this.smallImgWidth + Constants.LIST_ITEM_SPACE;
        this.selectedIndex = Math.round(((this.bigScroller.currentOffset().
xOffset as number) +
          smallWidth / Constants.DOUBLE_NUMBER) / this.imageWidth);
      if (type === scrollTypeEnum.SCROLL) {
            this.smallScroller.scrollTo({ xOffset: this.selectedIndex *
smallWidth, yOffset: 0 });
      } else {
            this.bigScroller.scrollTo({ xOffset: this.selectedIndex * this.
imageWidth, yOffset: 0 });
      }
    }

    build() {
      Navigation() {
        Stack({ alignContent: Alignment.Bottom }) {
          List({ scroller: this.bigScroller, initialIndex: this.selectedIndex }) {
            ForEach(this.photoArr, (img: Resource) => {
              ListItem() {
                Image(img)
                  .height(Constants.FULL_PERCENT)
                  .width(Constants.FULL_PERCENT)
                  .objectFit(ImageFit.Contain)
                  .gesture(PinchGesture({ fingers: Constants.DOUBLE_NUMBER })
                    .onActionStart(() => this.goXPage()))
                  .onClick(() => this.goXPage())
              }
              .padding({
                left: this.smallImgWidth / Constants.DOUBLE_NUMBER,
                right: this.smallImgWidth / Constants.DOUBLE_NUMBER
              })
              .width(this.imageWidth)
            }, (item: Resource) => JSON.stringify(item))
          }
          .onScroll((scrollOffset, scrollState) => {
            if (scrollState === ScrollState.Fling) {
              this.bigScrollAction(scrollTypeEnum.SCROLL);
            }
          })
          .onScrollStop(() => this.bigScrollAction(scrollTypeEnum.STOP))
          .width(Constants.FULL_PERCENT)
          .height(Constants.FULL_PERCENT)
          .padding({ bottom: this.smallImgWidth * Constants.DOUBLE_NUMBER })
```

```
          .listDirection(Axis.Horizontal)

          List({
            scroller: this.smallScroller,
            space: Constants.LIST_ITEM_SPACE,
            initialIndex: this.selectedIndex
          }) {
            ForEach(this.smallPhotoArr, (img: Resource, index?: number) => {
              ListItem() {
                this.SmallImgItemBuilder(img, index)
              }
              .width(this.smallImgWidth)
              .aspectRatio(1)
            }, (item: Resource) => JSON.stringify(item))
          }
          .listDirection(Axis.Horizontal)
          .onScroll((scrollOffset, scrollState) => {
            if (scrollState === ScrollState.Fling) {
              this.smallScrollAction(scrollTypeEnum.SCROLL);
            }
          })
          .onScrollStop(() => this.smallScrollAction(scrollTypeEnum.STOP))
          .margin({ top: $r('app.float.detail_list_margin'), bottom: $r ('app.float.detail_list_margin') })
          .height(this.smallImgWidth)
          .width(Constants.FULL_PERCENT)
        }
        .width(this.imageWidth)
        .height(Constants.FULL_PERCENT)
      }
      .title(Constants.PAGE_TITLE)
      .hideBackButton(false)
      .titleMode(NavigationTitleMode.Mini)
    }
  }
```

上述代码的实现流程如下。

- 首先，定义了一个名为 **XListPage** 的鸿蒙页面，该页面包含两个横向滚动的图片列表和底部导航栏。页面导入了显示（display）、路由模块和常量定义，并使用了枚举和状态变量。
- 其次，在页面结构体中，通过 @Entry 和 @Component 注解标明它是一个入口页面和组件。使用了两个滚动控制器 smallScroller 和 bigScroller 以及一些状态变量来管理页面布局和状态。
- 最后，在 build 方法中，通过 Navigation 和 Stack 创建了页面的导航栏和图片列表。页面包含两个水平方向滚动的列表，通过 List 和 ForEach 遍历图片数组，展示了大图和小图。设置了滚动事件和单击事件，实现了大图和小图的联动。

总之，上述代码实现了一个水平滚动的图片详情列表页面，用户可以通过滑动或点击图片进行交互，并支持手势缩放查看大图。执行效果，如图 6-4 所示。

（8）编写文件 src/main/ets/pages/XPage.ets，定义了一个名为 XPage 的鸿蒙页面，实现了水平滚动查看详细图片的功能。页面中包含了水平滚动的大图列表功能，也包含了通过手势缩放查看图片详细介绍的功能。通过路由获取图片数组，实现了大图的滚动浏览、手势缩放和交互功能。

执行效果，如图 6-5 所示。

图 6-4　图片详情列表页面

图 6-5　水平滚动查看详细图片

6.2　绘制几何图形

在 HarmonyOS 系统中，绘制几何图形（Shape）组件的功能是在页面中绘制图形。Shape 是绘制组件的父组件，内置了所有绘制组件均支持的通用属性。

6.2.1　Shape 基础

在 HarmonyOS 系统中，创建绘制组件有以下两种形式。
- 绘制组件使用 Shape 作为父组件，实现类似 SVG 的效果。此时调用接口 Shape 的语法格式如下。

```
Shape(value?: PixelMap)
```

接口 Shape 用于创建带有父组件的绘制组件，其中 value 用于设置绘制目标，可以将图形绘制在指定的 PixelMap 对象中，若未设置，则在当前绘制目标中进行绘制。例如，下面的代码创建了一个矩形（Rect），并设置了其宽度为 300，高度为 50。

```
Shape() {
```

```
      Rect().width(300).height(50)
}
```

- 单独使用绘制组件，用于在页面上绘制指定的图形。Shape 有 7 种绘制类型，分别为 Circle（圆形）、Ellipse（椭圆形）、Line（直线）、Polyline（折线）、Polygon（多边形）、Path（路径）、Rect（矩形）。例如，下面是绘制 Circle（圆形）的语法格式：

```
Circle(options?: {width?: string | number, height?: string | number}
```

该接口用于在页面绘制圆形，其中 width 用于设置圆形外接矩形的宽度，height 用于设置圆形外接矩形的高度，圆形直径由外接矩形的宽、高最小值确定。例如，下面的代码绘制了一个半径为 150 的圆形。

```
Circle({ width: 150, height: 150 })
```

1. 形状视口

在 HarmonyOS 系统中，形状视口（viewport）的功能是指定用户空间中的一个矩形区域，该区域映射到与关联的 SVG 元素建立的视区边界。其语法格式如下。

```
viewPort{ x?: number | string, y?: number | string, width?: number | string, height?: number | string }
```

在上述格式中，属性 viewport 的值包含 x、y、width 和 height 四个可选参数，其中 x 和 y 表示视区的左上角坐标，width 和 height 表示其尺寸。

2. 自定义样式

- 绘制组件 Shape 支持通过各种属性对组件样式进行更改。例如，在下面的代码中，通过 fill 设置了组件填充区域的颜色为 #E87361（橙色）。

```
Path()
  .width(100)
  .height(100)
  .commands('M150 0 L300 300 L0 300 Z')
  .fill("#E87361")
```

- 可以通过 fill 设置组件填充区域颜色，演示代码如下。

```
Path()
  .width(100)
  .height(100)
  .commands('M150 0 L300 300 L0 300 Z')
  .fill("#E87361")
```

- 可以通过 stroke 设置组件的边框颜色，演示代码如下。

```
Path()
  .width(100)
  .height(100)
  .fillOpacity(0)
  .commands('M150 0 L300 300 L0 300 Z')
  .stroke(Color.Red)
```

- 可以通过 strokeOpacity 设置边框的透明度，演示代码如下。

```
Path()
  .width(100)
  .height(100)
  .fillOpacity(0)
  .commands('M150 0 L300 300 L0 300 Z')
  .stroke(Color.Red)
  .strokeWidth(10)
  .strokeOpacity(0.2)
```

- 通过 strokeLineJoin 可以设置线条拐角绘制样式。拐角绘制样式分为 Bevel（使用斜角连接路径段）、Miter（使用尖角连接路径段）、Round（使用圆角连接路径段）。例如，下面的代码创建了一个折线，宽度和高度均为 100，填充透明度为 0，描边颜色为红色，描边宽度为 8，折线经过三个点（[20, 0]，[0, 100]，[100, 90]）。同时，设置折线拐角处为圆弧，使连接处呈现圆润效果。

```
Polyline()
  .width(100)
  .height(100)
  .fillOpacity(0)
  .stroke(Color.Red)
  .strokeWidth(8)
  .points([[20, 0], [0, 100], [100, 90]])
  // 设置折线拐角处为圆弧
  .strokeLineJoin(LineJoinStyle.Round)
```

- 通过 strokeMiterLimit 设置斜接长度与边框宽度比值的极限值。

斜接长度表示外边框外边交点到内边交点的距离，边框宽度即 strokeWidth 属性的值。strokeMiterLimit 取值需大于等于 1，且在 strokeLineJoin 属性取值 LineJoinStyle.Miter 时生效。例如，下面的代码设置了两个 Polyline（折线）组件的样式和形状，虽然这两个 Polyline 具有相似的属性，但是通过调整 strokeMiterLimit 和其他参数，可以影响拐角的形状和锐利度。

```
Polyline()
  .width(100)
  .height(100)
  .fillOpacity(0)
  .stroke(Color.Red)
  .strokeWidth(10)
  .points([[20, 0], [20, 100], [100, 100]])
  // 设置折线拐角处为尖角
  .strokeLineJoin(LineJoinStyle.Miter)
  // 设置斜接长度与线宽的比值
  .strokeMiterLimit(1/Math.sin(45))
Polyline()
  .width(100)
  .height(100)
  .fillOpacity(0)
  .stroke(Color.Red)
```

```
      .strokeWidth(10)
      .points([[20, 0], [20, 100], [100, 100]])
      .strokeLineJoin(LineJoinStyle.Miter)
      .strokeMiterLimit(1.42)
```

- 通过 antiAlias 设置是否开启抗锯齿，默认值为 true（开启抗锯齿）。例如，在下面的代码中，在绘制圆时开启了抗锯齿。

```
// 开启抗锯齿
Circle()
  .width(150)
  .height(200)
  .fillOpacity(0)
  .strokeWidth(5)
  .stroke(Color.Black)
```

6.2.2 Shape 实战：绘制各种各样的图形

实例 6-2 演示了使用 Shape 组件绘制各种样式图形的过程。首先，创建了一个包含基本形状（矩形、椭圆、直线路径）的界面，其中各种形状具有不同的边框、颜色和样式。其次，在例子中使用 Shape 组件绘制了不同位置和样式的矩形、椭圆和直线，同时演示了视口的使用来调整形状的位置和大小。最后，展示了调整线条样式的方法，包括线宽、颜色、间隙、两端样式及拐角样式，同时还演示了开启抗锯齿和透明度功能的用法。

实例 6-2：绘制各种各样的图形（源码路径 :codes\6\Shape）

编写文件 src/main/ets/pages/Index.ets，功能是使用 Shape 组件绘制各种样式的不同类型图形，具体实现代码如下。

扫码看视频

```
@Entry
@Component
struct ShapeExample {
  build() {
    Column({ space: 10 }) {
      Text('basic').fontSize(11).fontColor(0xCCCCCC).width(320)
      // 在 Shape 的 (-2, -2) 点绘制一个 300 * 50 带边框的矩形，颜色 0x317AF7，边
框颜色黑色，边框宽度4，边框间隙20，向左偏移10，线条两端样式为半圆，拐角样式圆角，抗锯齿（默
认开启）
      // 在 Shape 的 (-2, 58) 点绘制一个 300 * 50 带边框的椭圆，颜色 0x317AF7，边
框颜色黑色，边框宽度4，边框间隙20，向左偏移10，线条两端样式为半圆，拐角样式圆角，抗锯齿（默
认开启）
      // 在 Shape 的 (-2, 118) 点绘制一个 300 * 10 直线路径，颜色 0x317AF7，边框
颜色黑色，宽度4，间隙20，向左偏移10，线条两端样式为半圆，拐角样式圆角，抗锯齿（默认开启）
      Shape() {
        Rect().width(300).height(50)
        Ellipse().width(300).height(50).offset({ x: 0, y: 60 })
        Path().width(300).height(10).commands('M0 0 L900 0').offset({ x: 0,
y: 120 })
      }
      .viewPort({ x: -2, y: -2, width: 304, height: 130 })
      .fill(0x317AF7)
```

```
      .stroke(Color.Black)
      .strokeWidth(4)
      .strokeDashArray([20])
      .strokeDashOffset(10)
      .strokeLineCap(LineCapStyle.Round)
      .strokeLineJoin(LineJoinStyle.Round)
      .antiAlias(true)
      // 分别在 Shape 的 (0, 0)、(-5, -5) 点绘制一个 300 × 50 带边框的矩形，可以
看出之所以将视口的起始位置坐标设为负值，是因为绘制的起点默认为线宽的中点位置，因此要让边框
完全显示，则需要让视口偏移半个线宽
      Shape() {
        Rect().width(300).height(50)
      }
      .viewPort({ x: 0, y: 0, width: 320, height: 70 })
      .fill(0x317AF7)
      .stroke(Color.Black)
      .strokeWidth(10)

      Shape() {
        Rect().width(300).height(50)
      }
      .viewPort({ x: -5, y: -5, width: 320, height: 70 })
      .fill(0x317AF7)
      .stroke(Color.Black)
      .strokeWidth(10)

      Text('path').fontSize(11).fontColor(0xCCCCCC).width(320)
      // 在 Shape 的 (0, -5) 点绘制一条直线路径，颜色为 0xEE8443，线条宽度为 10，线条间
隙为 20
      Shape() {
        Path().width(300).height(10).commands('M0 0 L900 0')
      }
      .viewPort({ x: 0, y: -5, width: 300, height: 20 })
      .stroke(0xEE8443)
      .strokeWidth(10)
      .strokeDashArray([20])
      // 在 Shape 的 (0, -5) 点绘制一条直线路径，颜色为 0xEE8443，线条宽度为 10，线
条间隙为 20，向左偏移为 10
      Shape() {
        Path().width(300).height(10).commands('M0 0 L900 0')
      }
      .viewPort({ x: 0, y: -5, width: 300, height: 20 })
      .stroke(0xEE8443)
      .strokeWidth(10)
      .strokeDashArray([20])
      .strokeDashOffset(10)
      // 在 Shape 的 (0, -5) 点绘制一条直线路径，颜色：0xEE8443，线条宽度为 10，透
明度为 0.5
      Shape() {
        Path().width(300).height(10).commands('M0 0 L900 0')
      }
      .viewPort({ x: 0, y: -5, width: 300, height: 20 })
```

```
          .stroke(0xEE8443)
          .strokeWidth(10)
          .strokeOpacity(0.5)
          // 在 Shape 的 (0, -5) 点绘制一条直线路径，颜色为 0xEE8443，线条宽度为 10，线
条间隙为 20，线条两端样式为半圆
        Shape() {
          Path().width(300).height(10).commands('M0 0 L900 0')
        }
        .viewPort({ x: 0, y: -5, width: 300, height: 20 })
        .stroke(0xEE8443)
        .strokeWidth(10)
        .strokeDashArray([20])
        .strokeLineCap(LineCapStyle.Round)
          // 在 Shape 的 (-80, -5) 点绘制一个封闭路径，颜色为 0x317AF7，线条宽度为 10，
边框颜色为 0xEE8443，拐角样式为锐角（默认值）
        Shape() {
          Path().width(200).height(60).commands('M0 0 L400 0 L400 150 Z')
        }
        .viewPort({ x: -80, y: -5, width: 310, height: 90 })
        .fill(0x317AF7)
        .stroke(0xEE8443)
        .strokeWidth(10)
        .strokeLineJoin(LineJoinStyle.Miter)
        .strokeMiterLimit(5)
    }.width('100%').margin({ top: 15 })
  }
}
```

上述代码的实现流程如下。

- 首先，定义了一个名为 ShapeExample 的入口组件，通过 Shape 组件嵌套使用 Rect、Ellipse 和 Path 组件来绘制矩形、椭圆和直线。
- 其次，通过设置不同的视口和样式属性，调整了形状的位置、大小和样式。例如，通过设置 viewPort 调整了形状的起始点，通过 fill 和 stroke 设置了填充和描边颜色，通过 strokeWidth 设置了线宽，通过 strokeDashArray 设置了虚线的间隙，通过 strokeDashOffset 调整了虚线的起始位置。
- 最后，通过多个 Shape 组件的嵌套，展示了在不同位置和样式下创建多个形状的能力，包括封闭路径的绘制，以及对拐角样式的调整。

执行效果，如图 6-6 所示。

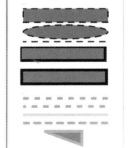

图 6-6 绘制各种样式的图形

6.3 画布

在 HarmonyOS 系统中，画布（Canvas）组件用于绘制自定义图形。开发者可以通过使用 CanvasRenderingContext2D 对象和 OffscreenCanvasRenderingContext2D 对象，在 Canvas 组件上绘制不同样式、不同类型的图像。绘制对象可以是基础形状、文本、图片等。

6.3.1 Canvas 绘制自定义图形

在 HarmonyOS 系统中,可以通过以下三种形式在画布上绘制自定义图形。

(1)使用 CanvasRenderingContext2D 对象在 Canvas 画布上绘图:例如,在下面的代码中,通过配置 CanvasRenderingContext2D 对象的参数,包括设置抗锯齿参数(RenderingContextSettings(true)),在 Canvas 上绘制了一个具有背景色的边框矩形。

```
@Entry
@Component
struct CanvasExample1 {
  // 用来配置 CanvasRenderingContext2D 对象的参数,包括是否开启抗锯齿,true 表明开启抗锯齿。
  private settings: RenderingContextSettings = new RenderingContextSettings(true)
  // 用来创建 CanvasRenderingContext2D 对象,通过在 canvas 中调用 CanvasRenderingContext2D 对象来绘制。
  private context: CanvasRenderingContext2D = new CanvasRenderingContext2D(this.settings)

  build() {
    Flex({ direction: FlexDirection.Column, alignItems: ItemAlign.Center, justifyContent: FlexAlign.Center }) {
      // 在 canvas 中调用 CanvasRenderingContext2D 对象。
      Canvas(this.context)
        .width('100%')
        .height('100%')
        .backgroundColor('#F5DC62')
        .onReady(() => {
          // 可以在这里绘制内容。
          this.context.strokeRect(50, 50, 200, 150);
        })
    }
    .width('100%')
    .height('100%')
  }
}
```

(2)离屏绘制:离屏绘制是指将需要绘制的内容先绘制在缓存区,再将其转换成图片,一次性绘制到 Canvas 上,从而加快了绘制速度。具体绘制过程如下。

- 通过 transferToImageBitmap 方法将离屏画布最近渲染的图像创建为一个 ImageBitmap 对象。
- 通过 CanvasRenderingContext2D 对象的 transferFromImageBitmap 方法显示给定的 ImageBitmap 对象。

例如,在以下代码中,通过配置 CanvasRenderingContext2D 和 OffscreenCanvasRenderingContext2D 的参数,绘制了一个边框带有背景颜色的矩形,然后将离屏绘制的图像显示在画布中。

```
@Entry
@Component
struct CanvasExample2 {
```

```
    // 用来配置 CanvasRenderingContext2D 对象和 OffscreenCanvasRenderingContext
2D 对象的参数，包括是否开启抗锯齿。true 表明开启抗锯齿
    private settings: RenderingContextSettings = new RenderingContextSettings(true)
    private context: CanvasRenderingContext2D = new CanvasRenderingContext2D
(this.settings)
    // 用来创建 OffscreenCanvasRenderingContext2D 对象，width 为离屏画布的宽度，
height 为离屏画布的高度。通过在 canvas 中调用 OffscreenCanvasRenderingContext2D 对象
来绘制。
    private offContext: OffscreenCanvasRenderingContext2D = new OffscreenC
anvasRenderingContext2D(600, 600, this.settings)

    build() {
        Flex({ direction: FlexDirection.Column, alignItems: ItemAlign.
Center, justifyContent: FlexAlign.Center }) {
            Canvas(this.context)
              .width('100%')
              .height('100%')
              .backgroundColor('#F5DC62')
              .onReady(() =>{
                // 可以在这里绘制内容
                this.offContext.strokeRect(50, 50, 200, 150);
                // 将离屏绘制的图像在普通画布上显示
                let image = this.offContext.transferToImageBitmap();
                this.context.transferFromImageBitmap(image);
              })
        }
        .width('100%')
        .height('100%')
    }
}
```

（3）在 Canvas 上加载 Lottie 动画时，需要先按照如下方式下载 Lottie。

```
import lottie from '@ohos/lottie'
```

6.3.2　Canvas 的常用绘图方法

在 HarmonyOS 系统中，OffscreenCanvasRenderingContext2D 对象和 CanvasRenderingContext2D 对象提供了大量的属性和方法，可以用来绘制文本、图形，处理像素等。这些方法是 Canvas 组件的核心。其中常用的绘图接口包括对封闭路径进行填充（fill）、设置当前路径为剪切路径（clip）、进行边框绘制操作（stroke）等。同时提供了指定绘制的填充色（fillStyle）、设置透明度（globalAlpha）与设置描边的颜色（strokeStyle）等属性，用以修改绘制内容的样式。

1. 绘制基础形状

通过绘制弧线路径（arc）、绘制一个椭圆（ellipse）、创建矩形路径（rect）等接口绘制基础形状。例如，在下面的代码中，在 Canvas 上绘制了一个具有背景色的矩形，并通过 onReady 回调中的绘图操作绘制了一个矩形、一个圆形和一个椭圆。矩形的起始点为 (100, 50)，宽度为 100，高度为 100；圆形的中心坐标为 (150, 250)，半径为 50；椭圆的中心坐标

为 (150, 450)，横轴半径为 50，纵轴半径为 100，旋转角度为 Math.PI * 0.25。绘制的基础形状效果，如图 6-7 所示。

```
Canvas(this.context)
    .width('100%')
    .height('100%')
    .backgroundColor('#F5DC62')
    .onReady(() =>{
       // 绘制矩形
       this.context.beginPath();
       this.context.rect(100, 50, 100, 100);
       this.context.stroke();
       // 绘制圆形
       this.context.beginPath();
       this.context.arc(150, 250, 50, 0, 6.28);
       this.context.stroke();
       // 绘制椭圆
       this.context.beginPath();
       this.context.ellipse(150, 450, 50, 100, Math.PI * 0.25, Math.PI * 0, Math.PI * 2);
       this.context.stroke();
    })
```

图 6-7 绘制的基础形状效果

2. 绘制文本

通过 fillText（绘制填充类文本）、strokeText（绘制描边类文本）等接口绘制文本。例如，在下面的代码中，在 Canvas 上绘制了一个具有背景色的文本。通过 onReady 回调中的绘图操作，首先绘制了一行填充类文本"Hello World!"，设置字体大小为 50 像素，起始点坐标为 (50, 100)；然后，绘制了一行描边类文本"Hello World!"，设置字体大小为 55 像素，起始点坐标为 (50, 150)。绘制的文本效果，如图 6-8 所示。

图 6-8 绘制的文本效果

```
Canvas(this.context)
    .width('100%')
    .height('100%')
    .backgroundColor('#F5DC62')
    .onReady(() =>{
       // 绘制填充类文本
       this.context.font = '50px sans-serif';
       this.context.fillText("Hello World!", 50, 100);
       // 绘制描边类文本
       this.context.font = '55px sans-serif';
       this.context.strokeText("Hello World!", 50, 150);
    })
```

3. 绘制图片和处理图像的像素信息

在 HarmonyOS 系统中，通过图像绘制（drawImage）、使用 ImageData 数据填充新的矩形区域（putImageData）等接口绘制图片，通过创建新的 ImageData 对象（createImageData）、以当前 canvas 指定区域内的像素创建 PixelMap 对象（getPixelMap）、以当前 canvas 指定区域内的像素创建 ImageData 对象（getImageData）等接口处理图像的像素信息。例如，在下面的演示代码中，在 Canvas 上绘制了图片，获取了指定区域的图像数据（ImageData），然后将得到的 ImageData 重新绘制在不同的位置上，最终呈现在主 Canvas 上。绘制的图片效果，如图 6-9 所示。

图 6-9　绘制的图片效果

```
@Entry
@Component
struct GetImageData {
  private settings: RenderingContextSettings = new RenderingContextSettings(true)
  private context: CanvasRenderingContext2D = new CanvasRenderingContext2D(this.settings)
  private offContext: OffscreenCanvasRenderingContext2D = new OffscreenCanvasRenderingContext2D(600, 600, this.settings)
  private img:ImageBitmap = new ImageBitmap("/common/images/1234.png")

  build() {
    Flex({ direction: FlexDirection.Column, alignItems: ItemAlign.Center, justifyContent: FlexAlign.Center }) {
      Canvas(this.context)
        .width('100%')
        .height('100%')
        .backgroundColor('#F5DC62')
        .onReady(() =>{
          // 使用 drawImage 接口将图片画在（0, 0）为起点，宽高 130 的区域
          this.offContext.drawImage(this.img,0,0,130,130);
          // 使用 getImageData 接口，获得 canvas 组件区域中，（50, 50）为起点，宽高 130 范围内的绘制内容
          let imagedata = this.offContext.getImageData(50,50,130,130);
          // 使用 putImageData 接口将得到的 ImageData 画在起点为（150, 150）的区域中
          this.offContext.putImageData(imagedata,150,150);
          // 将离屏绘制的内容画到 canvas 组件上
          let image = this.offContext.transferToImageBitmap();
          this.context.transferFromImageBitmap(image);
        })
    }
    .width('100%')
    .height('100%')
  }
}
```

4. 其他绘图方法

除了基本的绘图方法外，Canvas 还提供了其他类型的绘图方法，例如，与渐变相关的方法，包括 createLinearGradient（创建线性渐变）和 createRadialGradient（创建径向渐变）等。例如，在下面的演示代码中，在 Canvas 上绘制了一个具有渐变色的矩形。通过 onReady 回调中的绘图操作，创建了一个渐变色的 CanvasGradient 对象，设置了渐变的起点、终点和颜色断点，然后用该渐变对象填充一个矩形，实现矩形区域内的径向渐变效果。绘制径向渐变图像效果，如图 6-10 所示。

图 6-10　绘制径向渐变图像效果

```
Canvas(this.context)
  .width('100%')
  .height('100%')
  .backgroundColor('#F5DC62')
  .onReady(() =>{
    // 创建一个径向渐变色的 CanvasGradient 对象
    let grad = this.context.createRadialGradient(200,200,50, 200,200,200)
    // 为 CanvasGradient 对象设置渐变断点值，包括偏移和颜色
    grad.addColorStop(0.0, '#E87361');
    grad.addColorStop(0.5, '#FFFFF0');
    grad.addColorStop(1.0, '#BDDB69');
    // 用 CanvasGradient 对象填充矩形
    this.context.fillStyle = grad;
    this.context.fillRect(0, 0, 400, 400);
})
```

6.4 动画

在 HarmonyOS 系统中，实现动画的原理是在一个时间段内多次改变 UI 外观。由于人眼会产生视觉暂留现象，所以最终看到的就是一个"连续"的动画。UI 的一次改变称为一个动画帧，对应一次屏幕刷新，而决定动画流畅度的一个重要指标就是帧率 FPS（Frames Per Second），即每秒的动画帧数。帧率越高，动画就会越流畅。

6.4.1 ArkUI 动画的分类

在 ArkUI 中，产生动画的方式是改变属性值并指定动画参数。动画参数包括动画时长、变化规律（即曲线）等。当属性值发生变化后，按照动画参数从原来的状态过渡到新的状态，即可形成一个动画。

按照页面的分类方式，可以将 ArkUI 中的动画分为页面内的动画和页面间的动画，如图 6-11 所示，页面内的动画指在一个页面内即可发生的动画，页面间的动画指两个页面跳转时才会发生的动画。

如果按照基础能力进行划分，可以将 ArkUI 中的动画分为属性动画、显式动画和转场动画，如图 6-12 所示。

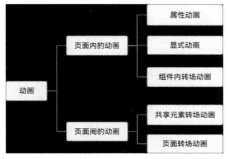

图 6-11　按照页面分类的方式

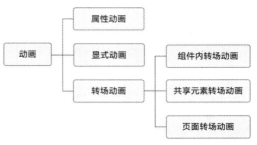

图 6-12　按照基础能力分类的方式

6.4.2 布局更新动画

在 ArkUI 中，显式动画（animateTo）和属性动画（animation）是最常用的动画功能。在布局属性（如尺寸属性、位置属性）发生变化时，可以通过属性动画或显式动画，按照动画参数过渡到新的布局参数状态。这两种动画的具体特点如下。

- 显式动画：闭包内的变化均会触发动画，包括由数据变化引起的组件的增删、组件属性的变化等，可以制作较为复杂的动画。
- 属性动画：动画设置简单，属性变化时自动触发动画。

1. 使用显式动画产生布局更新动画

我们可以使用显式动画产生布局更新动画。使用显式动画接口的语法格式如下：

```
animateTo(value: AnimateParam, event: () => void): void
```

第一个参数指定动画参数，第二个参数为动画的闭包函数。

实例 6-3 是一个使用显式动画产生布局更新动画的例子。当 Column 组件的 alignItems 属性改变后，其子组件的布局位置结果发生变化。只要该属性是在 animateTo 的闭包函数中修改的，那么，由其引起的所有变化都会按照 animateTo 的动画参数执行动画过渡到终点值。

实例 6-3：使用显式动画（源码路径 :codes\6\Xian）

编写文件 src/main/ets/pages/Index.ets，通过按钮点击触发动画，逐步改变嵌套的 Column 容器内按钮的布局方式（alignItems 属性），呈现水平对齐方式在开始、中间和末尾之间的平滑切换效果。

扫码看视频

```
@Entry
@Component
struct LayoutChange {
  // 用于控制 Column 的 alignItems 属性
  @State itemAlign: HorizontalAlign = HorizontalAlign.Start;
  allAlign: HorizontalAlign[] = [HorizontalAlign.Start, HorizontalAlign.Center, HorizontalAlign.End];
  alignIndex: number = 0;

  build() {
    Column() {
      Column({ space: 10 }) {
        Button("1").width(100).height(50)
        Button("2").width(100).height(50)
        Button("3").width(100).height(50)
      }
      .margin(20)
      .alignItems(this.itemAlign)
      .borderWidth(2)
      .width("90%")
      .height(200)

      Button("click").onClick(() => {
        // 动画时长为 1000ms，曲线为 EaseInOut
```

```
                animateTo({ duration: 1000, curve: Curve.EaseInOut }, () => {
                    this.alignIndex = (this.alignIndex + 1) % this.allAlign.length;
                    // 在闭包函数中修改 this.itemAlign 参数，使 Column 容器内部孩子的布局方
式变化，使用动画过渡到新位置
                    this.itemAlign = this.allAlign[this.alignIndex];
                });
            })
        }
        .width("100%")
        .height("100%")
    }
}
```

对上述代码的具体说明如下。
- 使用 Column 嵌套 Button 组件，设置按钮垂直方向排列，并设置了宽度、高度、边距等样式。
- 使用 @State 注解创建状态变量 itemAlign，控制 Column 的 alignItems 属性，初始值为 HorizontalAlign.Start。
- 使用数组 allAlign 保存所有可能的 HorizontalAlign 值，以及利用 alignIndex 记录当前选用的对齐方式的索引。
- 当 click 按钮被点击时，通过 animateTo 函数设置动画，切换 itemAlign 的值，使按钮所在的 Column 容器内部的布局方式发生变化，实现平滑的动画过渡效果。

执行效果如图 6-13 所示。

图 6-13　使用显示动画效果

2. 使用属性动画产生布局更新动画

使用属性动画可以产生布局更新动画。与显式动画不同，属性动画不需要使用闭包，只需要将 animation 属性加在要进行属性动画的组件属性后即可。属性动画的接口为：

```
animation(value: AnimateParam)
```

要指定组件随某个属性值的变化而产生动画，此属性需要加在 animation 属性之前。如果某些属性变化不希望通过 animation 产生属性动画，可以放在 animation 属性之后。实例 6-3 显式动画的示例很容易改成属性动画实现。例如，在下面的实例中，第一个 button 上的 animation 属性，只对写在 animation 之前的 type、width、height 属性生效，而对写在

animation 之后的 backgroundColor、margin 属性无效。运行结果是 width、height 属性会按照 animation 的动画参数执行动画，而 backgroundColor 会直接跳变，不会产生动画。

实例 6-4：使用属性动画产生布局更新动画（源码路径：codes\6\Bu）

编写文件 src/main/ets/pages/Index.ets，通过按钮点击触发属性动画，实现按钮的宽度、高度和背景颜色在不同状态下的平滑过渡。其中，text 按钮的宽高会在点击时根据 flag 的状态切换，而颜色会通过属性动画进行平滑过渡。 area: click me 按钮作为触发器，点击后改变属性值，触发 text 按钮的属性动画。

扫码看视频

```
@Entry
@Component
struct LayoutChange2 {
  @State myWidth: number = 100;
  @State myHeight: number = 50;
  @State flag: boolean = false;
  @State myColor: Color = Color.Blue;

  build() {
    Column({ space: 10 }) {
      Button("text")
        .type(ButtonType.Normal)
        .width(this.myWidth)
        .height(this.myHeight)
        // animation 只对其上面的 type、width、height 属性生效，时长为1000ms，曲线为Ease
        .animation({ duration: 1000, curve: Curve.Ease })
        // animation 对下面的 backgroundColor、margin 属性不生效
        .backgroundColor(this.myColor)
        .margin(20)

      Button("area: click me")
        .fontSize(12)
        .onClick(() => {
          // 改变属性值，配置了属性动画的属性会进行动画过渡
          if (this.flag) {
            this.myWidth = 100;
            this.myHeight = 50;
            this.myColor = Color.Blue;
          } else {
            this.myWidth = 200;
            this.myHeight = 100;
            this.myColor = Color.Pink;
          }
          this.flag = !this.flag;
        })
    }
  }
}
```

执行效果如图 6-14 所示。

图 6-14　使用属性动画效果

6.4.3　组件内转场动画

在 HarmonyOS 系统中，组件的插入和删除过程即为组件本身的转场过程，这个过程的动画称为组件内转场动画。通过组件内转场动画，可以定义组件出现和消失的效果。组件内转场动画的接口为：

```
transition(value: TransitionOptions)
```

在上述格式中，transition 函数的参数 Value 用于定义组件内转场的效果，可以指定平移、透明度、旋转、缩放等单个或组合的转场效果。注意，transition 函数必须与 animateTo 一起使用才能产生组件的转场效果。

1. transition 函数的常见用法

在使用 transition 函数时，type 属性用于指定当前的 transition 动画生效在组件的变化场景，类型为 TransitionType。

（1）组件的插入和删除使用同一个动画效果：当 type 属性为 TransitionType.All 时，表示指定的转场动画生效于组件的所有变化场景（插入和删除）。此时，删除动画和插入动画是相反的过程。例如，在下面的代码中定义了一个 Button 控件，在插入时，组件从 scale 的 x、y 均为 0 的状态变化到 scale 的 x、y 均为 1 的默认状态，以逐渐放大的方式出现。在删除时，组件从 scale 的 x、y 均为 1 的默认状态变化到 scale 的 x、y 均为 0 的状态，逐渐缩小至尺寸为 0。

```
Button()
    .transition({ type: TransitionType.All, scale: { x: 0, y: 0 } })
```

（2）组件的插入和删除使用不同的动画效果：当组件的插入和删除需要实现不同的转场动画效果时，可以调用两次 transition 函数，分别设置 type 属性为 TransitionType.Insert 和 TransitionType.Delete。例如，在下面的代码中定义了一个 Button 控件，在插入时，组件从相对于组件正常布局位置 x 方向平移 200vp、y 方向平移 -200vp 的位置、透明度为 0 的初始状态变化到 x、y 方向平移量为 0、透明度为 1 的默认状态。在删除时，组件从旋转角为 0 的默认状态变化到绕 z 轴旋转 360 度的终止状态。

```
Button()
    .transition({ type: TransitionType.Insert, translate: { x: 200, y: -200 }, opacity: 0 })
    .transition({ type: TransitionType.Delete, rotate: { x: 0, y: 0, z: 1, angle: 360 } })
```

（3）只定义组件的插入或删除其中一种动画效果：当只需要定义组件插入或删除时

的转场动画效果时，只需将过渡效果的类型（type 属性）设置为 TransitionType.Insert 或 TransitionType.Delete。例如，在下面的代码中定义了一个 Button 控件，当该组件被删除时，会从默认位置开始，没有任何平移，然后移到相对于正常布局位置向右平移 200 个视口单位（vp），向下平移 -200 个视口单位（vp）的位置。如果该组件被插入时不会触发转场动画。

```
Button()
  .transition({ type: TransitionType.Delete, translate: { x: 200, y: -200 } })
```

2. if/else 实现组件内转场动画

在 HarmonyOS 系统中，可以使用 if/else 语句控制组件的插入和删除操作。实例 6-5 中，通过按钮点击触发动画和条件语句控制，实现了图片的平滑出现和消失效果，展示了在不同状态下应用不同的过渡效果。

实例 6-5：使用 if/else 语句实现组件内转场动画（源码路径 :codes\6\Tiao）

编写文件 src/main/ets/pages/Index.ets，首先通过 Button 的单击事件控制 if 的条件是否满足，来控制 if 下的 Image 组件是否显示。然后使用 transition 参数以指定组件内转场的具体效果：在插入时加上平移效果，在删除时加上缩放和透明度效果。

扫码看视频

```
@Entry
@Component
struct IfElseTransition {
  @State flag: boolean = true;
  @State show: string = 'show';

  build() {
    Column() {
      Button(this.show).width(80).height(30).margin(30)
        .onClick(() => {
          if (this.flag) {
            this.show = 'hide';
          } else {
            this.show = 'show';
          }

          animateTo({ duration: 1000 }, () => {
            // 动画闭包内控制 Image 组件的出现和消失
            this.flag = !this.flag;
          })
        })
      if (this.flag) {
        // Image 的出现和消失配置为不同的过渡效果
        Image($r('app.media.shui')).width(200).height(200)
          .transition({ type: TransitionType.Insert, translate: { x: 200, y: -200 } })
          .transition({ type: TransitionType.Delete, opacity: 0, scale: { x: 0, y: 0 } })
      }
```

```
        }.height('100%').width('100%')
    }
}
```

执行效果如图 6-15 所示。

3. ForEach 产生组件内转场动画

与 if/else 类似，使用 ForEach 控制数组中的元素个数，可以控制组件的插入和删除操作。要通过 ForEach 产生组件内转场动画，需要以下两个条件。

- ForEach 里的组件配置了 transition 效果。
- 在 animateTo 的闭包中控制组件的插入或删除，即控制数组的元素添加和删除。

实例 6-6 演示了使用 ForEach 语句产生组件内转场动画的过程。

图 6-15 if/else 语句实现组件内转场动画效果

实例 6-6：使用 ForEach 语句产生组件内转场动画（源码路径 :codes\6\ForEach）

编写文件 src/main/ets/pages/Index.ets，通过 ForEach 遍历字符串数组，为每个元素创建带过渡效果的文本组件，包括位移和缩放效果。通过按钮点击触发数组的动态修改，向列表头尾添加元素和删除元素的平滑过渡效果，展示了在不同操作下列表的动态变化效果。

扫码看视频

```
@Entry
@Component
struct ForEachTransition {
  @State numbers: string[] = ["1", "2", "3", "4", "5"]
  startNumber: number = 6;

  build() {
    Column({ space: 10 }) {
      Column() {
        ForEach(this.numbers, (item) => {
          // ForEach 下的直接组件需配置 transition 效果
          Text(item)
            .width(240)
            .height(60)
            .fontSize(18)
            .borderWidth(1)
            .backgroundColor(Color.Orange)
            .textAlign(TextAlign.Center)
            .transition({ type: TransitionType.All, translate: { x: 200 },
scale: { x: 0, y: 0 } })
        }, item => item)
      }
      .margin(10)
      .justifyContent(FlexAlign.Start)
      .alignItems(HorizontalAlign.Center)
      .width("90%")
      .height("70%")
```

```
      Button(' 向头部添加元素 ')
        .fontSize(16)
        .width(160)
        .onClick(() => {
          animateTo({ duration: 1000 }, () => {
            // 往数组头部插入一个元素,导致 ForEach 在头部增加对应的组件
            this.numbers.unshift(this.startNumber.toString());
            this.startNumber++;
          })
        })
      Button(' 向尾部添加元素 ')
        .width(160)
        .fontSize(16)
        .onClick(() => {
          animateTo({ duration: 1000 }, () => {
            // 往数组尾部插入一个元素,导致 ForEach 在尾部增加对应的组件
            this.numbers.push(this.startNumber.toString());
            this.startNumber++;
          })
        })
      Button(' 删除头部元素 ')
        .width(160)
        .fontSize(16)
        .onClick(() => {
          animateTo({ duration: 1000 }, () => {
            // 删除数组的头部元素,导致 ForEach 删除头部的组件
            this.numbers.shift();
          })
        })
      Button(' 删除尾部元素 ')
        .width(160)
        .fontSize(16)
        .onClick(() => {
          animateTo({ duration: 1000 }, () => {
            // 删除数组的尾部元素,导致 ForEach 删除尾部的组件
            this.numbers.pop();
          })
        })
    }
    .width('100%')
    .height('100%')
  }
}
```

对上述代码的具体说明如下。
- 使用 ForEach 遍历字符串数组 numbers 中的元素,为每个元素创建一个带有过渡效果的文本组件。设置了位移和缩放效果。
- 通过按钮点击("向头部添加元素""向尾部添加元素""删除头部元素""删除尾部元素"),利用 animateTo 函数在动画过渡中改变数组的内容,触发 ForEach 动态更新列表,并实现平滑的过渡效果。

- 设置按钮的单击事件，分别在数组头部和尾部添加元素以及删除头部和尾部的元素，演示了列表在动态变化时的平滑过渡效果。执行效果，如图 6-16 所示。

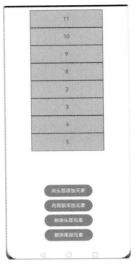

图 6-16　ForEach 语句产生的组件内转场动画效果

第 7 章
多媒体开发

多媒体开发是一个综合性领域,要求开发者具备音频、视频、图形等多个方向的技能。在 HarmonyOS 等分布式操作系统中,多媒体开发与多设备互联互通紧密相关,能够为用户提供一体化的多媒体体验。通过深入了解这一领域的技术和工具,开发者可以创造出引人入胜、功能丰富的多媒体应用,从而丰富用户的数字生活。本章将详细讲解在 HarmonyOS 系统中开发多媒体应用程序的知识。

7.1 HarmonyOS 多媒体开发架构

在实际应用中，多媒体开发涉及处理和展示音频、视频、图形等多种形式的媒体内容，同时充分利用硬件和软件的优势，为用户提供出色的互动性和视听感受。在 HarmonyOS 系统中，多媒体系统提供了用户视觉、听觉信息的处理能力，如音视频信息的采集、压缩存储、解压播放等。在操作系统实现中，通常基于不同的媒体信息处理内容，将媒体划分为不同的模块，包括音频、视频（播放与录制）、图片等。如图 7-1 所示，多媒体系统为开发者提供了音视频应用、图库应用的编程框架接口，为设备开发者提供了对接不同硬件芯片适配加速功能，为中间服务者提供了媒体核心功能和管理机制。

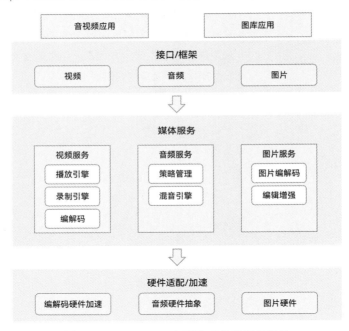

图 7-1 HarmonyOS 多媒体系统的框架结构

HarmonyOS 多媒体系统框架由以下三部分组成。
- 视频（media）：提供音视频解压播放、压缩录制接口与服务。
- 音频（audio）：提供音量管理、音频路由管理、混音管理接口与服务。
- 图片（image）：提供图片编解码、图片处理接口与服务。

7.2 AVPlayer 和 AVRecorder

在 HarmonyOS 系统中，音频和视频功能是由 AVPlayer 和 AVRecorder 接口实现的。本节将详细讲解 AVPlayer 和 AVRecorder 的知识。

7.2.1 AVPlayer

在 HarmonyOS 系统中，AVPlayer 的功能是将 Audio/Video 媒体资源（如 mp4/mp3/mkv/mpeg-ts 等）转码为可供渲染的图像和可听见的音频模拟信号，并通过输出设备进行播放。AVPlayer 提供了一体化的媒体播放能力，应用程序只需要提供多媒体资源，不负责数据解析和解码即可达成播放效果。

1. 音频播放

当使用 AVPlayer 开发音频播放程序时，整个交互过程如图 7-2 所示。

图 7-2　音频播放外部模块交互过程

当音频类应用程序调用 AVPlayer 接口（JS 接口层提供的）实现相应功能时，框架层会通过播放服务（Player Framework）将资源解析成音频数据流（PCM），音频数据流经过软件解码后输出至音频服务（Audio Framework），由音频服务输出至音频驱动渲染，实现音频播放功能。由此可见，完整的音频播放功能需要应用程序、Player Framework、Audio Framework 和音频 HDI 共同实现。

注意：图 7-2 中的数字标记表示数据与外部模块的传递。

（1）音乐应用将媒体资源传递给 AVPlayer 接口。

（2）Player Framework 将音频 PCM 数据流输出给 Audio Framework，再由 Audio Framework 输出给音频 HDI。

2. 视频播放

当使用 AVPlayer 开发视频播放程序时，整个交互过程如图 7-3 所示。

图 7-3　视频播放外部模块交互过程

当 HarmonyOS 应用程序通过调用 AVPlayer 接口（JS 接口层提供的）实现相应功能时，框架层会通过播放服务（Player Framework）解析成单独的音频数据流和视频数据流，音频数据流经过软件解码后输出至音频服务（Audio Framework），再至硬件接口层的音频 HDI，实现音频播放功能。视频数据流经过硬件（推荐）/ 软件解码后输出至图形渲染服务（Graphic Framework），再输出至硬件接口层的显示 HDI，完成图形渲染。由此可见，要实现完整的视频播放功能，需要经过应用程序、XComponent、Player Framework、Graphic Framework、Audio Framework、显示 HDI 和音频 HDI 共同实现。

注意：在图 7-3 中，数字标注表示数据与外部模块的传递。

（1）应用从 XComponent 组件获取窗口 SurfaceID，获取方式参考 XComponent。

（2）应用把媒体资源、SurfaceID 传递给 AVPlayer 接口。

（3）Player Framework 把视频 ES 数据流输出给解码 HDI，解码获得视频帧（NV12/NV21/RGBA）。

（4）Player Framework 把音频 PCM 数据流输出给 Audio Framework，Audio Framework 输出给音频 HDI。

（5）Player Framework 把视频帧（NV12/NV21/RGBA）输出给 Graphic Framework，Graphic Framework 输出给显示 HDI。

7.2.2　AVRecorder

在 HarmonyOS 系统中，AVRecorder 的功能是捕获音频信号和视频信号，完成音视频编码工作并保存到文件中，帮助开发者实现音视频录制功能，包括开始录制、暂停录制、恢复录制、停止录制、释放资源等功能控制。在操作过程中，允许开发者设置录制的编码格式、封装格式、文件路径等参数。整个交互过程如图 7-4 所示。

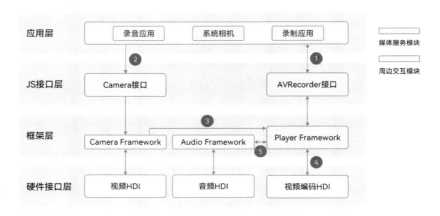

图 7-4　视频录制外部模块交互过程

注意：在图 7-4 中，数字标注表示数据与外部模块的传递。

（1）应用通过 AVRecorder 接口从录制服务获取 SurfaceID。

（2）应用将 SurfaceID 设置给相机服务，相机服务可以通过 SurfaceID 获取到 Surface。相机服务通过视频 HDI 捕获图像数据送至框架层的录制服务。

（3）相机服务通过 Surface 将视频数据传递给录制服务。

（4）录制服务通过视频编码 HDI 模块将视频数据编码。

（5）录制服务将音频参数设置给音频服务，并从音频服务获取到音频数据。

在 HarmonyOS 系统中，通过音、视频录制组合功能，可以分别实现音频录制、视频录制功能，具体说明如下。

- 音频录制：当应用程序通过调用 AVRecorder 接口（JS 接口层提供的）实现音频录制时，框架层会通过录制服务（Player Framework），调用音频服务（Audio Framework）通过音频 HDI 捕获音频数据，通过软件编码封装后保存至文件中，实现音频录制功能。
- 视频录制：当应用程序通过调用 AVRecorder 接口（JS 接口层提供的）实现视频录制时，先通过 Camera 接口调用相机服务（Camera Framework）通过视频 HDI 捕获图像数据送至框架层的录制服务，录制服务将图像数据通过视频编码 HDI 编码，再将编码后的图像数据封装至文件中，实现视频录制功能。

注意：AVRecorder 只负责视频数据的处理，需要与视频数据采集模块配合才能完成视频录制。视频数据采集模块需要通过 Surface 将视频数据传递给 AVRecorder 进行数据处理。当前常用的数据采集模块为相机模块，相关说明以相机为例，相机模块目前仅对系统应用开放。

7.3 音频播放

本节将介绍在 HarmonyOS 系统中开发音频播放程序的知识，指导开发者使用系统提供的音视频 API。例如，使用 AVPlayer 实现音乐播放器，循环播放一首音乐。

7.3.1 使用 AVPlayer 开发音频播放程序

在 HarmonyOS 系统中，可以使用 AVPlayer 实现端到端播放原始媒体资源的功能。整个播放过程包括：创建 AVPlayer，设置播放资源，设置播放参数（音量/倍速/焦点模式），播放控制（播放/暂停/跳转/停止），重置，销毁资源。

使用 AVPlayer 开发音频播放程序的步骤如下。

（1）创建实例 createAVPlayer()，AVPlayer 初始化为 idle 状态。

（2）设置业务需要的监听事件，搭配全流程场景使用。支持的监听事件，如表 7-1 所示。

表 7-1 AVPlayer 音频播放支持的监听事件

事件类型	说明
stateChange	必要事件，监听播放器的 state 属性改变
error	必要事件，监听播放器的错误信息
durationUpdate	用于进度条，监听进度条长度，刷新资源时长
timeUpdate	用于进度条，监听进度条当前位置，刷新当前时间
seekDone	响应 API 调用，监听 seek() 请求完成情况。当使用 seek() 跳转到指定播放位置后，如果 seek 操作成功，将上报该事件
speedDone	响应 API 调用，监听 setSpeed() 请求完成情况。当使用 setSpeed() 设置播放倍速后，如果 setSpeed 操作成功，将上报该事件

续表

事件类型	说明
volumeChange	响应 API 调用，监听 setVolume() 请求完成情况。当使用 setVolume() 调节播放音量后，如果 setVolume 操作成功，将上报该事件
bufferingUpdate	用于网络播放，监听网络播放缓冲信息，用于上报缓冲百分比以及缓存播放进度
audioInterrupt	监听音频焦点切换信息，搭配属性 audioInterruptMode 使用。如果当前设备存在多个音频正在播放，音频焦点被切换（即播放其他媒体如通话等）时将上报该事件，应用可以及时处理

（3）设置资源：设置属性 url，AVPlayer 进入 initialized 状态。

（4）准备播放：调用 prepare()，AVPlayer 进入 prepared 状态，此时可以获取 duration，设置音量。

（5）音频播控：播放 play()，暂停 pause()，跳转 seek()，停止 stop()，等操作。

（6）（可选）更换资源：调用 reset() 重置资源，AVPlayer 重新进入 idle 状态，允许更换资源 url。

（7）退出播放：调用 release() 销毁实例，AVPlayer 进入 released 状态，退出播放。

请看下面的代码，展示了使用 AVPlayer 进行音频播放的过程，包括音频初始化、准备、播放、暂停、停止等操作，并通过状态机变化回调函数清晰地展示了整个播放过程中的状态变化。通过这个例子，开发者可以学习在 HarmonyOS 中使用 AVPlayer 进行多媒体开发，实现音频文件播放功能的方法。

```
import media from '@ohos.multimedia.media';
import fs from '@ohos.file.fs';
import common from '@ohos.app.ability.common';

export class AVPlayerDemo {
  private avPlayer;
  private count: number = 0;

  // 注册 avplayer 回调函数
  setAVPlayerCallback() {
    // seek 操作结果回调函数
    this.avPlayer.on('seekDone', (seekDoneTime) => {
      console.info(`AVPlayer seek succeeded, seek time is ${seekDoneTime}`);
    })
    // error 回调监听函数，当 avPlayer 在操作过程中出现错误时调用 reset 接口触发重置流程
    this.avPlayer.on('error', (err) => {
      console.error(`Invoke avPlayer failed, code is ${err.code}, message is ${err.message}`);
      this.avPlayer.reset(); // 调用 reset 重置资源，触发 idle 状态
    })
    // 状态机变化回调函数
    this.avPlayer.on('stateChange', async (state, reason) => {
      switch (state) {
        case 'idle': // 成功调用 reset 接口后触发该状态机上报
          console.info('AVPlayer state idle called.');
          this.avPlayer.release(); // 调用 release 接口销毁实例对象
          break;
```

```
        case 'initialized': // avplayer 设置播放源后触发该状态上报
          console.info('AVPlayerstate initialized called.');
          this.avPlayer.prepare().then(() => {
            console.info('AVPlayer prepare succeeded.');
          }, (err) => {
            console.error(`Invoke prepare failed, code is ${err.code}, message is ${err.message}`);
          });
          break;
        case 'prepared': // prepare 调用成功后上报该状态机
          console.info('AVPlayer state prepared called.');
          this.avPlayer.play(); // 调用播放接口开始播放
          break;
        case 'playing': // play 成功调用后触发该状态机上报
          console.info('AVPlayer state playing called.');
          if (this.count !== 0) {
            console.info('AVPlayer start to seek.');
            this.avPlayer.seek(this.avPlayer.duration); //seek 到音频末尾
          } else {
            this.avPlayer.pause(); // 调用暂停接口暂停播放
          }
          this.count++;
          break;
        case 'paused': // pause 成功调用后触发该状态机上报
          console.info('AVPlayer state paused called.');
          this.avPlayer.play(); // 再次播放接口开始播放
          break;
        case 'completed': // 播放结束后触发该状态机上报
          console.info('AVPlayer state completed called.');
          this.avPlayer.stop(); //调用播放结束接口
          break;
        case 'stopped': // stop 接口成功调用后触发该状态机上报
          console.info('AVPlayer state stopped called.');
          this.avPlayer.reset(); // 调用 reset 接口初始化 avplayer 状态
          break;
        case 'released':
          console.info('AVPlayer state released called.');
          break;
        default:
          console.info('AVPlayer state unknown called.');
          break;
      }
    })
  }

  // 以下 demo 为使用 fs 文件系统打开沙箱地址获取媒体文件地址并通过 url 属性进行播放示例
  async avPlayerUrlDemo() {
    // 创建 avPlayer 实例对象
    this.avPlayer = await media.createAVPlayer();
    // 创建状态机变化回调函数
    this.setAVPlayerCallback();
    let fdPath = 'fd://';
```

```
        // 通过 UIAbilityContext 获取沙箱地址 filesDir,以下为 Stage 模型获方式,如需在
FA 模型上获取请参考《访问应用沙箱》获取地址
        let context = getContext(this) as common.UIAbilityContext;
        let pathDir = context.filesDir;
        let path = pathDir + '/01.mp3';
        // 打开相应的资源文件地址获取 fd,并为 url 赋值触发 initialized 状态机上报
        let file = await fs.open(path);
        fdPath = fdPath + '' + file.fd;
        this.avPlayer.url = fdPath;
    }

        // 以下 demo 为使用资源管理接口获取打包在 HAP 内的媒体资源文件并通过 fdSrc 属性进
行播放示例
    async avPlayerFdSrcDemo() {
        // 创建 avPlayer 实例对象
        this.avPlayer = await media.createAVPlayer();
        // 创建状态机变化回调函数
        this.setAVPlayerCallback();
        // 通过 UIAbilityContext 的 resourceManager 成员的 getRawFd 接口获取媒体资源
播放地址
        // 返回类型为{fd,offset,length},fd 为 HAP 包 fd 地址,offset 为媒体资源偏移量,
length 为播放长度
        let context = getContext(this) as common.UIAbilityContext;
        let fileDescriptor = await context.resourceManager.getRawFd('01.mp3');
        // 为 fdSrc 赋值触发 initialized 状态机上报
        this.avPlayer.fdSrc = fileDescriptor;
    }
}
```

上述代码的实现流程如下。
- 首先,创建了一个 AVPlayerDemo 类,其中包含了 AVPlayer 的实例和一些状态变化的回调函数。通过两个不同的方法演示了如何使用 AVPlayer 播放媒体文件。
- 其次,avPlayerUrlDemo 方法通过文件系统(fs)打开沙箱地址,获取媒体文件的地址,并通过 AVPlayer 的 url 属性进行播放。这一过程中,使用了状态机变化回调函数,监控了 AVPlayer 的各个状态,如初始化、准备、播放等。
- 最后,通过 avPlayerFdSrcDemo 方法,使用资源管理接口获取打包在 HAP 内的媒体资源文件,并通过 AVPlayer 的 fdSrc 属性进行播放。同样,也使用了状态机变化回调函数,跟踪 AVPlayer 的状态变化。

7.3.2　使用 AudioRenderer 开发音频播放程序

在 HarmonyOS 系统中,AudioRenderer 是音频渲染器,用于播放 PCM(Pulse Code Modulation)音频数据,与 AVPlayer 相比,AudioRenderer 可以在输入前添加数据预处理,更适合有音频开发经验的开发者,以实现更灵活的播放功能。

使用 AudioRenderer 播放音频涉及 AudioRenderer 实例的创建、音频渲染参数的配置、渲染的开始与停止、资源的释放等。图 7-5 所示为 AudioRenderer 的状态变化,在创建实例后,调用对应的方法可以进入指定的状态实现对应的行为。需要注意的是在确定的状态执

行不合适的方法可能导致 AudioRenderer 发生错误,建议开发者在调用状态转换的方法前进行状态检查,避免程序运行产生预期以外的结果。

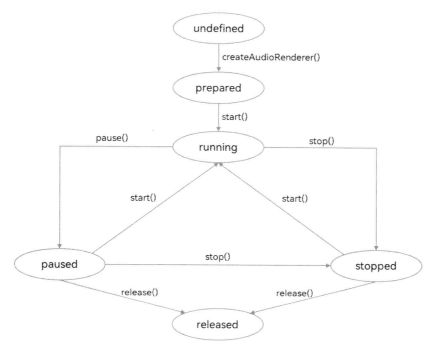

图 7-5　AudioRenderer 状态变化

为保证 UI 线程不被阻塞,大部分 AudioRenderer 调用都是异步的。对于每个 API 均提供了 callback 函数和 Promise 函数,在实际开发中,建议使用 callback 函数。

在开发 HarmonyOS 应用程序的过程中,建议开发者通过 on('stateChange') 方法订阅 AudioRenderer 的状态变更。因为针对 AudioRenderer 的某些操作,仅在音频播放器在固定状态时才能执行。如果应用在音频播放器处于错误状态时执行操作,系统可能会抛出异常或生成其他未定义的行为。

- prepared 状态:通过调用 createAudioRenderer() 方法进入该状态。
- running 状态:正在进行音频数据播放,可以在 prepared 状态通过调用 start() 方法进入此状态,也可以在 paused 状态和 stopped 状态通过调用 start() 方法进入此状态。
- paused 状态:在 running 状态可以通过调用 pause() 方法暂停音频数据的播放并进入 paused 状态,暂停播放之后可以通过调用 start() 方法继续音频数据播放。
- stopped 状态:在 paused/running 状态可以通过 stop() 方法停止音频数据的播放。
- released 状态:在 prepared、paused、stopped 等状态,用户均可通过 release() 方法释放掉所有占用的硬件和软件资源,并且不会再进入到其他的任何一种状态了。

例如,下面的代码展示了使用 AudioRenderer 创建音频渲染器、进行音频文件的读取和渲染的过程,以及如何管理音频渲染器的状态和资源释放的方法。通过这个例子,开发者可以学习如何在 HarmonyOS 中实现音频渲染功能,为应用添加音频播放的支持。

```
import audio from '@ohos.multimedia.audio';
import fs from '@ohos.file.fs';
```

```
  const TAG = 'AudioRendererDemo';

export default class AudioRendererDemo {
  private renderModel = undefined;
  private audioStreamInfo = {
    samplingRate: audio.AudioSamplingRate.SAMPLE_RATE_48000, // 采样率
    channels: audio.AudioChannel.CHANNEL_2, // 通道
    sampleFormat: audio.AudioSampleFormat.SAMPLE_FORMAT_S16LE, // 采样格式
    encodingType: audio.AudioEncodingType.ENCODING_TYPE_RAW // 编码格式
  }
  private audioRendererInfo = {
    content: audio.ContentType.CONTENT_TYPE_MUSIC, // 媒体类型
    usage: audio.StreamUsage.STREAM_USAGE_MEDIA, // 音频流使用类型
    rendererFlags: 0 // 音频渲染器标志
  }
  private audioRendererOptions = {
    streamInfo: this.audioStreamInfo,
    rendererInfo: this.audioRendererInfo
  }

  // 初始化，创建实例，设置监听事件
  init() {
    audio.createAudioRenderer(this.audioRendererOptions, (err, renderer) => { // 创建 AudioRenderer 实例
      if (!err) {
        console.info(`${TAG}: creating AudioRenderer success`);
        this.renderModel = renderer;
        this.renderModel.on('stateChange', (state) => { // 设置监听事件，当转换到指定的状态时触发回调
          if (state == 2) {
            console.info('audio renderer state is: STATE_RUNNING');
          }
        });
        this.renderModel.on('markReach', 1000, (position) => { // 订阅 markReach 事件，当渲染的帧数达到 1000 帧时触发回调
          if (position == 1000) {
            console.info('ON Triggered successfully');
          }
        });
      } else {
        console.info(`${TAG}: creating AudioRenderer failed, error: ${err.message}`);
      }
    });
  }

  // 开始一次音频渲染
  async start() {
    let stateGroup = [audio.AudioState.STATE_PREPARED, audio.AudioState.STATE_PAUSED, audio.AudioState.STATE_STOPPED];
    if (stateGroup.indexOf(this.renderModel.state) === -1) { // 当且仅当状态为 prepared、paused 和 stopped 之一时才能启动渲染
```

```
      console.error(TAG + 'start failed');
      return;
    }
    await this.renderModel.start(); // 启动渲染

    const bufferSize = await this.renderModel.getBufferSize();
    let context = getContext(this);
    let path = context.filesDir;
    const filePath = path + '/test.wav'; // 使用沙箱路径获取文件,实际路径为 /
data/ storage/el2/base/haps/entry/files/test.wav

    let file = fs.openSync(filePath, fs.OpenMode.READ_ONLY);
    let stat = await fs.stat(filePath);
    let buf = new ArrayBuffer(bufferSize);
    let len = stat.size % bufferSize === 0 ? Math.floor(stat.size / bufferSize) :
Math.floor(stat.size / bufferSize + 1);
    for (let i = 0; i < len; i++) {
      let options = {
        offset: i * bufferSize,
        length: bufferSize
      };
      let readsize = await fs.read(file.fd, buf, options);

      // buf 是要写入缓冲区的音频数据,在调用 AudioRenderer.write() 方法前可以进
行音频数据的预处理,实现个性化的音频播放功能,AudioRenderer 会读出写入缓冲区的音频数据进
行渲染

      let writeSize = await new Promise((resolve, reject) => {
        this.renderModel.write(buf, (err, writeSize) => {
          if (err) {
            reject(err);
          } else {
            resolve(writeSize);
          }
        });
      });
      if (this.renderModel.state === audio.AudioState.STATE_RELEASED) {
// 如果渲染器状态为 released,停止渲染
        fs.close(file);
        await this.renderModel.stop();
      }
      if (this.renderModel.state === audio.AudioState.STATE_RUNNING) {
        if (i === len - 1) { // 如果音频文件已经被读取完,停止渲染
          fs.close(file);
          await this.renderModel.stop();
        }
      }
    }
  }

  // 暂停渲染
  async pause() {
```

```
    // 只有渲染器状态为 running 的时候才能暂停
    if (this.renderModel.state !== audio.AudioState.STATE_RUNNING) {
      console.info('Renderer is not running');
      return;
    }
    await this.renderModel.pause(); // 暂停渲染
    if (this.renderModel.state === audio.AudioState.STATE_PAUSED) {
      console.info('Renderer is paused.');
    } else {
      console.error('Pausing renderer failed.');
    }
  }

  // 停止渲染
  async stop() {
    // 只有渲染器状态为 running 或 paused 的时候才可以停止
    if (this.renderModel.state !== audio.AudioState.STATE_RUNNING &&
this.renderModel.state !== audio.AudioState.STATE_PAUSED) {
      console.info('Renderer is not running or paused.');
      return;
    }
    await this.renderModel.stop(); // 停止渲染
    if (this.renderModel.state === audio.AudioState.STATE_STOPPED) {
      console.info('Renderer stopped.');
    } else {
      console.error('Stopping renderer failed.');
    }
  }

  // 销毁实例，释放资源
  async release() {
    // 渲染器状态不是 released 状态，才能 release
    if (this.renderModel.state === audio.AudioState.STATE_RELEASED) {
      console.info('Renderer already released');
      return;
    }
    await this.renderModel.release(); // 释放资源
    if (this.renderModel.state === audio.AudioState.STATE_RELEASED) {
      console.info('Renderer released');
    } else {
      console.error('Renderer release failed.');
    }
  }
}
```

上述代码的实现流程如下。

- 首先，创建了一个 AudioRendererDemo 类，其中包含了一个 AudioRenderer 实例和一些用于配置音频流信息、音频渲染信息以及音频渲染选项的属性。
- 其次，通过 init 方法初始化了音频渲染器实例，设置了监听事件，包括状态变化和帧数达到特定值的事件。在成功创建渲染器后，将音频文件的路径设置为沙箱地址，并使用沙箱路径打开文件。
- 再次，通过 start 方法启动音频渲染器，同时读取音频文件的数据，对数据进行预处理，然后通过 AudioRenderer.write 方法写入音频渲染器进行播放。在每次写入前，代码检查渲染器的状态，如果状态为 STATE_RELEASED，则停止渲染。在完成音频文件的读取后，也停止渲染。
- 复次，还提供了 pause 和 stop 方法，用于暂停和停止渲染器的播放。
- 最后，通过 release 方法释放音频渲染器的资源。在释放资源前会检查渲染器的状态，确保只有在非 STATE_RELEASED 状态时才执行释放操作。

7.3.3 使用 OpenSL ES 开发音频播放程序

OpenSL ES 全称为 Open Sound Library for Embedded Systems，是一个嵌入式、跨平台、免费的音频处理库。它为嵌入式移动多媒体设备上的应用开发者提供标准化、高性能、低延迟的 API。HarmonyOS 的 Native API 基于 Khronos Group 开发的 OpenSL ES 1.0.1 API 规范实现，开发者可以通过 <OpenSLES.h> 和 <OpenSLES_OpenHarmony.h> 在 HarmonyOS 上使用相关 API。

目前，HarmonyOS 仅实现了部分 OpenSL ES 接口，可以实现音频播放的基础功能，以下为 HarmonyOS 系统支持的接口。

1）HarmonyOS 上支持的 Engine 接口
- SLresult (*CreateAudioPlayer) (SLEngineItf self, SLObjectItf * pPlayer, SLDataSource *pAudioSrc, SLDataSink *pAudioSnk, SLuint32 numInterfaces, const SLInterfaceID * pInterfaceIds, const SLboolean * pInterfaceRequired)
- SLresult (*CreateAudioRecorder) (SLEngineItf self, SLObjectItf * pRecorder, SLDataSource *pAudioSrc, SLDataSink *pAudioSnk, SLuint32 numInterfaces, const SLInterfaceID * pInterfaceIds, const SLboolean * pInterfaceRequired)
- SLresult (*CreateOutputMix) (SLEngineItf self, SLObjectItf * pMix, SLuint32 numInterfaces, const SLInterfaceID * pInterfaceIds, const SLboolean * pInterfaceRequired)

2）HarmonyOS 上支持的 Object 接口
- SLresult (*Realize) (SLObjectItf self, SLboolean async)
- SLresult (*GetState) (SLObjectItf self, SLuint32 * pState)
- SLresult (*GetInterface) (SLObjectItf self, const SLInterfaceID iid, void * pInterface)
- void (*Destroy) (SLObjectItf self)

3）HarmonyOS 上支持的 Playback 接口
- SLresult (*SetPlayState) (SLPlayItf self, SLuint32 state)
- SLresult (*GetPlayState) (SLPlayItf self, SLuint32 *pState)

4）HarmonyOS 上支持的 Volume 控制接口
- SLresult (*SetVolumeLevel) (SLVolumeItf self, SLmillibel level)
- SLresult (*GetVolumeLevel) (SLVolumeItf self, SLmillibel *pLevel)
- SLresult (*GetMaxVolumeLevel) (SLVolumeItf self, SLmillibel *pMaxLevel)

在 HarmonyOS 系统中，使用 OpenSL ES 开发音频播放程序的流程如下。

（1）添加头文件，代码如下。

```
#include <OpenSLES.h>
#include <OpenSLES_OpenHarmony.h>
#include <OpenSLES_Platform.h>
```

（2）使用 slCreateEngine 接口和获取 engine 实例，代码如下。

```
SLObjectItf engineObject = nullptr;
slCreateEngine(&engineObject, 0, nullptr, 0, nullptr, nullptr);
(*engineObject)->Realize(engineObject, SL_BOOLEAN_FALSE);
```

（3）获取接口 SL_IID_ENGINE 的 engineEngine 实例，代码如下。

```
SLEngineItf engineEngine = nullptr;
(*engineObject)->GetInterface(engineObject, SL_IID_ENGINE, &engineEngine);
```

（4）配置播放器信息，创建 AudioPlayer。代码如下。

```
SLDataLocator_BufferQueue slBufferQueue = {
    SL_DATALOCATOR_BUFFERQUEUE,
    0
};

// 具体参数需要根据音频文件格式进行适配
SLDataFormat_PCM pcmFormat = {
    SL_DATAFORMAT_PCM,
    2,                             // 通道数
    SL_SAMPLINGRATE_48,            // 采样率
    SL_PCMSAMPLEFORMAT_FIXED_16,   // 音频采样格式
    0,
    0,
    0
};
SLDataSource slSource = {&slBufferQueue, &pcmFormat};
SLObjectItf pcmPlayerObject = nullptr;
(*engineEngine)->CreateAudioPlayer(engineEngine, &pcmPlayerObject,
&slSource, nullptr, 0, nullptr, nullptr);
(*pcmPlayerObject)->Realize(pcmPlayerObject, SL_BOOLEAN_FALSE);
```

（5）获取接口 SL_IID_OH_BUFFERQUEUE 的 bufferQueueItf 实例，代码如下。

```
SLOHBufferQueueItf bufferQueueItf;
(*pcmPlayerObject)->GetInterface(pcmPlayerObject, SL_IID_OH_BUFFERQUEUE,
&bufferQueueItf);
```

（6）打开音频文件，注册 BufferQueueCallback 回调。代码如下。

```
static void BufferQueueCallback (SLOHBufferQueueItf bufferQueueItf, void
*pContext, SLuint32 size)
{
    SLuint8 *buffer = nullptr;
    SLuint32 pSize;
```

```
            (*bufferQueueItf)->GetBuffer(bufferQueueItf, &buffer, &pSize);
            // 将待播放音频数据写入buffer
            (*bufferQueueItf)->Enqueue(bufferQueueItf, buffer, size);
    }
    void *pContext; // 可传入自定义的上下文信息，会在Callback内收到
    (*bufferQueueItf)->RegisterCallback(bufferQueueItf, BufferQueueCallback,
pContext);
```

（7）获取接口 SL_PLAYSTATE_PLAYING 的 playItf 实例，开始播放。代码如下。

```
SLPlayItf playItf = nullptr;
(*pcmPlayerObject)->GetInterface(pcmPlayerObject, SL_IID_PLAY, &playItf);
(*playItf)->SetPlayState(playItf, SL_PLAYSTATE_PLAYING);
```

（8）结束音频播放，代码如下。

```
(*playItf)->SetPlayState(playItf, SL_PLAYSTATE_STOPPED);
(*pcmPlayerObject)->Destroy(pcmPlayerObject);
(*engineObject)->Destroy(engineObject);
```

7.3.4　音频播放实战：多功能音乐播放器

在实例 7-1 中，将通过一个具体实例展示使用 AVPlayer 实现一个多功能音乐播放器的过程。实例中 AVPlayer 的主要工作是将 Audio/Video 媒体资源转码为可供渲染的图像和可听见的音频模拟信号，并通过输出设备进行播放，同时对播放任务进行管理，包括开始播放、暂停播放、停止播放、释放资源、设置音量、跳转播放位置、获取轨道信息等功能控制。

实例 7-1：多功能音乐播放器（源码路径 :codes\7\MusicPlay）

（1）编写文件 src/main/resources/base/element/string.json，定义一个鸿蒙项目中的字符串资源文件，包含了多语言支持的字符串及其中文翻译。这些字符串涵盖了音乐播放器模块的描述、表单信息、后台运行提示、歌单信息以及相关操作按钮的标签和描述。通过这些字符串的国际化支持，代码实现了在多语言环境中展示合适的文本信息，使音乐播放器界面更具可读性和友好性。

（2）编写文件 src/main/ets/constants/SongConstants.ets，该 TypeScript 类定义一组与歌曲列表和页面过渡相关的常量，包括歌曲列表索引、切片操作的起始和结束位置、表单 ID 的不存在情况以及页面过渡的持续时间。这些常量用于在代码中统一管理和使用，提高代码的可读性和维护性。

（3）编写文件 src/main/ets/constants/PlayConstants.ets，定义一个名为 PlayConstants 的类，包含了常用播放组件的常量。这些常量包括播放控制索引、滑动步进值、网格行列参数、页面转换持续时间等。通过定义这些常量，本项目实现了可重用的数值和文本信息，以便在应用程序中进行一致的管理和调整。

（4）编写文件 src/main/ets/pages/MainPage.ets，定义一个名为 MainPage 的主页面组件，包含音乐播放器和内容展示。通过引用其他组件、工具类和常量，实现了页面生命周期的管理、布局的构建以及页面过渡效果的配置。

（5）编写文件 src/main/ets/pages/MusicComment.ets，定义一个名为 MusicComment 的音乐评论页面组件，包含了精彩评论和最新评论的展示。通过引用其他组件、常量和视图模型，实现了页面布局的构建、标题的显示以及评论列表的展示。页面还响应断点变化，在不同屏幕尺寸下进行样式和布局的调整。

```
import { StyleConstants } from '../constants/StyleConstants';
import { BreakpointConstants } from '../constants/BreakpointConstants';
import { Comment } from '../model/Comment';
import CommentViewModel from '../viewmodel/CommentViewModel';
import { ListItemComponent } from '../components/ListItemComponent';
import { CommentMusicComponent } from '../components/CommentMusicComponent';
import { CommonConstants } from '../constants/CommonConstants';

@Entry
@Component
struct MusicComment {
  @State currentBp: string = BreakpointConstants.CURRENT_BREAKPOINT;
  @State wonderfulComment: Comment[] = CommentViewModel.getWonderfulReview();
  @State newComment: Comment[] = CommentViewModel.getNewComment();

  @Builder ShowTitle(title: ResourceStr) {
    Row() {
      Text(title)
        .fontSize($r('app.float.comment_title_size'))
        .fontColor($r('app.color.comment_title_color'))
        .lineHeight($r('app.float.title_line_height'))
        .fontWeight(FontWeight.Medium)
        .margin({
          top: $r('app.float.title_margin_top'),
          bottom: $r('app.float.title_margin_bottom'),
          left: this.currentBp === BreakpointConstants.BREAKPOINT_SM ?
            $r('app.float.margin_left_sm') : $r('app.float.margin_left'),
          right: this.currentBp === BreakpointConstants.BREAKPOINT_SM ?
            $r('app.float.margin_right_sm') : $r('app.float.margin_right')
        })
    }
    .justifyContent(FlexAlign.Start)
    .width(StyleConstants.FULL_WIDTH)
  }

  build() {
    GridRow({
      breakpoints: {
        value: BreakpointConstants.BREAKPOINT_VALUE,
        reference: BreakpointsReference.WindowSize
      },
      columns: {
        sm: BreakpointConstants.COLUMN_SM,
        md: BreakpointConstants.COLUMN_MD,
        lg: BreakpointConstants.COLUMN_LG
      },
```

```
        gutter: { x: BreakpointConstants.GUTTER_X }
}) {
  GridCol({
    span: {
      sm: BreakpointConstants.COLUMN_SM,
      md: BreakpointConstants.COLUMN_MD,
      lg: BreakpointConstants.COLUMN_LG
    }
  }) {
    Column() {
      CommentMusicComponent()
        .margin({
          left: this.currentBp === BreakpointConstants.BREAKPOINT_SM ?
          $r('app.float.margin_left_sm') : $r('app.float.margin_left'),
          right: this.currentBp === BreakpointConstants.BREAKPOINT_SM ?
          $r('app.float.margin_right_sm') : $r('app.float.margin_right')
        })

      this.ShowTitle($r('app.string.wonderful_comment'))

      List() {
        ForEach(this.wonderfulComment, (comment: Comment, index?:
        number) => {
          if (this.currentBp === BreakpointConstants.BREAKPOINT_SM ||
            this.currentBp === BreakpointConstants.BREAKPOINT_MD) {
            if (index && index < CommonConstants.LIST_COUNT) {
              ListItem() {
                ListItemComponent({ item: comment })
                  .margin({
                    left: this.currentBp === BreakpointConstants.
                    BREAKPOINT_SM ?
                      0 : $r('app.float.margin_left_list'),
                    right: this.currentBp === BreakpointConstants.
                    BREAKPOINT_SM ?
                      0 : $r('app.float.margin_right_list')
                  })
              }
              .width(StyleConstants.FULL_WIDTH)
              .padding({
                bottom: $r('app.float.padding_bottom')
              })
            }
          } else {
            ListItem() {
              ListItemComponent({ item: comment })
                .margin({
                  left: this.currentBp === BreakpointConstants.
                  BREAKPOINT_SM ?
                    0 : $r('app.float.margin_left_list'),
                  right: this.currentBp === BreakpointConstants.
                  BREAKPOINT_SM ?
                    0 : $r('app.float.margin_right_list')
```

```
          })
        }
        .width(StyleConstants.FULL_WIDTH)
        .padding({
          bottom: $r('app.float.padding_bottom')
        })
      }
    }, (item: Comment, index?: number) => index + JSON.stringify(item))
}
.lanes(this.currentBp === BreakpointConstants.BREAKPOINT_LG ? 2 : 1)
.scrollBar(BarState.Off)
.divider({
  color: $r('app.color.list_divider'),
  strokeWidth: $r('app.float.stroke_width'),
  startMargin: this.currentBp === BreakpointConstants.BREAKPOINT_SM ?
  $r('app.float.start_margin') : $r('app.float.start_margin_lg'),
  endMargin: this.currentBp === BreakpointConstants.BREAKPOINT_SM ?
    0 : $r('app.float.divider_margin_left')
})
.margin({
  left: this.currentBp === BreakpointConstants.BREAKPOINT_SM ?
  $r('app.float.margin_left_sm') : $r('app.float.margin_left_list'),
  right: this.currentBp === BreakpointConstants.BREAKPOINT_SM ?
  $r('app.float.margin_right_sm') : $r('app.float.margin_right_list')
})

this.ShowTitle($r('app.string.new_comment'))

List() {
  ForEach(this.newComment, (comment: Comment) => {
    ListItem() {
      ListItemComponent({ item: comment })
        .margin({
          left: this.currentBp === BreakpointConstants.
          BREAKPOINT_SM ?
            0 : $r('app.float.margin_left_list'),
          right: this.currentBp === BreakpointConstants.
          BREAKPOINT_SM ?
            0 : $r('app.float.margin_right_list')
        })
    }
    .width(StyleConstants.FULL_WIDTH)
    .padding({
      bottom: $r('app.float.padding_bottom')
    })
  }, (item: Comment, index?: number) => index + JSON.stringify(item))
}
.layoutWeight(1)
.lanes(this.currentBp === BreakpointConstants.BREAKPOINT_LG ? 2 : 1)
.scrollBar(BarState.Off)
.margin({
  left: this.currentBp === BreakpointConstants.BREAKPOINT_SM ?
```

```
            $r('app.float.margin_left_sm') : $r('app.float.margin_left_list'),
          right: this.currentBp === BreakpointConstants.BREAKPOINT_SM ?
            $r('app.float.margin_right_sm') : $r('app.float.margin_right_list')
        })
        .divider({
          color: $r('app.color.list_divider'),
          strokeWidth: $r('app.float.stroke_width'),
          startMargin: this.currentBp === BreakpointConstants.BREAKPOINT_SM ?
            $r('app.float.start_margin') : $r('app.float.start_margin_lg'),
          endMargin: this.currentBp === BreakpointConstants.BREAKPOINT_SM ?
            0 : $r('app.float.divider_margin_left')
        })
      }
      .height(StyleConstants.FULL_HEIGHT)
    }
  }
  .backgroundColor(Color.White)
  .onBreakpointChange((breakpoint) => {
    this.currentBp = breakpoint;
  })
 }
}
```

上述代码的实现流程如下。

- 首先，在 MusicComment 组件中引入了多个常量、模型、视图模型和组件，包括样式常量、断点常量、评论模型、评论视图模型以及列表项和音乐评论组件。
- 其次，在组件定义的代码中，使用 @State 注解声明了一些状态变量，其中包括当前断点尺寸（currentBp）、精彩评论列表（wonderfulComment）和最新评论列表（newComment）。
- 再次，通过 @Builder 注解定义了 ShowTitle 方法，用于构建显示标题的部分，包括标题文字和样式设置。
- 复次，在 build 方法中，使用 GridRow 和 GridCol 构建了页面的网格布局，包含了精彩评论和最新评论的展示区域。在展示区域中，通过 CommentMusicComponent 显示音乐评论的相关组件，并通过 List 和 ForEach 构建了评论列表，根据不同断点尺寸显示不同数量的评论项。同时，使用 divider 和其他样式常量进行布局和样式设置。
- 最后，通过 onBreakpointChange 方法监听断点变化，实现了在不同屏幕尺寸下的风格和布局调整。整体而言，该组件实现了音乐评论页面的布局构建、评论列表展示，并通过断点变化实现了响应式设计。

（6）编写文件 src/main/ets/pages/PlayPage.ets，实现音乐播放页面的布局和交互。页面能够根据不同屏幕尺寸，以垂直或标签页形式展示歌曲信息、歌词和控制组件。页面背景通过背景图、模糊效果、线性渐变呈现，并添加了返回按钮。页面切换时具有滑动过渡效果。

```
import router from '@ohos.router';
import { StyleConstants } from '../constants/StyleConstants';
import { BreakpointType } from '../utils/BreakpointSystem';
import { SongItem } from '../model/SongData';
import { MusicList } from '../viewmodel/MusicList';
```

```
import { ControlComponent } from '../components/ControlComponent';
import { LyricsComponent } from '../components/LyricsComponent';
import { MusicInfoComponent } from '../components/MusicInfoComponent';
import { PlayConstants } from '../constants/PlayConstants';

@Entry
@Component
struct PlayPage {
  @State currentTabIndex: number = 0;
  @StorageProp('currentBreakpoint') currentBreakpoint: string = 'sm';
  songList: SongItem[] = MusicList;
  @StorageProp('selectIndex') selectIndex: number = 0;

  build() {
    Stack({ alignContent: Alignment.TopStart }) {
      if (this.currentBreakpoint === PlayConstants.LG) {
        Stack({ alignContent: Alignment.Top }) {
          LyricsComponent()

          Flex({ direction: FlexDirection.Column }) {
            MusicInfoComponent()
            ControlComponent()
          }
        }
      } else {
        Flex({ direction: FlexDirection.Column }) {
          Tabs({ barPosition: BarPosition.Start, index: this.currentTabIndex }) {
            TabContent() {
              MusicInfoComponent()
            }
            .tabBar(this.TabTitle(PlayConstants.TAB_SONG, 0))

            TabContent() {
              LyricsComponent()
            }
            .tabBar(this.TabTitle(PlayConstants.TAB_LYRICS, 1))
          }
          .onChange(index => this.currentTabIndex = index)
          .vertical(false)
          .barHeight($r('app.float.fifty_six'))
          .barWidth(PlayConstants.TAB_WIDTH)

          ControlComponent()
        }
      }

      Image($r('app.media.ic_back_down'))
        .width($r('app.float.image_back_size'))
        .height($r('app.float.image_back_size'))
        .margin({ left: $r('app.float.twenty_four'), top: $r('app.float.image_back_margin_top') })
        .onClick(() => router.back())
```

```
      }
      .backgroundImage(this.songList[this.selectIndex].label)
      .backgroundImageSize(ImageSize.Cover)
      .backdropBlur(PlayConstants.BLUR)
      .linearGradient({
        direction: GradientDirection.Bottom,
        colors: [
          [PlayConstants.EIGHTY_WHITE_COLOR, PlayConstants.EIGHTY_WHITE],
          [PlayConstants.NINETY_WHITE_COLOR, PlayConstants.NINETY_WHITE]
        ]
      })
      .height(StyleConstants.FULL_HEIGHT)
      .width(StyleConstants.FULL_WIDTH)
    }

    @Builder TabTitle(title: string, index: number) {
      Text(title)
        .fontColor(this.currentTabIndex === index ? $r('app.color.text_color') : $r('app.color.text_forty_color'))
        .fontWeight(this.currentTabIndex === index ? PlayConstants.FIVE_HUNDRED : PlayConstants.FOUR_HUNDRED)
        .fontSize(new BreakpointType({
          sm: $r('app.float.font_sixteen'),
          md: $r('app.float.font_twenty')
        }).getValue(this.currentBreakpoint))
        .border({
          width: { bottom: this.currentTabIndex === index ? $r('app.float.tab_border_width') : 0 },
          color: $r('app.color.text_color')
        })
        .padding({ bottom: $r('app.float.tab_text_padding_bottom') })
    }

    pageTransition() {
      PageTransitionEnter({ duration: PlayConstants.FIVE_HUNDRED, curve: Curve.Smooth }).slide(SlideEffect.Bottom);
      PageTransitionExit({ duration: PlayConstants.FIVE_HUNDRED, curve: Curve.Smooth }).slide(SlideEffect.Bottom);
    }
  }
```

上述代码的实现流程如下。

- 首先，在 PlayPage 组件中引入了不同模块和组件，包括路由模块、样式常量、断点类型、歌曲数据模型、音乐列表视图模型以及控制组件、歌词组件和音乐信息组件。
- 其次，在组件定义代码中使用 @State 注解声明了一些状态变量，其中包括当前标签页索引（currentTabIndex）、当前断点尺寸（currentBreakpoint）、歌曲列表（songList）和选中的索引（selectIndex）。
- 再次，在 build 方法中根据当前断点尺寸选择性地构建了页面布局。如果断点为大屏（PlayConstants.LG），则呈现歌词组件、音乐信息组件和控制组件的垂直布局；否则，在小屏幕上使用标签页（Tabs）实现了歌曲信息和歌词的切换，并添加了控制组件。
- 复次，通过 Image 添加了返回按钮，并通过 router.back() 方法实现了返回功能。整体页面通过背景图、模糊效果、线性渐变和样式设置，营造了音乐播放页面的视觉效果。
- 又次，在 @Builder 注解中定义了 TabTitle 方法，用于构建标签页标题的样式和设置，包括字体颜色、粗细、大小、底部边框等。
- 最后，通过 pageTransition 方法定义了页面的进入和退出过渡效果，使用了滑动效果并设置了过渡的持续时间和曲线。

执行效果如图 7-6 所示。

图 7-6　多功能音乐播放器执行效果

7.4　开发音频录制程序

在 HarmonyOS 系统中，有多种 API 接口提供了音频录制开发的支持，不同的 API 适用于不同的录音输出格式、音频使用场景或不同的开发语言。因此，选择合适的音频录制 API，有助于降低开发工作量，实现更佳的音频录制效果。

- AVRecorder：功能较完善的音频、视频录制 ArkTS/JS API，集成了音频输入录制、音频编码和媒体封装的功能。开发者可以直接调用设备硬件如麦克风录音，并生成 m4a 音频文件。

- **AudioCapturer**：用于音频输入的 ArkTS/JS API，仅支持 PCM 格式，需要应用持续读取音频数据进行工作。应用可以在音频输出后添加数据处理，要求开发者具备音频处理的基础知识，适用于更专业、更多样化的媒体录制应用开发。
- **OpenSL ES**：一套跨平台标准化的音频 Native API，目前阶段唯一的音频类 Native API，同样提供音频输入原子能力，仅支持 PCM 格式，适用于从其他嵌入式平台迁移，或依赖在 Native 层实现音频输入功能的录音应用使用。

7.4.1 使用 AVRecorder 开发音频录制程序

在 HarmonyOS 系统中，使用 AVRecorder 可以实现音频录制功能，具体过程包括"开始录制—暂停录制—恢复录制—停止录制"。

使用 AVRecorder 开发音频录制程序的步骤如下。

（1）创建 AVRecorder 实例，实例创建完成后进入 idle 状态。

```
import media from '@ohos.multimedia.media';

let avRecorder = undefined;
media.createAVRecorder().then((recorder) => {
  avRecorder = recorder;
}, (err) => {
  console.error(`Invoke createAVRecorder failed, code is ${err.code}, message is ${err.message}`);
})
```

（2）设置业务需要的监听事件，监听状态变化及错误上报。涉及的事件如下。

- **stateChange**：必要事件，监听 AVRecorder 的 state 属性改变。
- **error**：必要事件，监听 AVRecorder 的错误信息。

```
// 状态上报回调函数
avRecorder.on('stateChange', (state, reason) => {
  console.log(`current state is ${state}`);
  // 用户可以在此补充状态发生切换后想要进行的动作
})

// 错误上报回调函数
avRecorder.on('error', (err) => {
  console.error(`avRecorder failed, code is ${err.code}, message is ${err.message}`);
})
```

（3）配置音频录制参数，调用 prepare() 接口，此时进入 prepared 状态。

```
let avProfile = {
  audioBitrate: 100000, // 音频比特率
  audioChannels: 2, // 音频声道数
  audioCodec: media.CodecMimeType.AUDIO_AAC, // 音频编码格式，当前只支持 aac
  audioSampleRate: 48000, // 音频采样率
  fileFormat: media.ContainerFormatType.CFT_MPEG_4A, // 封装格式，当前只支持 m4a
}
let avConfig = {
```

```
      audioSourceType: media.AudioSourceType.AUDIO_SOURCE_TYPE_MIC, // 音频输
入源,这里设置为麦克风
      profile: avProfile,
      url: 'fd://35', // 参考应用文件访问与管理中的开发示例获取创建的音频文件 fd 填入此处
    }
    avRecorder.prepare(avConfig).then(() => {
      console.log('Invoke prepare succeeded.');
    }, (err) => {
      console.error(`Invoke prepare failed, code is ${err.code}, message is ${err.message}`);
    })
```

（4）开始录制，调用 start() 接口，此时进入 started 状态。
（5）暂停录制，调用 pause() 接口，此时进入 paused 状态。
（6）恢复录制，调用 resume() 接口，此时再次进入 started 状态。
（7）停止录制，调用 stop() 接口，此时进入 stopped 状态。
（8）重置资源，调用 reset() 重新进入 idle 状态，允许重新配置录制参数。
（9）销毁实例，调用 release() 进入 released 状态，退出录制。

以下代码实现了完整的录音过程，并且实现了"开始录制—暂停录制—恢复录制—停止录制"完整流程。

```
import media from '@ohos.multimedia.media';

export class AudioRecorderDemo {
  private avRecorder;
  private avProfile = {
    audioBitrate: 100000, // 音频比特率
    audioChannels: 2, // 音频声道数
    audioCodec: media.CodecMimeType.AUDIO_AAC, // 音频编码格式,当前只支持 aac
    audioSampleRate: 48000, // 音频采样率
    fileFormat: media.ContainerFormatType.CFT_MPEG_4A, // 封装格式,当前只支持 m4a
  };
  private avConfig = {
    audioSourceType: media.AudioSourceType.AUDIO_SOURCE_TYPE_MIC, // 音频输
入源,这里设置为麦克风
    profile: this.avProfile,
    url: 'fd://35', // 参考应用文件访问与管理开发示例新建并读写一个文件
  };

  // 注册 audioRecorder 回调函数
  setAudioRecorderCallback() {
    // 状态机变化回调函数
    this.avRecorder.on('stateChange', (state, reason) => {
      console.log(`AudioRecorder current state is ${state}`);
    })
    // 错误上报回调函数
    this.avRecorder.on('error', (err) => {
      console.error(`AudioRecorder failed, code is ${err.code}, message is ${err.message}`);
    })
```

```
  }
  // 开始录制对应的流程
  async startRecordingProcess() {
    // 1.创建录制实例
    this.avRecorder = await media.createAVRecorder();
    this.setAudioRecorderCallback();
    // 2.获取录制文件 fd 赋予 avConfig 里的 url;参考 FilePicker 文档
    // 3.配置录制参数完成准备工作
    await this.avRecorder.prepare(this.avConfig);
    // 4.开始录制
    await this.avRecorder.start();
  }

  // 暂停录制对应的流程
  async pauseRecordingProcess() {
      if (this.avRecorder.state === 'started') { // 仅在 started 状态下调用 pause 为合理状态切换
      await this.avRecorder.pause();
    }
  }

  // 恢复录制对应的流程
  async resumeRecordingProcess() {
      if (this.avRecorder.state === 'paused') { // 仅在 paused 状态下调用 resume 为合理状态切换
      await this.avRecorder.resume();
    }
  }

  // 停止录制对应的流程
  async stopRecordingProcess() {
    // 1.停止录制
    if (this.avRecorder.state === 'started'
      || this.avRecorder.state === 'paused') { // 仅在 started 或者 paused 状态下调用 stop 为合理状态切换
      await this.avRecorder.stop();
    }
    // 2.重置
    await this.avRecorder.reset();
    // 3.释放录制实例
    await this.avRecorder.release();
    // 4.关闭录制文件 fd
  }

  // 一个完整的【开始录制—暂停录制—恢复录制—停止录制】示例
  async audioRecorderDemo() {
    await this.startRecordingProcess(); // 开始录制
    // 用户此处可以自行设置录制时长,例如,通过设置休眠阻止代码执行
    await this.pauseRecordingProcess(); // 暂停录制
    await this.resumeRecordingProcess(); // 恢复录制
    await this.stopRecordingProcess(); // 停止录制
```

```
    }
  }
```

7.4.2 使用 AudioCapturer 开发音频录制程序

在 HarmonyOS 系统中，AudioCapturer 是一个音频采集器，用于录制 PCM（Pulse Code Modulation）音频数据，适合有音频开发经验的开发者实现更灵活的录制功能。使用 AudioCapturer 开发音频录制程序，涉及 AudioCapturer 实例的创建、音频采集参数的配置、采集的开始与停止、资源的释放等工作。

以下代码演示了使用 AudioCapturer 录制音频的过程，通过初始化、启动、停止和释放操作，使用麦克风采集音频数据，并将捕获的音频以 WAV 格式写入文件。并且在代码中包括了事件监听操作，当捕获帧数达到特定阈值时触发回调。

```
import audio from '@ohos.multimedia.audio';
import fs from '@ohos.file.fs';

const TAG = 'AudioCapturerDemo';

export default class AudioCapturerDemo {
  private audioCapturer = undefined;
  private audioStreamInfo = {
    samplingRate: audio.AudioSamplingRate.SAMPLE_RATE_44100,
    channels: audio.AudioChannel.CHANNEL_1,
    sampleFormat: audio.AudioSampleFormat.SAMPLE_FORMAT_S16LE,
    encodingType: audio.AudioEncodingType.ENCODING_TYPE_RAW
  }
  private audioCapturerInfo = {
    source: audio.SourceType.SOURCE_TYPE_MIC, // 音源类型
    capturerFlags: 0 // 音频采集器标志
  }
  private audioCapturerOptions = {
    streamInfo: this.audioStreamInfo,
    capturerInfo: this.audioCapturerInfo
  }

  // 初始化，创建实例，设置监听事件
  init() {
    audio.createAudioCapturer(this.audioCapturerOptions, (err, capturer) => { // 创建 AudioCapturer 实例
      if (err) {
        console.error(`Invoke createAudioCapturer failed, code is ${err.code}, message is ${err.message}`);
        return;
      }

      console.info(`${TAG}: create AudioCapturer success`);
      this.audioCapturer = capturer;
      this.audioCapturer.on('markReach', 1000, (position) => { // 订阅 markReach 事件，当采集的帧数达到 1000 时触发回调
        if (position === 1000) {
```

```
                console.info('ON Triggered successfully');
            }
        });
        this.audioCapturer.on('periodReach', 2000, (position) => {  // 订阅
periodReach 事件，当采集的帧数达到 2000 时触发回调
            if (position === 2000) {
                console.info('ON Triggered successfully');
            }
        });

    });
}

// 开始一次音频采集
async start() {
    let stateGroup = [audio.AudioState.STATE_PREPARED, audio.AudioState.
STATE_PAUSED, audio.AudioState.STATE_STOPPED];
        if (stateGroup.indexOf(this.audioCapturer.state) === -1) {  // 当且仅
当状态为 STATE_PREPARED、STATE_PAUSED 和 STATE_STOPPED 之一时才能启动采集
        console.error(`${TAG}: start failed`);
        return;
    }
    await this.audioCapturer.start();  // 启动采集

    let context = getContext(this);
    const path = context.filesDir + '/test.wav';  // 采集到的音频文件存储路径

    let file = fs.openSync(path, 0o2 | 0o100);  // 如果文件不存在则创建文件
    let fd = file.fd;
    let numBuffersToCapture = 150;  // 循环写入 150 次
    let count = 0;
    while (numBuffersToCapture) {
        let bufferSize = await this.audioCapturer.getBufferSize();
        let buffer = await this.audioCapturer.read(bufferSize, true);
        let options = {
            offset: count * bufferSize,
            length: bufferSize
        };
        if (buffer === undefined) {
            console.error(`${TAG}: read buffer failed`);
        } else {
            let number = fs.writeSync(fd, buffer, options);
            console.info(`${TAG}: write date: ${number}`);
        }
        numBuffersToCapture--;
        count++;
    }
}

// 停止采集
async stop() {
    // 只有采集器状态为 STATE_RUNNING 或 STATE_PAUSED 的时候才可以停止
```

```
        if (this.audioCapturer.state !== audio.AudioState.STATE_RUNNING && this.
audioCapturer.state !== audio.AudioState.STATE_PAUSED) {
            console.info('Capturer is not running or paused');
            return;
        }
        await this.audioCapturer.stop(); // 停止采集
        if (this.audioCapturer.state === audio.AudioState.STATE_STOPPED) {
            console.info('Capturer stopped');
        } else {
            console.error('Capturer stop failed');
        }
    }

    // 销毁实例,释放资源
    async release() {
        // 采集器状态不是 STATE_RELEASED 或 STATE_NEW 状态,才能 release
        if (this.audioCapturer.state === audio.AudioState.STATE_RELEASED || this.
audioCapturer.state === audio.AudioState.STATE_NEW) {
            console.info('Capturer already released');
            return;
        }
        await this.audioCapturer.release(); // 释放资源
        if (this.audioCapturer.state == audio.AudioState.STATE_RELEASED) {
            console.info('Capturer released');
        } else {
            console.error('Capturer release failed');
        }
    }
}
```

上述代码的实现流程如下。

- 首先,通过 audio.createAudioCapturer 初始化了一个音频捕获器实例,设置了采样率、通道数等参数,并订阅了帧数达到特定阈值时触发的事件。
- 其次,在启动方法中检查音频捕获器的状态,如果是预备、暂停或停止状态,则启动捕获器并进入循环,将采集到的音频数据写入文件,循环次数由 numBuffersToCapture 控制。
- 最后,在运行或暂停状态下通过停止方法停止音频捕获器。并通过释放方法释放音频捕获器的资源,确保其不再可用。

7.4.3 使用 OpenSL ES 开发音频录制程序

在 HarmonyOS 系统中,使用 OpenSL ES 实现音频录制的步骤如下。
(1)添加头文件,代码如下。

```
#include <OpenSLES.h>
#include <OpenSLES_OpenHarmony.h>
#include <OpenSLES_Platform.h>
```

(2)使用 slCreateEngine 接口创建引擎对象,并实例化引擎对象 engine,代码如下。

```
SLObjectItf engineObject = nullptr;
```

```
slCreateEngine(&engineObject, 0, nullptr, 0, nullptr, nullptr);
(*engineObject)->Realize(engineObject, SL_BOOLEAN_FALSE);
```

(3) 获取接口 SL_IID_ENGINE 的引擎接口 engineEngine 实例,代码如下。

```
SLEngineItf engineItf = nullptr;
(*engineObject)->GetInterface(engineObject, SL_IID_ENGINE, &engineItf);
```

(4) 配置录音器信息(配置输入源 audiosource、输出源 audiosink),创建录音对象 pcmCapturerObject。代码如下。

```
SLDataLocator_IODevice io_device = {
    SL_DATALOCATOR_IODEVICE,
    SL_IODEVICE_AUDIOINPUT,
    SL_DEFAULTDEVICEID_AUDIOINPUT,
    NULL
};
SLDataSource audioSource = {
    &io_device,
    NULL
};
SLDataLocator_BufferQueue buffer_queue = {
    SL_DATALOCATOR_BUFFERQUEUE,
    3
};
// 具体参数需要根据音频文件格式进行适配
SLDataFormat_PCM format_pcm = {
    SL_DATAFORMAT_PCM,              // 输入的音频格式
    1,                                                          // 单声道
    SL_SAMPLINGRATE_44_1,           // 采样率:44100HZ
    SL_PCMSAMPLEFORMAT_FIXED_16,    // 音频采样格式,小尾数,带符号的16位整数
    0,
    0,
    0
};
SLDataSink audioSink = {
    &buffer_queue,
    &format_pcm
};

SLObjectItf pcmCapturerObject = nullptr;
(*engineItf)->CreateAudioRecorder(engineItf, &pcmCapturerObject,
    &audioSource, &audioSink, 0, nullptr, nullptr);
(*pcmCapturerObject)->Realize(pcmCapturerObject, SL_BOOLEAN_FALSE);
```

(5) 获取录音接口 SL_IID_RECORD 的 recordItf 接口实例,代码如下。

```
SLRecordItf  recordItf;
(*pcmCapturerObject)->GetInterface(pcmCapturerObject, SL_IID_RECORD,
&recordItf);
```

(6) 获取接口 SL_IID_OH_BUFFERQUEUE 的 bufferQueueItf 实例,代码如下。

```
SLOHBufferQueueItf bufferQueueItf;
```

```
(*pcmCapturerObject)->GetInterface(pcmCapturerObject, SL_IID_OH_
BUFFERQUEUE, &bufferQueueItf);
```

(7) 注册 BufferQueueCallback 回调,代码如下。

```
static void BufferQueueCallback(SLOHBufferQueueItf bufferQueueItf, void
*pContext, SLuint32 size)
{
    // 可从 pContext 获取注册时传入的使用者信息
    SLuint8 *buffer = nullptr;
    SLuint32 pSize = 0;
    (*bufferQueueItf)->GetBuffer(bufferQueueItf, &buffer, &pSize);
    if (buffer != nullptr) {
        // 可从 buffer 内读取录音数据进行后续处理
        (*bufferQueueItf)->Enqueue(bufferQueueItf, buffer, size);
    }
}
void *pContext; // 可传入自定义的上下文信息,会在 Callback 内收到
(*bufferQueueItf)->RegisterCallback(bufferQueueItf, BufferQueueCallback,
pContext);
```

(8) 开始录音,代码如下。

```
(*recordItf)->SetRecordState(recordItf, SL_RECORDSTATE_RECORDING);
```

(9) 结束音频录制,代码如下。

```
(*recordItf)->SetRecordState(recordItf, SL_RECORDSTATE_STOPPED);
(*pcmCapturerObject)->Destroy(pcmCapturerObject);
```

7.4.4 管理麦克风

在 HarmonyOS 系统中,录制音频的过程中需要使用麦克风录制相关音频数据,因此建议开发者在调用录制接口前查询麦克风状态,并在录制过程中监听麦克风的状态变化,避免影响录制效果。

需要注意的是,在音频录制过程中将麦克风静音,此时录音过程正常进行,录制生成的数据文件的大小随录制时长递增,但写入文件的数据均为 0,即无声数据(空白数据)。

在 AudioVolumeGroupManager 中提供了管理麦克风状态的方法,例如,以下代码演示了使用 AudioVolumeGroupManager 设置麦克风静音和取消麦克风静音的过程。

```
import audio from '@ohos.multimedia.audio';

@Entry
@Component
struct AudioVolumeGroup {
 private audioVolumeGroupManager: audio.AudioVolumeGroupManager;

    async loadVolumeGroupManager() {
      const groupid = audio.DEFAULT_VOLUME_GROUP_ID;
      this.audioVolumeGroupManager = await audio.getAudioManager().
getVolumeManager().getVolumeGroupManager(groupid);
```

```
      console.info('audioVolumeGroupManager------create-------success.');
  }
  async on() {     // 监听麦克风状态变化
    await this.loadVolumeGroupManager();
    this.audioVolumeGroupManager.on('micStateChange', (micStateChange) => {
      console.info(`Current microphone status is: ${micStateChange.mute} `);
    });
  }
  async isMicrophoneMute() { // 查询麦克风是否静音
    await this.audioVolumeGroupManager.isMicrophoneMute().then((value) => {
      console.info(`isMicrophoneMute is: ${value}.`);
    });
  }
  async setMicrophoneMuteTrue() { // 设置麦克风静音
    await this.loadVolumeGroupManager();
    await this.audioVolumeGroupManager.setMicrophoneMute(true).then(() => {
      console.info('setMicrophoneMute to mute.');
    });
  }
  async setMicrophoneMuteFalse() { // 取消麦克风静音
    await this.loadVolumeGroupManager();
    await this.audioVolumeGroupManager.setMicrophoneMute(false).then(() => {
      console.info('setMicrophoneMute to not mute.');
    });
  }
  async test(){
    await this.on();
    await this.isMicrophoneMute();
    await this.setMicrophoneMuteTrue();
    await this.isMicrophoneMute();
    await this.setMicrophoneMuteFalse();
    await this.isMicrophoneMute();
    await this.setMicrophoneMuteTrue();
    await this.isMicrophoneMute();
  }
}
```

上述代码定义了一个名为 AudioVolumeGroup 的结构体，通过 audio 模块实现了在 OpenHarmony 环境中控制麦克风音量和静音状态的功能。上述代码的实现流程如下。

- 首先，通过 audio.getAudioManager().getVolumeManager().getVolumeGroupManager 方法获取默认音量组的管理器，并在 loadVolumeGroupManager 方法中进行初始化操作。
- 其次，通过 on 方法监听麦克风状态变化的事件 micStateChange，并在麦克风状态发生变化时输出显示当前麦克风的静音状态信息。
- 再次，通过 isMicrophoneMute 方法查询当前麦克风是否处于静音状态，并打印输出查询结果。
- 复次，在接下来的 setMicrophoneMuteTrue 方法和 setMicrophoneMuteFalse 方法中，分别设置麦克风静音为 true 和 false，并输出相应的状态信息，true 和 false 分别表示设置静音和取消静音。
- 最后，通过 test 方法依次执行了一系列操作：开启监听麦克风状态、查询麦克风是否静音、将麦克风静音、再次查询麦克风状态、取消麦克风静音、最终查询麦克风状态。这样的测试流程验证了整个功能的正确性。

7.5 音频通话

现实应用中，常用的音频通话模式包括 VOIP 通话和蜂窝通话，具体说明如下。
- VOIP（Voice over Internet Protocol）通话：是指基于互联网协议（IP）进行通信的一种语音通话技术。VOIP 通话会将通话信息打包成数据包，通过网络进行传输，因此 VOIP 通话对网络要求较高，通话质量与网络连接速度紧密相关。
- 蜂窝通话（仅对系统应用开放）：是指传统的电话功能，由运营商提供服务，目前仅对系统应用开放，未向三方应用提供开发接口。

在开发音频通话相关功能时，开发者可以根据实际情况，检查当前的音频场景模式和铃声模式，以使用相应的音频处理策略。

7.5.1 音频通话基础

在使用音频通话的相关功能时，系统会切换至与通话相关的音频场景模式（AudioScene）。在 HarmonyOS 系统中预置了多种音频场景，包括响铃、通话、语音聊天等，在不同的场景下，系统会采用不同的策略来处理音频。当前在 HarmonyOS 系统中预置的音频场景有以下几点。
- AUDIO_SCENE_DEFAULT：默认音频场景，音频通话之外的场景均可使用。
- AUDIO_SCENE_VOICE_CHAT：语音聊天音频场景，VOIP 通话时使用。

HarmonyOS 应用程序可通过 AudioManager 的 getAudioScene 来获取当前的音频场景模式，当应用开始或结束使用音频通话相关功能时，可通过此方法检查系统是否已切换为合适的音频场景模式。

在用户进入音频通话时，可以使用铃声或振动来提示用户。系统通过调整铃声模式（AudioRingMode），实现便捷地管理铃声音量，并调整设备的振动模式。当前 HarmonyOS 预置了如下三种铃声模式。
- RINGER_MODE_SILENT：静音模式，此模式下铃声音量为零（即静音）。
- RINGER_MODE_VIBRATE：振动模式，此模式下铃声音量为零，设备振动开启（即响铃时静音，触发振动）。
- RINGER_MODE_NORMAL：响铃模式，此模式下铃声音量正常。

HarmonyOS 应用程序可以调用 AudioVolumeGroupManager 中的 getRingerMode 获取当前的铃声模式，以便采取合适的提示策略。

如果希望及时获取铃声模式的变化情况，可以通过 AudioVolumeGroupManager 中的 on('ringerModeChange') 监听铃声模式变化事件，使应用在铃声模式发生变化时及时收到通知，方便应用作出相应的调整。

7.5.2 开发音频通话功能

在音频通话场景下，音频输出（播放对端声音）和音频输入（录制本端声音）会同时进行。在 HarmonyOS 应用程序中，可以使用 AudioRenderer 实现音频输出功能，使用 AudioCapturer 实现音频输入功能，同时使用 AudioRenderer 和 AudioCapturer 实现音频通话功能。在音频通话开始和结束时，应用程序可以自行检查当前的音频场景模式和铃声模式，以便采取合适的音频管理及提示策略。

1. 使用 AudioRenderer 播放对端的通话声音

在 HarmonyOS 系统中，使用 AudioRenderer 播放对端的通话声音的过程与使用 AudioRenderer 开发音频播放功能的过程相似，关键区别在于 audioRenderInfo 参数和音频数据来源。在 audioRenderInfo 参数中，音频内容类型需设置为语音，CONTENT_TYPE_SPEECH，音频流使用类型需设置为语音通信，STREAM_USAGE_VOICE_COMMUNICATION。

在以下代码中，演示了使用 AudioRenderer 实现音频通话的过程，其中未包含音频通话数据的传输过程。在实际开发中，需要将网络传输来的对端通话数据解码播放，此处仅以读取音频文件的数据代替；同时需要将本端录制的通话数据编码打包，通过网络发送给对端，此处仅以将数据写入音频文件代替。

```
import audio from '@ohos.multimedia.audio';
import fs from '@ohos.file.fs';
const TAG = 'VoiceCallDemoForAudioRenderer';
// 与使用 AudioRenderer 开发音频播放功能过程相似，关键区别在于 audioRendererInfo
参数和音频数据来源
export default class VoiceCallDemoForAudioRenderer {
  private renderModel = undefined;
  private audioStreamInfo = {
    samplingRate: audio.AudioSamplingRate.SAMPLE_RATE_48000, // 采样率
    channels: audio.AudioChannel.CHANNEL_2, // 通道
    sampleFormat: audio.AudioSampleFormat.SAMPLE_FORMAT_S16LE, // 采样格式
    encodingType: audio.AudioEncodingType.ENCODING_TYPE_RAW // 编码格式
  }
  private audioRendererInfo = {
    // 需使用通话场景相应的参数
    content: audio.ContentType.CONTENT_TYPE_SPEECH, // 音频内容类型：语音
    usage: audio.StreamUsage.STREAM_USAGE_VOICE_COMMUNICATION, // 音频流使用
类型：语音通信
    rendererFlags: 0 // 音频渲染器标志：默认为 0 即可
  }
  private audioRendererOptions = {
    streamInfo: this.audioStreamInfo,
    rendererInfo: this.audioRendererInfo
  }
  // 初始化，创建实例，设置监听事件
  init() {
    audio.createAudioRenderer(this.audioRendererOptions, (err, renderer) => { // 创建 AudioRenderer 实例
      if (!err) {
        console.info(`${TAG}: creating AudioRenderer success`);
```

```
            this.renderModel = renderer;
            this.renderModel.on('stateChange', (state) => { // 设置监听事件，当转
换到指定的状态时触发回调
                if (state == 1) {
                    console.info('audio renderer state is: STATE_PREPARED');
                }
                if (state == 2) {
                    console.info('audio renderer state is: STATE_RUNNING');
                }
            });
            this.renderModel.on('markReach', 1000, (position) => { // 订阅markReach
事件，当渲染的帧数达到1000帧时触发回调
                if (position == 1000) {
                    console.info('ON Triggered successfully');
                }
            });
        } else {
            console.info(`${TAG}: creating AudioRenderer failed, error: ${err.message}`);
        }
    });
}
// 开始一次音频渲染
async start() {
    let stateGroup = [audio.AudioState.STATE_PREPARED, audio.AudioState.STATE_PAUSED, audio.AudioState.STATE_STOPPED];
        if (stateGroup.indexOf(this.renderModel.state) === -1) { // 当且仅当
状态为STATE_PREPARED、STATE_PAUSED和STATE_STOPPED之一时才能启动渲染
        console.error(TAG + 'start failed');
        return;
    }
    await this.renderModel.start(); // 启动渲染
    const bufferSize = await this.renderModel.getBufferSize();
    // 此处仅以读取音频文件的数据举例，实际音频通话开发中，需要读取的是通话对端传输来
的音频数据
    let context = getContext(this);
    let path = context.filesDir;
    const filePath = path + '/voice_call_data.wav'; // 沙箱路径，实际路径为/
data/storage/el2/base/haps/entry/files/voice_call_data.wav
    let file = fs.openSync(filePath, fs.OpenMode.READ_ONLY);
    let stat = await fs.stat(filePath);
    let buf = new ArrayBuffer(bufferSize);
    let len = stat.size % bufferSize === 0 ? Math.floor(stat.size / bufferSize) : Math.floor(stat.size / bufferSize + 1);
    for (let i = 0; i < len; i++) {
        let options = {
            offset: i * bufferSize,
            length: bufferSize
        };
        let readsize = await fs.read(file.fd, buf, options);
        // buf 是要写入缓冲区的音频数据，在调用AudioRenderer.write()方法前可以进行音
频数据的预处理，实现个性化的音频播放功能，AudioRenderer会读出写入缓冲区的音频数据进行渲染
```

```
          let writeSize = await new Promise((resolve, reject) => {
            this.renderModel.write(buf, (err, writeSize) => {
              if (err) {
                reject(err);
              } else {
                resolve(writeSize);
              }
            });
          });
          if (this.renderModel.state === audio.AudioState.STATE_RELEASED) {
// 如果渲染器状态为 STATE_RELEASED, 停止渲染
            fs.close(file);
            await this.renderModel.stop();
          }
          if (this.renderModel.state === audio.AudioState.STATE_RUNNING) {
            if (i === len - 1) { // 如果音频文件已经被读取完, 停止渲染
              fs.close(file);
              await this.renderModel.stop();
            }
          }
        }
    }
    // 暂停渲染
    async pause() {
      // 只有渲染器状态为 STATE_RUNNING 的时候才能暂停
      if (this.renderModel.state !== audio.AudioState.STATE_RUNNING) {
        console.info('Renderer is not running');
        return;
      }
      await this.renderModel.pause(); // 暂停渲染
      if (this.renderModel.state === audio.AudioState.STATE_PAUSED) {
        console.info('Renderer is paused.');
      } else {
        console.error('Pausing renderer failed.');
      }
    }
    // 停止渲染
    async stop() {
      // 只有渲染器状态为 STATE_RUNNING 或 STATE_PAUSED 的时候才可以停止
      if (this.renderModel.state !== audio.AudioState.STATE_RUNNING && this.renderModel.state !== audio.AudioState.STATE_PAUSED) {
        console.info('Renderer is not running or paused.');
        return;
      }
      await this.renderModel.stop(); // 停止渲染
      if (this.renderModel.state === audio.AudioState.STATE_STOPPED) {
        console.info('Renderer stopped.');
      } else {
        console.error('Stopping renderer failed.');
      }
    }
    // 销毁实例, 释放资源
```

```
  async release() {
    // 渲染器状态不是 STATE_RELEASED 状态,才能 release
    if (this.renderModel.state === audio.AudioState.STATE_RELEASED) {
      console.info('Renderer already released');
      return;
    }
    await this.renderModel.release(); // 释放资源
    if (this.renderModel.state === audio.AudioState.STATE_RELEASED) {
      console.info('Renderer released');
    } else {
      console.error('Renderer release failed.');
    }
  }
}
```

对上述代码的具体说明如下。

- 首先,通过 audio.createAudioRenderer 创建一个 AudioRenderer 实例,配置了音频流信息、渲染器信息,以及相关的监听事件。在初始化过程中,通过监听 stateChange 事件和 markReach 事件,实时获取渲染器的状态变化和帧数达到指定阈值的信息。
- 其次,在 start 方法中判断渲染器的状态,仅当状态为 STATE_PREPARED、STATE_PAUSED 或 STATE_STOPPED 时才启动渲染。随后,通过读取音频文件数据,以缓冲区为单位进行循环写入,实现音频渲染功能。在每次写入之前,可以进行音频数据的预处理,以实现个性化的音频播放功能。
- 再次,通过 pause 方法暂停渲染,仅在渲染器状态为 STATE_RUNNING 时才可执行。同样,通过 stop 方法停止渲染,条件是渲染器状态为 STATE_RUNNING 或 STATE_PAUSED。
- 最后,通过 release 方法释放渲染器的资源,仅当渲染器状态不为 STATE_RELEASED 时才执行释放工作。

2. 使用 AudioCapturer 录制本端的通话声音

在 HarmonyOS 系统中,使用 AudioCapturer 录制本端的通话声音的过程与使用 AudioCapturer 开发音频录制功能的过程相似,主要区别在于参数 audioCapturerInfo 和音频数据流向。需要将音源类型参数 audioCapturerInfo 设置为使用语音通话:SOURCE_TYPE_VOICE_COMMUNICATION。例如,在以下代码中使用了 @ohos.multimedia.audio 模块,通过类 AudioCapturer 进行音频采集,以及通过类 AudioRenderer 进行音频渲染,从而实现了语音通话的录制和播放功能。在语音通话过程中,这段代码还考虑了一些关键参数,如采样率、通道数、采样格式等,以满足语音通话场景的需求。

```
import audio from '@ohos.multimedia.audio';
import fs from '@ohos.file.fs';
const TAG = 'VoiceCallDemoForAudioCapturer';
// 与使用 AudioCapturer 开发音频录制功能过程相似,关键区别在于 audioCapturerInfo
参数和音频数据流向
export default class VoiceCallDemoForAudioCapturer {
  private audioCapturer = undefined;
  private audioStreamInfo = {
```

```
      samplingRate: audio.AudioSamplingRate.SAMPLE_RATE_44100, // 采样率
      channels: audio.AudioChannel.CHANNEL_1, // 通道
      sampleFormat: audio.AudioSampleFormat.SAMPLE_FORMAT_S16LE, // 采样格式
      encodingType: audio.AudioEncodingType.ENCODING_TYPE_RAW // 编码格式
    }
    private audioCapturerInfo = {
      // 需使用通话场景相应的参数
      source: audio.SourceType.SOURCE_TYPE_VOICE_COMMUNICATION, // 音源类型：语音通话
      capturerFlags: 0 // 音频采集器标志：默认为 0 即可
    }
    private audioCapturerOptions = {
      streamInfo: this.audioStreamInfo,
      capturerInfo: this.audioCapturerInfo
    }
    // 初始化，创建实例，设置监听事件
    init() {
      audio.createAudioCapturer(this.audioCapturerOptions, (err, capturer) => { // 创建 AudioCapturer 实例
        if (err) {
          console.error(`Invoke createAudioCapturer failed, code is ${err.code}, message is ${err.message}`);
          return;
        }
        console.info(`${TAG}: create AudioCapturer success`);
        this.audioCapturer = capturer;
        this.audioCapturer.on('markReach', 1000, (position) => { // 订阅 markReach 事件，当采集的帧数达到 1000 时触发回调
          if (position === 1000) {
            console.info('ON Triggered successfully');
          }
        });
        this.audioCapturer.on('periodReach', 2000, (position) => { // 订阅 periodReach 事件，当采集的帧数达到 2000 时触发回调
          if (position === 2000) {
            console.info('ON Triggered successfully');
          }
        });
      });
    }
    // 开始一次音频采集
    async start() {
      let stateGroup = [audio.AudioState.STATE_PREPARED, audio.AudioState.STATE_PAUSED, audio.AudioState.STATE_STOPPED];
      if (stateGroup.indexOf(this.audioCapturer.state) === -1) { // 当且仅当状态为 STATE_PREPARED、STATE_PAUSED 和 STATE_STOPPED 之一时才能启动采集
        console.error(`${TAG}: start failed`);
        return;
      }
      await this.audioCapturer.start(); // 启动采集
      // 此处仅以将音频数据写入文件举例，实际音频通话开发中，需要将本端采集的音频数据编码打包，通过网络发送给通话对端
```

```
      let context = getContext(this);
      const path = context.filesDir + '/voice_call_data.wav'; // 采集到的音频
文件存储路径
      let file = fs.openSync(path, 0o2 | 0o100); // 如果文件不存在则创建文件
      let fd = file.fd;
      let numBuffersToCapture = 150; // 循环写入 150 次
      let count = 0;
      while (numBuffersToCapture) {
        let bufferSize = await this.audioCapturer.getBufferSize();
        let buffer = await this.audioCapturer.read(bufferSize, true);
        let options = {
          offset: count * bufferSize,
          length: bufferSize
        };
        if (buffer === undefined) {
          console.error(`${TAG}: read buffer failed`);
        } else {
          let number = fs.writeSync(fd, buffer, options);
          console.info(`${TAG}: write date: ${number}`);
        }
        numBuffersToCapture--;
        count++;
      }
    }
    // 停止采集
    async stop() {
      // 只有采集器状态为 STATE_RUNNING 或 STATE_PAUSED 的时候才可以停止
      if (this.audioCapturer.state !== audio.AudioState.STATE_RUNNING &&
this.audioCapturer.state !== audio.AudioState.STATE_PAUSED) {
        console.info('Capturer is not running or paused');
        return;
      }
      await this.audioCapturer.stop(); // 停止采集
      if (this.audioCapturer.state === audio.AudioState.STATE_STOPPED) {
        console.info('Capturer stopped');
      } else {
        console.error('Capturer stop failed');
      }
    }
    // 销毁实例，释放资源
    async release() {
      // 采集器状态不是 STATE_RELEASED 或 STATE_NEW 状态，才能 release
      if (this.audioCapturer.state === audio.AudioState.STATE_RELEASED ||
this.audioCapturer.state === audio.AudioState.STATE_NEW) {
        console.info('Capturer already released');
        return;
      }
      await this.audioCapturer.release(); // 释放资源
      if (this.audioCapturer.state == audio.AudioState.STATE_RELEASED) {
        console.info('Capturer released');
      } else {
        console.error('Capturer release failed');
```

        }
    }
}

7.6 视频播放

在 HarmonyOS 系统中，提供了如下两种开发视频播放程序的方案。
- AVPlayer：这是一种功能较完善的音视频播放 ArkTS/JS API，集成了流媒体和本地资源解析、媒体资源解封装、视频解码和渲染功能，适用于对媒体资源进行端到端播放的场景，可直接播放 mp4、mkv 等格式的视频文件。
- Video 组件：封装了视频播放的基础能力，需要设置数据源以及基础信息即可播放视频，但相对扩展能力较弱。Video 组件由 ArkUI 提供能力，相关指导请参考 UI 开发文档 -Video 组件。

HarmonyOS 官方建议使用 AVPlayer。在 HarmonyOS 系统中，使用 AVPlayer 播放视频的方法与前面 7.3.1 介绍的播放音频类似，AVPlayer 的有关内置属性、内置方法和播放步骤请参考本书 7.3.1 中的内容。在接下来的内容中，将讲解使用 AVPlayer 开发一个视频播放器的过程。

在实例 7-2 中，将讲解使用 AVPlayer 开发一个视频播放器的过程。

实例 7-2：AVPlayer 视频播放器（源码路径 :codes\7\AVPlayer）

（1）编写文件 src/main/module.json5，这是一个典型的鸿蒙项目 module.json5 文件，它设置了应用程序的模块信息。这个文件主要定义了应用程序的入口模块，设置入口文件的路径是 "./ets/entryability/EntryAbility.ts"。

扫码看视频

（2）编写文件 src/main/ets/common/util/DateFormatUtil.ets，定义了类 DateFormatUtil，其中包含了两个方法。
- 第一个 secondToTime 方法接受一个表示秒数的参数，将其转换为 HH:mm:ss 格式的时间字符串。
- 第二个 padding 方法用于对数字进行零填充，确保保留两位。

由此可见，这个工具类 DateFormatUtil 的主要功能是处理时间格式，尤其是将秒数转换为可读的时间表示形式。

（3）编写文件 src/main/ets/pages/HomePage.ets，定义了一个名为 HomePage 的 HarmonyOS 页面结构，用于展示两个选项卡（本地视频和网络视频）。在页面中包含了选项卡构建、样式设置以及切换选项卡时更新当前索引等功能，具体包括设置屏幕大小、构建选项卡内容、设置选项卡样式和行为等。

（4）编写文件 src/main/ets/pages/PlayPage.ets，定义了一个名为 PlayPage 的 HarmonyOS 页面，用于展示视频播放相关的界面和控制，实现了视频播放的相关功能，包括亮度调节和音量调节的手势操作。

```
import router from '@ohos.router';
import { PlayTitle } from '../view/PlayTitle';
import { PlayPlayer } from '../view/PlayPlayer';
import { PlayControl } from '../view/PlayControl';
```

```
import { PlayProgress } from '../view/PlayProgress';
import { VideoController } from '../controller/VideoController';
import { CommonConstants } from '../common/constants/CommonConstants';
import { PlayConstants } from '../common/constants/PlayConstants';
import resourceManager from '@ohos.resourceManager';

@Entry
@Component
struct PlayPage {
  @State videoHeight: string = PlayConstants.PLAY_PLAYER_HEIGHT;
  @State videoWidth: string = CommonConstants.FULL_PERCENT;
  @State videoMargin: string = PlayConstants.MARGIN_ZERO;
  @State videoPosition: FlexAlign = FlexAlign.Center;
  private playVideoModel: VideoController = new VideoController();
  @Provide src: resourceManager.RawFileDescriptor = {} as resourceManager.RawFileDescriptor;
  @Provide iSrc: string = '';
  @Provide index: number = 0;
  @Provide type: number = 0;
  @Provide status: number = CommonConstants.STATUS_START;
  private panOptionBright: PanGestureOptions = new PanGestureOptions({ direction: PanDirection.Vertical });
  private panOptionVolume: PanGestureOptions = new PanGestureOptions({ direction: PanDirection.Horizontal });

  aboutToAppear() {
    // 获取页面跳转参数并初始化视频控制器
    let params = router.getParams() as Record<string, Object>;
    this.src = params.src as resourceManager.RawFileDescriptor;
    this.iSrc = params.iSrc as string;
    this.index = params.index as number;
    this.type = params.type as number;
    this.playVideoModel.initPlayPageThis(this);
  }

  aboutToDisappear() {
    // 页面即将消失时释放资源
    this.playVideoModel.release();
  }

  onPageHide() {
    // 页面隐藏时暂停播放
    this.status = CommonConstants.STATUS_PAUSE;
    this.playVideoModel.pause();
  }

  build() {
    // 页面结构
    Stack() {
      Column() {
        Column() {
        }
```

```
        .height(this.videoMargin)

      // 播放器视图
      PlayPlayer({ playVideoModel: this.playVideoModel })
        .width(this.videoWidth)
        .height(this.videoHeight)
  }
  .height(CommonConstants.FULL_PERCENT)
  .width(CommonConstants.FULL_PERCENT)
  .justifyContent(this.videoPosition)
  .zIndex(0)

  Column() {
    // 播放标题
    PlayTitle({ playVideoModel: this.playVideoModel })
      .width(CommonConstants.FULL_PERCENT)
      .height(PlayConstants.HEIGHT)
    Column()
      .width(CommonConstants.FULL_PERCENT)
      .height(PlayConstants.COLUMN_HEIGHT_ONE)
      .gesture(
        PanGesture(this.panOptionBright)
          .onActionStart((event?: GestureEvent) => {
            this.playVideoModel.onBrightActionStart(event);
          })
          .onActionUpdate((event?: GestureEvent) => {
            this.playVideoModel.onBrightActionUpdate(event);
          })
          .onActionEnd(() => {
            this.playVideoModel.onActionEnd();
          })
      )
    Column() {
    }
    .width(CommonConstants.FULL_PERCENT)
    .height(PlayConstants.PLAY_PLAYER_HEIGHT)

    Column()
      .width(CommonConstants.FULL_PERCENT)
      .height(PlayConstants.COLUMN_HEIGHT_TWO)
      .gesture(
        PanGesture(this.panOptionVolume)
          .onActionStart((event?: GestureEvent) => {
            this.playVideoModel.onVolumeActionStart(event);
          })
          .onActionUpdate((event?: GestureEvent) => {
            this.playVideoModel.onVolumeActionUpdate(event);
          })
          .onActionEnd(() => {
            this.playVideoModel.onActionEnd();
          })
      )
```

```
      // 播放控制器
      PlayControl({ playVideoModel: this.playVideoModel })
        .width(CommonConstants.FULL_PERCENT)
        .height(PlayConstants.HEIGHT)
      // 播放进度条
      PlayProgress({ playVideoModel: this.playVideoModel })
        .width(CommonConstants.FULL_PERCENT)
        .height(PlayConstants.PLAY_PROGRESS_HEIGHT)
    }
    .height(CommonConstants.FULL_PERCENT)
    .width(CommonConstants.FULL_PERCENT)
    .zIndex(1)
  }
  .height(CommonConstants.FULL_PERCENT)
  .width(CommonConstants.FULL_PERCENT)
  .backgroundColor(Color.Black)
  }
}
```

上述代码包含了视频播放器、播放控制、播放进度等组件，并通过 VideoController 控制视频的播放行为。其中，页面提供了一些状态信息，并通过页面生命周期函数来处理页面即将出现、即将消失和页面隐藏时的操作。执行效果如图 7-7 所示。

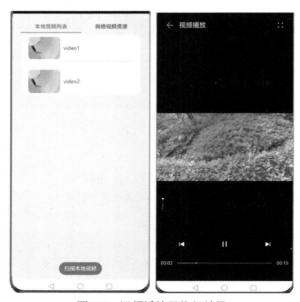

图 7-7　视频播放器执行效果

第 8 章
相机开发

相机程序开发在手机应用领域至关重要,因为用户普遍依赖手机相机来捕捉生活瞬间、分享照片和视频。一个优秀的相机应用程序不仅能提供良好的用户体验,还能增强社交互动,同时促进技术创新,为整个应用生态系统注入活力,提高应用的竞争力。本章将详细讲解在 HarmonyOS 系统中开发相机拍照应用程序的知识。

8.1 相机开发概述

HarmonyOS 相机模块支持相机业务的开发，开发者通过已开放的接口实现相机硬件的访问、操作和新功能开发。最常见的操作包括预览、拍照、连拍和录像等。与相机开发相关的概念如下。

- 相机静态能力：用于描述相机固有能力的一系列参数，如朝向、支持的分辨率等信息。
- 物理相机：独立的实体摄像头设备。物理相机 ID 用于标识每个物理摄像头的唯一字符串。
- 逻辑相机：多个物理相机组合成的抽象设备，逻辑相机通过同时控制多个物理相机设备来完成某些相机功能，如大光圈、变焦等。逻辑相机 ID 是一个唯一的字符串，标识多个物理相机的抽象能力。
- 帧捕获：相机启动后对帧的捕获动作统称为帧捕获，主要包含单帧捕获、多帧捕获和循环帧捕获。
- 单帧捕获：相机启动后，在帧数据流中捕获一帧数据，常用于普通拍照。
- 多帧捕获：相机启动后，在帧数据流中连续捕获多帧数据，常用于连拍。
- 循环帧捕获：相机启动后，在帧数据流中持续捕获帧数据，常用于预览和录像。

8.2 开发相机程序

在 HarmonyOS 系统中，相机模块的主要工作是为相机应用开发者提供基本的相机 API 接口，使用相机系统的功能，进行相机硬件的访问、操作和新功能开发。开发 HarmonyOS 相机程序的流程，如图 8-1 所示。

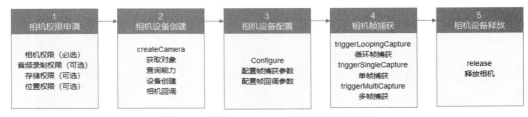

图 8-1 开发 HarmonyOS 相机程序的流程

8.2.1 相机接口

为了方便开发者更容易地实现相机功能，HarmonyOS 相机模块为相机应用开发者提供了 3 个包的内容，分别为方法、枚举以及常量 / 变量，具体说明如下。

- ohos.media.camera.CameraKit：相机功能入口类，用于获取当前支持的相机列表及其静态能力信息，创建相机对象。
- ohos.media.camera.device：相机设备操作类，提供相机能力查询、相机配置、相机帧捕获、相机状态回调等功能。
- ohos.media.camera.params：相机参数类，提供相机属性、参数和操作结果的定义。

在使用 HarmonyOS 相机模块之前，需要先申请相机权限以保证应用程序拥有相机硬件及其他功能权限。表 8-1 所示为 HarmonyOS 相机模块涉及的权限。

表 8-1　HarmonyOS 相机模块涉及的权限

权限名称	权限属性值	是否必选
相机权限	ohos.permission.CAMERA	必选
录音权限	ohos.permission.MICROPHONE	可选（需要录像时申请）
存储权限	ohos.permission.WRITE_MEDIA	可选（需要保存图像及视频到设备的外部存储时申请）
位置权限	ohos.permission.MEDIA_LOCATION	可选（需要保存图像及视频位置信息时申请）

8.2.2　创建相机设备

在 HarmonyOS 系统中，CameraKit 是相机的入口 API 类，用于实现获取相机设备特性、打开相机等功能。类 CameraKit 中包含以下接口。

- createCamera(String cameraId, CameraStateCallback callback, EventHandler handler)：创建相机对象。
- getCameraAbility(String cameraId)：获取指定逻辑相机或物理相机的静态能力。
- getCameraIds()：获取当前逻辑相机列表。
- getCameraInfo(String cameraId)：获取指定逻辑相机的信息。
- getInstance(Context context)：获取 CameraKit 实例。
- registerCameraDeviceCallback(CameraDeviceCallback callback, EventHandler handler)：注册相机使用状态回调。
- unregisterCameraDeviceCallback(CameraDeviceCallback callback)：注销相机使用状态回调。

在实现一个相机应用程序时，必须先创建一个独立的相机设备，然后才能继续实现相机的其他操作。HarmonyOS 官方建议创建相机设备的基本步骤如下。

（1）通过 CameraKit.getInstance(Context context) 方法获取唯一的 CameraKit 对象。例如，以下代码片段展示了打开相机的简单方法，使用的是类 CameraKit。

```
private void openCamera(){
    // 获取 CameraKit 对象
    CameraKit cameraKit = CameraKit.getInstance(getApplicationContext());
    if (cameraKit == null) {
        // 处理 cameraKit 获取失败的情况
    }
}
```

如果此步骤操作失败，相机可能被占用或无法使用。如果相机被占用，必须等到相机释放后才能重新获取 CameraKit 对象。

（2）通过 getCameraIds() 方法，获取当前使用的设备支持的逻辑相机列表。逻辑相机列表中存储了当前设备拥有的所有逻辑相机 ID。如果列表不为空，则列表中的每个 ID 都支持独立创建相机对象；否则，说明正在使用的设备无可用的相机，不能继续后续的操作。

```
try {
    // 获取当前设备的逻辑相机列表
    String[] cameraIds = cameraKit.getCameraIds();
    if (cameraIds.length <= 0) {
        HiLog.error(LABEL, "cameraIds size is 0");
    }
} catch (IllegalStateException e) {
    // 处理异常
}
```

接下来，还可以继续查询指定相机 ID 的静态信息，主要包括以下操作。
- 调用 getDeviceLinkType(String physicalId) 方法获取物理相机连接方式。
- 调用 getCameraInfo(String cameraId) 方法查询相机硬件朝向等信息。
- 调用 getCameraAbility(String cameraId) 方法查询相机能力信息（比如支持的分辨率列表等）。

在 HarmonyOS 操作系统中，CameraInfo 是用于描述相机设备信息的类，它提供了有关相机硬件和配置的信息，开发者可以使用这些信息来了解和配置相机的属性。类 CameraInfo 主要包括以下接口。
- getDeviceLinkType(String physicalId)：获取物理相机连接方式。
- getFacingType()：获取相机朝向信息。
- getLogicalId()：获取逻辑相机 ID。
- getPhysicalIdList()：获取对应的物理相机 ID 列表。

在 HarmonyOS 操作系统中，CameraAbility 用于描述相机设备能力的类，它提供了有关相机硬件支持的功能和性能信息，使开发者能够更好地了解相机设备的特性。类 CameraAbility 主要提供了以下接口。
- getSupportedSizes(int format)：根据格式查询输出图像的分辨率列表。
- getSupportedSizes(Class<T> clazz)：根据 Class 类型查询分辨率列表。
- getParameterRange(ParameterKey.Key<T> parameter)：获取指定参数能够设置的值范围。
- getPropertyValue(PropertyKey.Key<T> property)：获取指定属性对应的值。
- getSupportedAeMode()：获取当前相机支持的自动曝光模式。
- getSupportedAfMode()：获取当前相机支持的自动对焦模式。
- getSupportedFaceDetection()：获取相机支持的人脸检测类型范围。
- getSupportedFlashMode()：获取当前相机支持的闪光灯取值范围。
- getSupportedParameters()：获取当前相机支持的参数设置。
- getSupportedProperties()：获取当前相机的属性列表。
- getSupportedResults()：获取当前相机支持的参数设置可返回的结果列表。
- getSupportedZoom()：获取相机支持的变焦范围。

（3）通过 createCamera(String cameraId, CameraStateCallback callback, EventHandler handler)

方法，创建相机对象。如果此步骤执行成功，这意味着相机系统的硬件已经接通电源。

```
// 前置相机类型
int frontCamera = CameraInfo.FacingType.CAMERA_FACING_FRONT;
// 后置相机类型
int backCamera = CameraInfo.FacingType.CAMERA_FACING_BACK;
// 其他相机类型
int otherCamera = CameraInfo.FacingType.CAMERA_FACING_OTHERS;

// 选择想要创建的相机类型，如果不存在该类型相机，则返回false
boolean isCameraCreated = openCameraByFacingType(frontCamera);

// 根据类型创建相机的方法
private boolean openCameraByFacingType(int facingType) {
    CameraKit cameraKit = CameraKit.getInstance(getApplicationContext());
    for(String cameraId : cameraKit.getCameraIds()) {
        CameraInfo cameraInfo = cameraKit.getCameraInfo(cameraId);
        if(facingType == cameraInfo.getFacingType()) {
            cameraKit.createCamera(cameraId, cameraStateCallback, eventHandler);
            return true;
        }
    }
    return false;
}
```

在上述代码中，createCamera(cameraId, cameraStateCallback, eventHandler) 的参数 cameraId 可以是上一步获取的逻辑相机列表中的任何一个相机 ID。第二个参数和第三个参数负责相机创建和相机运行时的数据和状态检测，请务必保障在整个相机运行周期内有效。

```
private final class CameraStateCallbackImpl extends CameraStateCallback {
    @Override
    public void onCreated(Camera camera) {
        // 创建相机设备
    }

    @Override
    public void onConfigured(Camera camera) {
        // 配置相机设备
    }

    @Override
    public void onPartialConfigured(Camera camera) {
        // 当使用了addDeferredSurfaceSize配置了相机，会接到此回调
    }

    @Override
    public void onReleased(Camera camera) {
        // 释放相机设备
    }
}

// 相机创建和相机运行时的回调
```

```
CameraStateCallbackImpl cameraStateCallback = new CameraStateCallbackImpl();

import ohos.eventhandler.EventHandler;
import ohos.eventhandler.EventRunner;

// 执行回调的 EventHandler
EventHandler eventHandler = new EventHandler(EventRunner.create("CameraCb"));
```

至此，整个相机设备的创建工作已经完成。创建相机设备成功后会在CameraStateCallback中触发 onCreated(Camera camera) 回调。在进入相机设备配置前，请确保相机设备已经创建成功。否则会触发相机设备创建失败的回调并返回错误代码，在进行错误处理后重新执行相机设备的创建工作。

8.2.3 配置相机设备

在创建相机设备成功后，会在 CameraStateCallback 中触发 onCreated(Camera camera) 回调，并且带回 Camera 对象，用于执行相机设备的操作。当成功创建一个新的相机设备后，首先需要对相机进行配置，调用 configure(CameraConfig) 方法实现配置。相机配置主要是设置预览、拍照、录像用到的 Surface(详见 ohos.agp.graphics.Surface)，没有配置过 Surface，相应的功能不能使用。

为了实现相机帧捕获结果的数据和状态检测工作，还需要在配置相机时调用 setFrameStateCallback(FrameStateCallback, EventHandler) 方法设置帧回调。以下代码演示了使用 HarmonyOS 相机框架实现相机功能初始化和配置的过程。

```
// Surface 提供对象
private SurfaceProvider surfaceProvider;

private void initSurface() {
    surfaceProvider = new SurfaceProvider(this);
    DirectionalLayout.LayoutConfig params = new DirectionalLayout.LayoutConfig(
            ComponentContainer.LayoutConfig.MATCH_PARENT, ComponentContainer.LayoutConfig.MATCH_PARENT);
    surfaceProvider.setLayoutConfig(params);
    surfaceProvider.pinToZTop(false);
    surfaceProvider.getSurfaceOps().get().addCallback(new SurfaceCallBack());
    ((ComponentContainer)
            findComponentById(ResourceTable.Id_surface_container)).addComponent(surfaceProvider);
}

private FrameStateCallback frameStateCallbackImpl = new FrameStateCallback(){
    @Override
    public void onFrameStarted(Camera camera, FrameConfig frameConfig, long frameNumber, long timestamp) {
        ...
    }
    @Override
    public void onFrameProgressed(Camera camera, FrameConfig frameConfig, FrameResult frameResult) {
```

```java
            ...
        }
        @Override
         public void onFrameFinished(Camera camera, FrameConfig frameConfig,
FrameResult frameResult) {
            ...
        }
        @Override
         public void onFrameError(Camera camera, FrameConfig frameConfig, int
errorCode, FrameResult frameResult) {
            ...
        }
        @Override
        public void onCaptureTriggerStarted(Camera camera, int captureTriggerId,
long firstFrameNumber) {
            ...
        }
        @Override
        public void onCaptureTriggerFinished(Camera camera, int captureTriggerId,
long lastFrameNumber) {
            ...
        }
        @Override
        public void onCaptureTriggerInterrupted(Camera camera, int captureTriggerId) {
            ...
        }
    };

    // 相机设备
    private Camera cameraDevice;
    // 相机预览模板
    private Surface previewSurface;
    // 相机配置模板
    private CameraConfig.Builder cameraConfigBuilder;
    // 图像帧数据接收处理对象
    private ImageReceiver imageReceiver;

    private final class CameraStateCallbackImpl extends CameraStateCallback {
        @Override
        public void onCreated(Camera camera) {
            cameraDevice = camera;
            previewSurface = surfaceProvider.getSurfaceOps().get().getSurface();
            cameraConfigBuilder = camera.getCameraConfigBuilder();
            if (cameraConfigBuilder == null) {
                HiLog.error(LABEL, "onCreated cameraConfigBuilder is null");
                return;
            }
            // 配置预览的 Surface
            cameraConfigBuilder.addSurface(previewSurface);
            // 配置拍照的 Surface
            cameraConfigBuilder.addSurface(imageReceiver.getRecevingSurface());
            // 配置帧结果的回调
```

```
        cameraConfigBuilder.setFrameStateCallback(frameStateCallbackImpl, eventHandler);
        try {
            // 相机设备配置
            camera.configure(cameraConfigBuilder.build());
        } catch (IllegalArgumentException e) {
            HiLog.error(LABEL, "Argument Exception");
        } catch (IllegalStateException e) {
            HiLog.error(LABEL, "State Exception");
        }
    }
}
```

总体而言，上述代码的主要功能是初始化相机预览的 Surface，配置相机设备，设置帧结果的回调，并将相机预览和拍照的 Surface 添加到相机配置中。在 CameraStateCallbackImpl 中，创建相机后执行相应的配置，以便正常地进行相机预览和拍摄操作。此外，该代码使用了鸿蒙的异步事件处理机制，通过 eventHandler 处理相机事件的回调。

在 HarmonyOS 相机框架中，CameraConfig.Builder 用于配置相机参数的构建器类，它允许开发者通过链式调用方法来设置相机的各种配置，包括预览、拍照和录制视频等。CameraConfig.Builder 中的主要接口如下。

- addSurface(Surface surface)：在相机配置中增加 Surface。
- build()：相机配置的构建类。
- removeSurface(Surface surface)：移除先前添加的 Surface。
- setFrameStateCallback(FrameStateCallback callback, EventHandler handler)：设置用于相机帧结果返回的 FrameStateCallback 和 Handler。
- addDeferredSurfaceSize(Size surfaceSize, Class<T> clazz)：添加延迟 Surface 的尺寸和类型。
- addDeferredSurface(Surface surface)：设置延迟的 Surface，此 Surface 的尺寸和类型必须和使用 addDeferredSurfaceSize 配置的一致。

8.2.4 拍照

在 HarmonyOS 系统中，类 Camera 负责实现相机预览、录像、拍照等功能。类 Camera 中的主要接口如下。

- triggerSingleCapture(FrameConfig frameConfig)：启动相机帧的单帧捕获。
- triggerMultiCapture(List<FrameConfig> frameConfigs)：启动相机帧的多帧捕获。
- configure(CameraConfig config)：配置相机。
- flushCaptures()：停止并清除相机帧的捕获，包括循环帧 / 单帧 / 多帧捕获。
- getCameraConfigBuilder()：获取相机配置构造器对象。
- getCameraId()：获取当前相机的 ID。
- getFrameConfigBuilder(int type)：获取指定类型的相机帧配置构造器对象。
- release()：释放相机对象及资源。
- triggerLoopingCapture(FrameConfig frameConfig)：启动或更新相机帧的循环捕获。
- stopLoopingCapture()：停止当前相机帧的循环捕获。

在接下来的内容中，将详细讲解使用类 Camera 实现拍照功能的过程。

1. 启动预览（循环帧捕获）

在拍照时，用户一般都是先看见预览画面才执行拍照或者其他功能，所以对于一个普通的相机应用，预览是必不可少的。启动预览的基本步骤如下。

（1）通过 getFrameConfigBuilder(FRAME_CONFIG_PREVIEW) 方法获取预览配置模板，常用帧配置项如表 8-2 所示。

表 8-2 常用帧配置项

接口名	描述	是否必选
addSurface(Surface surface)	配置预览 surface 和帧的绑定	是
setAfMode(int afMode, Rect rect)	配置对焦模式	否
setAeMode(int aeMode, Rect rect)	配置曝光模式	否
setZoom(float value)	配置变焦值	否
setFlashMode(int flashMode)	配置闪光灯模式	否
setFaceDetection(int type, boolean isEnable)	配置人脸检测或者笑脸检测	否
setParameter(Key\<T\> key, T value)	配置其他属性（如自拍镜像等）	否
setMark(Object mark)	配置一个标签，后续可以从 FrameConfig 中通过 Object getMark() 拿到标签，判断两个是否相等，相等就说明是同一个配置	否
setCoordinateSurface(Surface surface)	配置坐标系基准 Surface，后续计算 Ae/Af 等区域都会基于此 Surface 为基本的中心坐标系，不设置默认使用添加的第一个 Surface	否

（2）通过 triggerLoopingCapture(FrameConfig) 方法实现循环帧捕获（如预览/录像），例如，以下代码在相机配置完成后，通过获取预览帧配置构建器，配置预览 Surface，构建帧配置对象，最后启动循环帧捕获，以实现持续的相机预览功能。异常处理部分用于捕获可能出现的配置异常，以确保代码的稳定性。

```
private final class CameraStateCallbackImpl extends CameraStateCallback {
    @Override
    public void onConfigured(Camera camera) {
        // 获取预览配置模板
        frameConfigBuilder = camera.getFrameConfigBuilder(FRAME_CONFIG_PREVIEW);
        // 配置预览 Surface
        frameConfigBuilder.addSurface(previewSurface);
        previewFrameConfig = frameConfigBuilder.build();
        try {
            // 启动循环帧捕获
            int triggerId = camera.triggerLoopingCapture(previewFrameConfig);
        } catch (IllegalArgumentException e) {
            HiLog.error(LABEL, "Argument Exception");
        } catch (IllegalStateException e) {
            HiLog.error(LABEL, "State Exception");
        }
    }
}
```

通过以上操作，现在相机应用程序已经可以正常进行实时预览了。在预览状态下，开发者还可以执行其他操作，比如，当预览帧配置更改时，可以通过 triggerLoopingCapture (FrameConfig) 方法实现预览帧配置的更新。

```
// 预览帧变焦值变更
frameConfigBuilder.setZoom(1.2f);
// 调用 triggerLoopingCapture 方法实现预览帧配置更新
triggerLoopingCapture(frameConfigBuilder.build());
```

通过 stopLoopingCapture() 方法停止循环帧捕获（停止预览）。

```
// 停止预览帧捕获
camera.stopLoopingCapture()
```

2. 实现拍照（单帧捕获）

拍照功能属于相机应用的最重要功能之一，而且照片质量对用户至关重要。相机模块基于相机复杂的逻辑，从应用接口层到器件驱动层都已经默认地做好了最适合用户的配置，这些默认配置尽可能地保证用户拍出的每张照片的质量。发起拍照的基本步骤如下。

（1）通过 getFrameConfigBuilder(FRAME_CONFIG_PICTURE) 方法获取拍照配置模板，并且设置拍照帧配置，如表 8-3 所示。

表 8-3　常用拍照帧配置的说明

接口名	描　　述	是否必选
FrameConfig.Builder addSurface(Surface)	实现拍照 Surface 和帧的绑定	必选
FrameConfig.Builder setImageRotation(int)	设置图片旋转角度	可选
FrameConfig.Builder setLocation(Location)	设置图片地理位置信息	可选
FrameConfig.Builder setParameter(Key<T>, T)	配置其他属性（如自拍镜像等）	可选

（2）在拍照前准备图像帧数据的接收实现，以下代码中实现了鸿蒙相机框架中拍照功能的初始化和处理工作。通过配置 ImageReceiver 对象接收 JPEG 图像帧数据，选择合适的拍照分辨率，以及通过类 ImageSaver 将图像数据保存为 JPEG 文件。整体流程包括初始化图像接收器、设置回调监听器，以及处理图像保存的线程任务。

```
    // 图像帧数据接收处理对象
    private ImageReceiver imageReceiver;
    // 执行回调的 EventHandler
    private EventHandler eventHandler = new EventHandler(EventRunner.create("CameraCb"));
    // 拍照支持分辨率
    private Size pictureSize;

    // 单帧捕获生成图像回调 Listener
    private final ImageReceiver.IImageArrivalListener imageArrivalListener = new ImageReceiver.IImageArrivalListener() {
        @Override
        public void onImageArrival(ImageReceiver imageReceiver) {
            StringBuffer fileName = new StringBuffer("picture_");
            fileName.append(UUID.randomUUID()).append(".jpg"); // 定义生成图片文件名
            File myFile = new File(dirFile, fileName.toString()); // 创建图片文件
            imageSaver = new ImageSaver(imageReceiver.readNextImage(), myFile); //
创建一个读写线程任务用于保存图片
```

```
            eventHandler.postTask(imageSaver); // 执行读写线程任务生成图片
        }
    };

    // 保存图片，图片数据读写，及图像生成见 run 方法
    class ImageSaver implements Runnable {
        private final Image myImage;
        private final File myFile;

        ImageSaver(Image image, File file) {
            myImage = image;
            myFile = file;
        }

        @Override
        public void run() {
            Image.Component component = myImage.getComponent(ImageFormat.
ComponentType.JPEG);
            byte[] bytes = new byte[component.remaining()];
            component.read(bytes);
            FileOutputStream output = null;
            try {
                output = new FileOutputStream(myFile);
                output.write(bytes); // 写图像数据
            } catch (IOException e) {
                HiLog.error(LABEL, "save picture occur exception!");
            } finally {
                if (output != null) {
                    try {
                        output.close(); // 关闭流
                    } catch (IOException e) {
                        HiLog.error(LABEL, "image release occur exception!");
                    }
                }
                myImage.release();
            }
        }
    }
    private void takePictureInit() {
        List<Size> pictureSizes = cameraAbility.getSupportedSizes
(ImageFormat.JPEG); // 获取拍照支持分辨率列表
        pictureSize = getPictureSize(pictureSizes) // 根据拍照要求选择合适的分辨率
        imageReceiver = ImageReceiver.create(Math.max(pictureSize.width, pictureSize.height),
            Math.min(pictureSize.width, pictureSize.height), ImageFormat.
JPEG, 5); // 创建 ImageReceiver 对象，注意 create 函数中宽度要大于高度；5 为最大支持的
图像数，请根据实际设置。
        imageReceiver.setImageArrivalListener(imageArrivalListener);
    }
```

（3）通过 triggerSingleCapture(FrameConfig) 方法实现单帧捕获（如拍照）功能。在以下代码中，通过配置拍照帧的参数，包括拍照时的 Surface、图像旋转角度等，然后触发单帧捕获，实现拍照功能。异常处理部分确保代码的稳定性。

```
private void capture() {
    // 获取拍照配置模板
    framePictureConfigBuilder = cameraDevice.getFrameConfigBuilder(FRAME_CONFIG_PICTURE);
    // 配置拍照 Surface
    framePictureConfigBuilder.addSurface(imageReceiver.getRecevingSurface());
    // 配置拍照其他参数
    framePictureConfigBuilder.setImageRotation(90);
    try {
        // 启动单帧捕获（拍照）
        cameraDevice.triggerSingleCapture(framePictureConfigBuilder.build());
    } catch (IllegalArgumentException e) {
        HiLog.error(LABEL, "Argument Exception");
    } catch (IllegalStateException e) {
        HiLog.error(LABEL, "State Exception");
    }
}
```

为了捕获到质量更高、效果更好的图片，需要在帧结果中，实时监测自动对焦和自动曝光的状态。一般而言，在自动对焦完成，自动曝光收敛后的瞬间是发起单帧捕获的最佳时机。

3. 实现连拍（多帧捕获）

连拍功能方便用户一次拍照获取多张照片，用于捕捉精彩瞬间。在 HarmonyOS 系统中，实现连拍功能的流程与普通拍照一致，但连拍需要使用 triggerMultiCapture(List<FrameConfig> frameConfigs) 方法。

4. 启动录像（循环帧捕获）

在 HarmonyOS 系统中，启动录像和启动预览类似，但需要另外配置录像 Surface 才能使用。

（1）在录像前需要进行音视频模块的配置工作，例如，以下代码实现了初始化录制音频、视频功能。通过配置音频和视频属性，设置录制参数，如比特率、方向、分辨率、编码方式等，创建录制文件对象，准备录制器，以便进行音视频录制。

```
private Source source; // 音视频源
private AudioProperty.Builder audioPropertyBuilder; // 音频属性构造器
private VideoProperty.Builder videoPropertyBuilder; // 视频属性构造器
private StorageProperty.Builder storagePropertyBuilder; // 音视频存储属性构造器
private Recorder mediaRecorder; // 录像操作对象
private String recordName; // 音视频文件名
private Size mRecordSize; // 录像分辨率

private void initMediaRecorder() {
    videoPropertyBuilder.setRecorderBitRate(10000000); // 设置录制比特率
    int rotation = DisplayManager.getInstance().getDefaultDisplay(this).get().getRotation();
    videoPropertyBuilder.setRecorderDegrees(getOrientation(rotation)); // 设置录像方向
    videoPropertyBuilder.setRecorderFps(30); // 设置录制采样率
    videoPropertyBuilder.setRecorderHeight(Math.min(recordSize.height, recordSize.width)); // 设置录像支持的分辨率，需保证 width > height
```

```
        videoPropertyBuilder.setRecorderWidth(Math.max(recordSize.height,
recordSize.width));
        videoPropertyBuilder.setRecorderVideoEncoder(Recorder.VideoEncoder.
H264); // 设置视频编码方式
        videoPropertyBuilder.setRecorderRate(30); // 设置录制帧率
        source.setRecorderAudioSource(Recorder.AudioSource.MIC); // 设置录制音频源
        source.setRecorderVideoSource(Recorder.VideoSource.SURFACE); // 设置视频窗口
        mediaRecorder.setSource(source); // 设置音视频源
        mediaRecorder.setOutputFormat(Recorder.OutputFormat.MPEG_4); // 设置音视频
输出格式
        StringBuffer fileName = new StringBuffer("record_"); // 生成随机文件名
        fileName.append(UUID.randomUUID()).append(".mp4");
        recordName = fileName.toString();
        File file = new File(dirFile, recordName); // 创建录像文件对象
        storagePropertyBuilder.setRecorderFile(file); // 设置存储音视频文件名
        mediaRecorder.setStorageProperty(storagePropertyBuilder.build());
        audioPropertyBuilder.setRecorderAudioEncoder(Recorder.AudioEncoder.
AAC); // 设置音频编码格式
        mediaRecorder.setAudioProperty(audioPropertyBuilder.build()); // 设置音
频属性
        mediaRecorder.setVideoProperty(videoPropertyBuilder.build()); // 设置
视频属性
        mediaRecorder.prepare(); // 准备录制
        HiLog.info(LABEL, "initMediaRecorder end");
    }
```

（2）配置录像帧，启动录像。例如，以下这段代码配置了录像帧，包括预览和录像的 Surface，并启动了循环帧捕获，以实现预览和录像功能的集成。异常处理部分用于捕获可能出现的配置异常，以确保代码的稳定性。

```
    private final class CameraStateCallbackImpl extends CameraStateCallback {
        @Override
        public void onConfigured(Camera camera) {
            // 获取录像配置模板
            frameConfigBuilder = camera.getFrameConfigBuilder(FRAME_CONFIG_RECORD);
            // 配置预览 Surface
            frameConfigBuilder.addSurface(previewSurface);
            // 配置录像的 Surface
            mRecorderSurface = mediaRecorder.getVideoSurface();
            frameConfigBuilder.addSurface(mRecorderSurface);
            previewFrameConfig = frameConfigBuilder.build();
            try {
                // 启动循环帧捕获
                int triggerId = camera.triggerLoopingCapture(previewFrameConfig);
            } catch (IllegalArgumentException e) {
                HiLog.error(LABEL, "Argument Exception");
            } catch (IllegalStateException e) {
                HiLog.error(LABEL, "State Exception");
            }
        }
    }
```

（3）通过 camera.stopLoopingCapture() 方法停止循环帧捕获（录像）功能，代码如下。

```
// 在需要停止录像功能的地方调用此方法
private void stopRecording() {
    try {
        // 停止循环帧捕获，即停止录像
        camera.stopLoopingCapture();
    } catch (IllegalArgumentException e) {
        HiLog.error(LABEL, "Argument Exception while stopping recording");
    } catch (IllegalStateException e) {
        HiLog.error(LABEL, "State Exception while stopping recording");
    }
}
```

5. 释放相机设备

使用完相机后，必须通过 release() 方法关闭相机和释放资源，否则可能导致其他相机应用无法启动。一旦相机被释放，它所提供的操作就不能再被调用，否则会导致不可预期的结果，或是会引发状态异常。

8.3 相机实战：多功能拍照程序

本节将介绍实现一个 HarmonyOS 相机应用程序的过程。这个项目实现了拍照和视频录制功能，主要包括主界面、拍照界面和视频录制界面，通过调用相机和媒体录制的 API 实现功能。界面使用 Ohos XML 布局，具备切换摄像头、拍摄照片、录制视频等用户友好的功能，展示了 HarmonyOS 应用开发的基本原理。

实例 8-1：多功能拍照程序（源码路径 :codes\8\Camera）

8.3.1 配置文件

扫码看视频

编写文件 src/main/config.json，这是一个 HarmonyOS 应用的配置文件，定义了应用的基本信息、模块配置和权限需求。其中包括应用包名、版本信息、模块配置（入口、能力、分发信息等），以及应用所需的权限（相机、存储、麦克风、定位等）。

8.3.2 布局文件

编写文件 base/layout/main_ability_slice.xml，定义了 HarmonyOS 应用程序的主界面布局。采用 DirectionalLayout 垂直方向布局，包含两个 DependentLayout，分别表示拍照和录像的选项。每个选项包括文本和箭头图标，采用相对定位进行布局。

编写文件 base/layout/main_camera_slice.xml，定义了 HarmonyOS 相机界面的布局。主要包括一个垂直方向的 DirectionalLayout，内部包含一个 DependentLayout，该布局包含两个子布局。

- surface_container：一个用于显示相机预览的 DirectionalLayout，占据整个父容器的高度和宽度。
- directionalLayout：一个包含三个按钮的水平方向的 DirectionalLayout，用于操作相机功能，包括退出、拍照和切换相机。这三个按钮是通过三个 Image 元素实现的。

分别表示退出按钮（ic_camera_back）、拍照按钮（ic_camera_photo）和切换相机按钮（ic_camera_switch），每个按钮的单击事件和图标来源均在 XML 中定义。

8.3.3 主界面逻辑

在本项目中，文件 MainAbility.java 和 MainAbilitySlice.java 分别代表应用的主能力和主能力片段。其中，文件 MainAbility.java 负责应用的启动和权限请求等初始化工作，而文件 MainAbilitySlice.java 处理了主界面的显示和用户交互逻辑，它们共同构成了应用的主要入口和界面。

（1）编写文件 src/main/java/ohos/samples/camera/MainAbility.java，通过 onStart 方法设置应用主界面路由，调用 requestPermissions 方法请求用户权限，包括写入用户存储、读取用户存储、相机、麦克风和定位权限。在 onRequestPermissionsFromUserResult 方法中处理用户权限请求结果，如果有权限未被授予，则终止应用。整体功能确保应用在启动时获取必要权限，以保障正常运行。

```java
public class MainAbility extends Ability {
    @Override
    public void onStart(Intent intent) {
        super.onStart(intent);
        super.setMainRoute(MainAbilitySlice.class.getName());
        requestPermissions();
    }

    private void requestPermissions() {
        String[] permissions = {
                SystemPermission.WRITE_USER_STORAGE, SystemPermission.READ_USER_STORAGE, SystemPermission.CAMERA,
                SystemPermission.MICROPHONE, SystemPermission.LOCATION
        };
        requestPermissionsFromUser(Arrays.stream(permissions)
                .filter(permission -> verifySelfPermission(permission) != IBundleManager.PERMISSION_GRANTED).toArray(String[]::new), 0);
    }

    @Override
    public void onRequestPermissionsFromUserResult(int requestCode, String[] permissions, int[] grantResults) {
            if (permissions == null || permissions.length == 0 || grantResults == null || grantResults.length == 0) {
                return;
            }
            for (int grantResult : grantResults) {
                if (grantResult != IBundleManager.PERMISSION_GRANTED) {
                    terminateAbility();
                    break;
                }
            }
    }
}
```

（2）编写文件 src/main/java/ohos/samples/camera/slice/MainAbilitySlice.java，定义了类 MainAbilitySlice，该类实现了主界面的逻辑，包括初始化 UI 内容和按钮单击事件，通过启动不同的能力实现拍照和录像功能。

```java
import ohos.samples.camera.ResourceTable;
import ohos.samples.camera.TakePhotoAbility;
import ohos.samples.camera.VideoRecordAbility;

import ohos.aafwk.ability.AbilitySlice;
import ohos.aafwk.content.Intent;
import ohos.aafwk.content.Operation;
import ohos.agp.components.Component;

/**
 * MainAbilitySlice
 */
public class MainAbilitySlice extends AbilitySlice {
    @Override
    public void onStart(Intent intent) {
        super.onStart(intent);
        super.setUIContent(ResourceTable.Layout_main_ability_slice);
        initComponents();
    }

    private void initComponents() {
        Component takePhoto = findComponentById(ResourceTable.Id_take_photo);
        Component videoRecord = findComponentById(ResourceTable.Id_video_record);
        takePhoto.setClickedListener((component) -> startAbility(TakePhotoAbility.class.getName()));
        videoRecord.setClickedListener((component) -> startAbility(VideoRecordAbility.class.getName()));
    }

    private void startAbility(String abilityName) {
        Operation operation = new Intent.OperationBuilder()
                .withDeviceId("")
                .withBundleName(getBundleName())
                .withAbilityName(abilityName)
                .build();
        Intent intent = new Intent();
        intent.setOperation(operation);
        startAbility(intent);
    }
}
```

对上述代码的具体说明如下。

- onStart 方法：在片段启动时调用，设置 UI 内容为 ResourceTable.Layout_main_ability_slice，即主界面的布局。其中调用了 initComponents 方法初始化界面组件。
- initComponents 方法：通过组件 ID 找到拍照和录像两个按钮，并为它们设置点击监听器。点击"拍照"按钮将启动 TakePhotoAbility 能力，点击"录像"按钮将启动 VideoRecordAbility 能力。
- startAbility 方法：根据给定的能力名称构建 Operation，然后创建 Intent 并启动相应能力。

8.3.4 拍照逻辑

编写文件 src/main/java/ohos/samples/camera/TakePhotoAbility.java，主要功能是在启动时设置与之关联的主界面片段，即 TakePhotoSlice。这样，当用户启动拍照功能时，将显示与之关联的界面。

```java
import ohos.aafwk.ability.Ability;
import ohos.aafwk.content.Intent;
import ohos.samples.camera.slice.TakePhotoSlice;

public class TakePhotoAbility extends Ability {

    @Override
    public void onStart (Intent intent) {
        super.onStart(intent);
        super.setMainRoute(TakePhotoSlice.class.getName());
    }
}
```

编写文件 src/main/java/ohos/samples/camera/slice/TakePhotoSlice.java，实现了相机预览、拍照、切换摄像头等基本功能，提供了用户友好的界面和操作体验。具体实现流程如下。

（1）onStart(Intent intent)：生命周期方法，设置界面布局并初始化界面组件，包括点击按钮和长按监听器，以及相机预览的 SurfaceProvider。

```java
@Override
public void onStart(Intent intent) {
    super.onStart(intent);
    super.setUIContent(ResourceTable.Layout_main_camera_slice);

    initComponents();
    initSurface();
}
```

（2）initSurface()：初始化相机预览的 SurfaceProvider，包括设置预览的 Surface 和添加相机回调。

```java
private void initSurface() {
    getWindow().setTransparent(true);
    DirectionalLayout.LayoutConfig params = new DirectionalLayout.LayoutConfig(
            ComponentContainer.LayoutConfig.MATCH_PARENT, ComponentContainer.LayoutConfig.MATCH_PARENT);
    surfaceProvider = new SurfaceProvider(this);
    surfaceProvider.setLayoutConfig(params);
    surfaceProvider.pinToZTop(false);
    if (surfaceProvider.getSurfaceOps().isPresent()) {
```

```java
            surfaceProvider.getSurfaceOps().get().addCallback(new
SurfaceCallBack());
        }
        surfaceContainer.addComponent(surfaceProvider);
    }
```

（3）initComponents()：初始化界面组件，包括拍照按钮、退出按钮、切换摄像头按钮等，并设置点击按钮和长按监听器。

```java
    private void initComponents() {
        buttonGroupLayout = findComponentById(ResourceTable.Id_directionalLayout);
        surfaceContainer = (ComponentContainer) findComponentById
(ResourceTable.Id_surface_container);
        Image takePhotoImage = (Image) findComponentById(ResourceTable.Id_
tack_picture_btn);
        Image exitImage = (Image) findComponentById(ResourceTable.Id_exit);
        Image switchCameraImage = (Image) findComponentById(ResourceTable.
Id_switch_camera_btn);
        exitImage.setClickedListener(component -> terminateAbility());
        takePhotoImage.setClickedListener(this::takeSingleCapture);
        takePhotoImage.setLongClickedListener(this::takeMultiCapture);
        switchCameraImage.setClickedListener(this::switchCamera);
    }
```

（4）openCamera()：打开相机，创建相机实例并配置预览和拍照的 Surface。

```java
    private void openCamera() {
        imageReceiver = ImageReceiver.create(SCREEN_WIDTH, SCREEN_HEIGHT,
ImageFormat.JPEG, IMAGE_RCV_CAPACITY);
        imageReceiver.setImageArrivalListener(this::saveImage);
        CameraKit cameraKit = CameraKit.getInstance(getApplicationContext());
        String[] cameraList = cameraKit.getCameraIds();
        String cameraId = "";
        for (String logicalCameraId : cameraList) {
            int faceType = cameraKit.getCameraInfo(logicalCameraId).
getFacingType();
            switch (faceType) {
                case CameraInfo.FacingType.CAMERA_FACING_FRONT:
                    if (isFrontCamera) {
                        cameraId = logicalCameraId;
                    }
                    break;
                case CameraInfo.FacingType.CAMERA_FACING_BACK:
                    if (!isFrontCamera) {
                        cameraId = logicalCameraId;
                    }
                    break;
                case CameraInfo.FacingType.CAMERA_FACING_OTHERS:
                default:
                    break;
            }
        }
```

```java
            if (cameraId != null && !cameraId.isEmpty()) {
                CameraStateCallbackImpl cameraStateCallback = new CameraStateCallbackImpl();
                cameraKit.createCamera(cameraId, cameraStateCallback, eventHandler);
            }
        }
```

(5) saveImage(ImageReceiver receiver)：保存拍摄的图片，将图片数据保存到文件，并显示保存成功。

```java
        private void saveImage(ImageReceiver receiver) {
            File saveFile = new File(getExternalFilesDir(null), "IMG_" + System.currentTimeMillis() + ".jpg");
            ohos.media.image.Image image = receiver.readNextImage();
            ohos.media.image.Image.Component component = image.getComponent(ImageFormat.ComponentType.JPEG);
            byte[] bytes = new byte[component.remaining()];
            component.read(bytes);
            try (FileOutputStream output = new FileOutputStream(saveFile)) {
                output.write(bytes);
                output.flush();
                String msg = "Take photo succeed, path=" + saveFile.getPath();
                showTips(this, msg);
            } catch (IOException e) {
                HiLog.error(LABEL_LOG, "%{public}s", "saveImage IOException");
            }
        }
```

(6) takeSingleCapture(Component component)：单帧捕获（拍照），触发相机拍照操作。

```java
        private void takeSingleCapture(Component component) {
            if (cameraDevice == null || imageReceiver == null) {
                return;
            }
            FrameConfig.Builder framePictureConfigBuilder = cameraDevice.getFrameConfigBuilder(FRAME_CONFIG_PICTURE);
            framePictureConfigBuilder.addSurface(imageReceiver.getRecevingSurface());
            FrameConfig pictureFrameConfig = framePictureConfigBuilder.build();
            cameraDevice.triggerSingleCapture(pictureFrameConfig);
        }
```

(7) takeMultiCapture(Component component)：多帧捕获，触发相机连续拍照操作。

```java
        private void takeMultiCapture(Component component) {
            FrameConfig.Builder framePictureConfigBuilder = cameraDevice.getFrameConfigBuilder(FRAME_CONFIG_PICTURE);
            framePictureConfigBuilder.addSurface(imageReceiver.getRecevingSurface());
            List<FrameConfig> frameConfigs = new ArrayList<>();
            FrameConfig firstFrameConfig = framePictureConfigBuilder.build();
            frameConfigs.add(firstFrameConfig);
            FrameConfig secondFrameConfig = framePictureConfigBuilder.build();
```

```
        frameConfigs.add(secondFrameConfig);
        cameraDevice.triggerMultiCapture(frameConfigs);
    }
```

（8）switchCamera(Component component)：切换摄像头，释放当前摄像头资源，切换到前/后摄像头，并重新打开相机。

```
    private void switchCamera(Component component) {
        isFrontCamera = !isFrontCamera;
        if (cameraDevice != null) {
            cameraDevice.release();
        }
        updateComponentVisible(false);
        openCamera();
    }
```

（9）CameraStateCallbackImpl：相机状态回调，在相机创建和配置完成时触发相应操作，如配置预览、拍照等。

```
    private class CameraStateCallbackImpl extends CameraStateCallback {
        CameraStateCallbackImpl() {
        }

        @Override
        public void onCreated(Camera camera) {
            if (surfaceProvider.getSurfaceOps().isPresent()) {
                previewSurface = surfaceProvider.getSurfaceOps().get().getSurface();
            }
            if (previewSurface == null) {
                HiLog.error(LABEL_LOG, "%{public}s", "Create camera filed, preview surface is null");
                return;
            }
            CameraConfig.Builder cameraConfigBuilder = camera.getCameraConfigBuilder();
            cameraConfigBuilder.addSurface(previewSurface);
            cameraConfigBuilder.addSurface(imageReceiver.getRecevingSurface());
            camera.configure(cameraConfigBuilder.build());
            cameraDevice = camera;
            updateComponentVisible(true);
        }

        @Override
        public void onConfigured(Camera camera) {
            FrameConfig.Builder framePreviewConfigBuilder = camera.getFrameConfigBuilder(FRAME_CONFIG_PREVIEW);
            framePreviewConfigBuilder.addSurface(previewSurface);
            camera.triggerLoopingCapture(framePreviewConfigBuilder.build());
        }
    }
```

（10）updateComponentVisible(boolean isVisible)：更新界面组件的可见性，根据参数决

定是否显示按钮组。

```
    private void updateComponentVisible(boolean isVisible) {
        buttonGroupLayout.setVisibility(isVisible ? Component.VISIBLE : Component.INVISIBLE);
    }
```

（11）SurfaceCallBack：相机预览 Surface 的回调，包括 surfaceCreated、surfaceChanged 和 surfaceDestroyed。

```
    private class SurfaceCallBack implements SurfaceOps.Callback {
        @Override
        public void surfaceCreated(SurfaceOps callbackSurfaceOps) {
            if (callbackSurfaceOps != null) {
                callbackSurfaceOps.setFixedSize(SCREEN_HEIGHT, SCREEN_WIDTH);
            }
            eventHandler.postTask(TakePhotoSlice.this::openCamera, 200);
        }

        @Override
        public void surfaceChanged(SurfaceOps callbackSurfaceOps, int format, int width, int height) {
        }

        @Override
        public void surfaceDestroyed(SurfaceOps callbackSurfaceOps) {
        }
    }
```

（12）showTips(Context context, String msg)：在界面上显示提示信息，使用 ToastDialog 弹出提示框。

```
    private void showTips(Context context, String msg) {
        getUITaskDispatcher().asyncDispatch(() -> new ToastDialog(context).setText(msg).show());
    }
```

（13）releaseCamera()：释放相机资源，包括释放相机实例和图片接收器。

```
    private void releaseCamera() {
        if (cameraDevice != null) {
            cameraDevice.release();
        }

        if (imageReceiver != null) {
            imageReceiver.release();
        }
    }
```

（14）onStop()：生命周期方法，在片段停止时释放相机资源。

```
    @Override
    protected void onStop() {
```

```
        releaseCamera();
    }
```

8.3.5 录制视频逻辑

编写文件 src/main/java/ohos/samples/camera/VideoRecordAbility.java，负责处理视频录制的功能。在启动时设置了主要路由为 VideoRecordSlice.class，指定了该能力对应的界面片段为视频录制界面。

```
import ohos.aafwk.ability.Ability;
import ohos.aafwk.content.Intent;
import ohos.samples.camera.slice.VideoRecordSlice;

public class VideoRecordAbility extends Ability {

    public void onStart (Intent intent) {
        super.onStart(intent);
        super.setMainRoute(VideoRecordSlice.class.getName());
    }
}
```

编写文件 src/main/java/ohos/samples/camera/slice/VideoRecordSlice.java，实现了一个使用相机录制视频的界面，其中包括摄像头切换、录制开始和停止等功能。通过与相机和 MediaRecorder 的交互，实现了用户交互和视频录制的控制。具体实现流程如下。

（1）initSurface()：初始化界面的 SurfaceProvider，设置 Surface 的大小和透明度，并为 Surface 添加回调。

```
        private void initSurface() {
            getWindow().setTransparent(true);
            DirectionalLayout.LayoutConfig params = new DirectionalLayout.LayoutConfig(
                    ComponentContainer.LayoutConfig.MATCH_PARENT, ComponentContainer.
LayoutConfig.MATCH_PARENT);
            surfaceProvider = new SurfaceProvider(this);
            surfaceProvider.setLayoutConfig(params);
            surfaceProvider.pinToZTop(false);
            if (surfaceProvider.getSurfaceOps().isPresent()) {
                surfaceProvider.getSurfaceOps().get().addCallback(new
SurfaceCallBack());
            }
            surfaceContainer.addComponent(surfaceProvider);
        }
```

（2）initComponents()：初始化界面组件，包括图标按钮，设置点击和长按事件。

```
        private void initComponents() {
            buttonGroupLayout = findComponentById(ResourceTable.Id_directionalLayout);
            surfaceContainer = (ComponentContainer) findComponentById (ResourceTable.
Id_surface_container);
            Image videoRecord = (Image) findComponentById(ResourceTable.Id_tack_
picture_btn);
            Image exitImage = (Image) findComponentById(ResourceTable.Id_exit);
```

```
            Image switchCameraImage = (Image) findComponentById(ResourceTable.
Id_switch_camera_btn);
            exitImage.setClickedListener(component -> terminateAbility());
            switchCameraImage.setClickedListener(this::switchCamera);

            videoRecord.setLongClickedListener(component -> {
                startRecord();
                isRecording = true;
                videoRecord.setPixelMap(ResourceTable.Media_ic_camera_video_
press);
            });

            videoRecord.setTouchEventListener((component, touchEvent) -> {
                    if (touchEvent != null && touchEvent.getAction() ==
TouchEvent.PRIMARY_POINT_UP && isRecording) {
                    stopRecord();
                    isRecording = false;
                    videoRecord.setPixelMap(ResourceTable.Media_ic_camera_video_
ready);
                }
                return true;
            });
    }
```

（3）initMediaRecorder()：初始化 MediaRecorder，设置视频的各项参数，包括视频尺寸、比特率、帧率等，并准备录制。

```
        private void initMediaRecorder() {
            mediaRecorder = new Recorder();
            VideoProperty.Builder videoPropertyBuilder = new VideoProperty.Builder();
            videoPropertyBuilder.setRecorderBitRate(10000000);
            videoPropertyBuilder.setRecorderDegrees(90);
            videoPropertyBuilder.setRecorderFps(30);
            videoPropertyBuilder.setRecorderHeight(Math.min(1440, 720));
            videoPropertyBuilder.setRecorderWidth(Math.max(1440, 720));
            videoPropertyBuilder.setRecorderVideoEncoder(Recorder.VideoEncoder.
H264);
            videoPropertyBuilder.setRecorderRate(30);

            Source source = new Source();
            source.setRecorderAudioSource(Recorder.AudioSource.MIC);
            source.setRecorderVideoSource(Recorder.VideoSource.SURFACE);
            mediaRecorder.setSource(source);
            mediaRecorder.setOutputFormat(Recorder.OutputFormat.MPEG_4);
            File file = new File(getExternalFilesDir(null), "VID_" + System.
currentTimeMillis() + ".mp4");
            videoPath = file.getPath();
            StorageProperty.Builder storagePropertyBuilder = new StorageProperty.
Builder();
            storagePropertyBuilder.setRecorderFile(file);
            mediaRecorder.setStorageProperty(storagePropertyBuilder.build());
```

```java
        AudioProperty.Builder audioPropertyBuilder = new AudioProperty.Builder();
        audioPropertyBuilder.setRecorderAudioEncoder(Recorder.AudioEncoder.AAC);
        mediaRecorder.setAudioProperty(audioPropertyBuilder.build());
        mediaRecorder.setVideoProperty(videoPropertyBuilder.build());
        mediaRecorder.prepare();
    }
```

（4）openCamera()：打开相机，根据摄像头类型选择前置摄像头或后置摄像头，并设置相机的预览 Surface。

```java
private void openCamera() {
    CameraKit cameraKit = CameraKit.getInstance(getApplicationContext());
    String[] cameraList = cameraKit.getCameraIds();
    String cameraId = "";
    for (String logicalCameraId : cameraList) {
        int faceType = cameraKit.getCameraInfo(logicalCameraId).getFacingType();
        switch (faceType) {
            case CameraInfo.FacingType.CAMERA_FACING_FRONT:
                if (isFrontCamera) {
                    cameraId = logicalCameraId;
                }
                break;
            case CameraInfo.FacingType.CAMERA_FACING_BACK:
                if (!isFrontCamera) {
                    cameraId = logicalCameraId;
                }
                break;
            case CameraInfo.FacingType.CAMERA_FACING_OTHERS:
            default:
                break;
        }
    }
    if (cameraId != null && !cameraId.isEmpty()) {
        CameraStateCallbackImpl cameraStateCallback = new CameraStateCallbackImpl();
        cameraKit.createCamera(cameraId, cameraStateCallback, eventHandler);
    }
}
```

（5）switchCamera(Component component)：切换摄像头，根据当前摄像头类型切换到另一摄像头。

```java
private void switchCamera(Component component) {
    isFrontCamera = !isFrontCamera;
    if (cameraDevice != null) {
        cameraDevice.release();
    }
    updateComponentVisible(false);
    openCamera();
}
```

（6）startRecord()：开始录制视频，初始化 MediaRecorder，配置相机，然后启动 MediaRecorder 开始录制视频。

```java
    private void startRecord() {
        if (cameraDevice == null) {
            HiLog.error(LABEL_LOG, "%{public}s", "startRecord failed, parameters is illegal");
            return;
        }
        synchronized (lock) {
            initMediaRecorder();
            recorderSurface = mediaRecorder.getVideoSurface();
            cameraConfigBuilder = cameraDevice.getCameraConfigBuilder();
            try {
                cameraConfigBuilder.addSurface(previewSurface);
                if (recorderSurface != null) {
                    cameraConfigBuilder.addSurface(recorderSurface);
                }
                cameraDevice.configure(cameraConfigBuilder.build());
            } catch (IllegalStateException | IllegalArgumentException e) {
                HiLog.error(LABEL_LOG, "%{public}s", "startRecord IllegalStateException | IllegalArgumentException");
            }
        }
        new ToastDialog(this).setText("Recording").show();
    }
```

（7）stopRecord()：停止录制视频，停止 MediaRecorder 的录制，更新相机配置。

```java
    private void stopRecord() {
        synchronized (lock) {
            try {
                eventHandler.postTask(() -> mediaRecorder.stop());
                if (cameraDevice == null || cameraDevice.getCameraConfigBuilder() == null) {
                    HiLog.error(LABEL_LOG, "%{public}s", "StopRecord cameraDevice or getCameraConfigBuilder is null");
                    return;
                }
                cameraConfigBuilder = cameraDevice.getCameraConfigBuilder();
                cameraConfigBuilder.addSurface(previewSurface);
                cameraConfigBuilder.removeSurface(recorderSurface);
                cameraDevice.configure(cameraConfigBuilder.build());
            } catch (IllegalStateException | IllegalArgumentException exception) {
                HiLog.error(LABEL_LOG, "%{public}s", "stopRecord occur exception");
            }
        }
        new ToastDialog(this).setText("video saved, path = " + videoPath).show();
    }
```

（8）updateComponentVisible(boolean isVisible)：更新界面组件的可见性。

```java
    private void updateComponentVisible(boolean isVisible) {
        buttonGroupLayout.setVisibility(isVisible ? Component.VISIBLE : Component.INVISIBLE);
    }
```

（9）SurfaceCallBack：Surface 回调，用于在 Surface 创建时延迟 200 毫秒打开相机。

```
private class SurfaceCallBack implements SurfaceOps.Callback {
    @Override
    public void surfaceCreated(SurfaceOps callbackSurfaceOps) {
        if (callbackSurfaceOps != null) {
            callbackSurfaceOps.setFixedSize(SCREEN_HEIGHT, SCREEN_WIDTH);
        }
        eventHandler.postTask(VideoRecordSlice.this::openCamera, 200);
    }

    @Override
    public void surfaceChanged(SurfaceOps callbackSurfaceOps, int format,
int width, int height) {
    }

    @Override
    public void surfaceDestroyed(SurfaceOps callbackSurfaceOps) {
    }
}
```

（10）releaseCamera()：释放相机资源，在 onStop() 时调用。

```
private void releaseCamera() {
    if (cameraDevice != null) {
        cameraDevice.release();
    }
}
```

为节省本书篇幅，视频录制的实现过程介绍于此。执行效果如图 8-2 所示。

图 8-2　视频录制执行效果

第 9 章
网络程序开发

鸿蒙系统致力于构建全场景智能生态系统，网络应用是实现跨设备、跨平台连接和互操作性的关键。通过开发鸿蒙系统的网络应用，可以实现更灵活、更高效的信息共享和互联互通，从而提升用户体验，并推动智能设备间的协同工作。因此，在鸿蒙系统中开发网络应用程序具有重要意义。本章将详细讲解在 HarmonyOS 系统中开发网络应用程序的知识。

9.1 网络管理开发

在 HarmonyOS 系统中,网络管理开发涉及管理设备间的网络连接和通信。这包括处理设备的网络状态、配置网络参数、管理网络设备的发现和通信等方面。HarmonyOS 旨在构建分布式的全场景智能生态系统,因此,网络管理开发对于确保设备间的无缝连接和协同工作至关重要。开发者可以利用 HarmonyOS 系统提供的网络管理 API,实现设备的自动发现、通信协议的制定及网络资源的优化利用,从而提升整体系统的性能和用户体验。网络管理开发在 HarmonyOS 系统中有助于构建更加智能、便捷且高效的设备互联互通体验。

9.1.1 HTTP 数据请求

HTTP(Hypertext Transfer Protocol)数据请求是指通过 HTTP 协议在客户端和服务器之间进行数据交换的过程。HTTP 是一种用于传输超文本(如 HTML)的应用层协议,通常用于在 Web 浏览器和 Web 服务器之间传递数据。

在 HTTP 数据请求中,客户端向服务器发送 HTTP 请求,请求可以包括获取资源(如网页、图片、文本文件等)或执行特定操作(如提交表单、发起搜索请求等)。HTTP 请求通常包含请求方法(GET、POST、PUT 等)、请求头部(包含一些关于客户端、请求和所需资源的信息)以及可选的请求体(对于 POST 等有请求体的请求)。

在 HarmonyOS 系统中,使用 http 模块实现 HTTP 数据请求功能,在使用该功能前需要申请 ohos.permission.INTERNET 权限。在 http 模块中,主要包含以下接口。

- createHttp():创建一个 HTTP 请求。
- request():根据 URL 地址,发起 HTTP 网络请求。
- destroy():中断请求任务。
- on(type: 'headersReceive'):订阅 HTTP Response Header 事件。
- off(type: 'headersReceive'):取消订阅 HTTP Response Header 事件。
- once('headersReceive'):订阅 HTTP Response Header 事件,但是只触发一次。

使用 http 模块开发应用程序的基本步骤如下。

(1)从 @ohos.net.http.d.ts 中导入 http 命名空间。

(2)调用 createHttp() 方法,创建一个 HttpRequest 对象。

(3)调用该对象的 on() 方法,订阅 HTTP 响应头事件,此接口会比 request 请求先返回。根据业务需要订阅此消息。

(4)调用该对象的 request() 方法,传入 HTTP 请求的 URL 地址和可选参数,发起网络请求。

(5)按照实际业务需要,解析返回结果。

(6)调用该对象的 off() 方法,取消订阅 HTTP 响应头事件。

(7)当该请求使用完毕时,调用 destroy() 方法主动销毁。

实例 9-1 使用 http 模块开发一个简单的 Web 浏览器应用程序,包括输入 URL 的文本框、加载页面的按钮和 Web 视图组件等功能。当用户输入 URL 后点击按钮,应用通过异步 HTTP 请求获取网络资源,成功则显示加载的 Web 页面,失败则提示错误信息。

实例 9-1:Web 浏览器程序(源码路径 :codes\9\HttpsRequest)

(1)编写文件 src/main/module.json5,请求使用 "ohos.permission.INTERNET" 权限,这是与网络访问相关的权限。

扫码看视频

```
"requestPermissions": [
  {
    "name": "ohos.permission.INTERNET"
  }
```

(2)编写文件 src/main/ets/common/utils/HttpUtil.ets,定义一个使用 HarmonyOS 的 httpGet 函数,用于发起异步的 HTTP GET 请求。该函数接受一个 URL 参数,配置请求选项包括请求方法、头部信息和超时设置,最后通过 @system.http 模块发起请求,并返回异步获取的 HTTP 响应结果。该函数通过导入通用常量确保了代码的可维护性。

```
/**
 * 发起 HTTP GET 请求的异步函数
 * @param {string} url - 请求的 URL 地址
 * @returns {Promise<any>} - 返回一个包含 HTTP 响应的 Promise 对象
 */
export default async function httpGet(url: string): Promise<any> {
  // 如果 URL 为空,则直接返回 undefined
  if (!url) {
    return undefined;
  }

  // 创建 HTTP 请求实例
  let request = http.createHttp();

  // 配置 HTTP 请求选项
  let options = {
    method: http.RequestMethod.GET,
    header: { 'Content-Type': 'application/json' },
    readTimeout: CommonConstant.READ_TIMEOUT,
    connectTimeout: CommonConstant.CONNECT_TIMEOUT
  } as http.HttpRequestOptions;

  // 发起 HTTP GET 请求,并等待响应
  let result = await request.request(url, options);

  // 返回 HTTP 响应结果
  return result;
}
```

(3)编写文件 src/main/ets/pages/WebPage.ets,定义一个 HarmonyOS 页面结构 WebPage,包含一个 Web 视图和一个按钮。用户输入 URL 后点击按钮,通过 httpGet 函数异步获取网络资源,若请求成功则显示 Web 视图加载对应页面,否则显示错误提示。页面具备响应式

设计,包括输入框、按钮和 Web 视图的交互,以及错误处理功能。

```
@Entry
@Component
struct WebPage {
  controller: webView.WebviewController = new webView.WebviewController();
  @State buttonName: Resource = $r('app.string.request_button_name');
  @State webVisibility: Visibility = Visibility.Hidden;
  @State webSrc: string = CommonConstant.SERVER;

  build() {
    Column() {
      Row() {
        Image($r('app.media.ic_network_global'))
          .height($r('app.float.image_height'))
          .width($r('app.float.image_width'))
        TextInput({ placeholder: $r('app.string.input_address'), text: this.webSrc })
          .height($r('app.float.text_input_height'))
          .layoutWeight(1)
          .backgroundColor(Color.White)
          .onChange((value: string) => {
            this.webSrc = value;
          })
      }
      .margin({
        top: $r('app.float.default_margin'),
        left: $r('app.float.default_margin'),
        right: $r('app.float.default_margin')
      })
      .height($r('app.float.default_row_height'))
      .backgroundColor(Color.White)
      .borderRadius($r('app.float.border_radius'))
      .padding({
        left: $r('app.float.default_padding'),
        right: $r('app.float.default_padding')
      })
      Row() {
        Web({ src: this.webSrc, controller: this.controller })
          .zoomAccess(true)
          .height(StyleConstant.FULL_HEIGHT)
          .width(StyleConstant.FULL_WIDTH)
      }
      .visibility(this.webVisibility)
      .height(StyleConstant.WEB_HEIGHT)
      .width(StyleConstant.FULL_WIDTH)
      .align(Alignment.Top)
      Row() {
        Button(this.buttonName)
          .fontSize($r('app.float.button_font_size'))
          .width(StyleConstant.BUTTON_WIDTH)
          .height($r('app.float.button_height'))
```

```
              .fontWeight(FontWeight.Bold)
              .onClick(() => {
                this.onRequest();
              })
          }
          .height($r('app.float.default_row_height'))
        }
        .width(StyleConstant.FULL_WIDTH)
        .height(StyleConstant.FULL_HEIGHT)
        .backgroundImage($r('app.media.ic_background_image', ImageRepeat.NoRepeat))
        .backgroundImageSize(ImageSize.Cover)
      }

      async onRequest() {
        if (this.webVisibility === Visibility.Hidden) {
          this.webVisibility = Visibility.Visible;
          try {
            let result = await httpGet(this.webSrc);
            if (result && result.responseCode === http.ResponseCode.OK) {
              this.controller.clearHistory();
              this.controller.loadUrl(this.webSrc);
            }
          } catch (error) {
            promptAction.showToast({
              message: $r('app.string.http_response_error')
            })
          }
        } else {
          this.webVisibility = Visibility.Hidden;
        }
      }
    }
```

执行以上代码后可以实现网页浏览器功能，效果如图 9-1 所示。

图 9-1 网页浏览器功能

9.1.2 WebSocket 连接

在 HarmonyOS 系统中，可以使用 WebSocket 建立服务器与客户端的双向连接。使用 WebSocket 时，需要先通过 createWebSocket() 方法创建 WebSocket 对象，然后通过 connect() 方法连接到服务器。当连接成功后，客户端会收到 open 事件的回调，之后客户端就可以通过 send() 方法与服务器进行通信。当服务器发送信息给客户端时，客户端会收到 message 事件的回调。当客户端不需要此连接时，可以通过调用 close() 方法主动断开连接，之后客户端会收到 close 事件的回调。

在 HarmonyOS 系统中，WebSocket 连接功能主要由 webSocket 模块实现，在使用 webSocket 模块前需要先申请 ohos.permission.INTERNET 权限。webSocket 模块包含以下接口。

- createWebSocket()：创建一个 WebSocket 连接。
- connect()：根据 URL 地址，建立一个 WebSocket 连接。
- send()：通过 WebSocket 连接发送数据。
- close()：关闭 WebSocket 连接。
- on(type: 'open')：订阅 WebSocket 的打开事件。
- off(type: 'open')：取消订阅 WebSocket 的打开事件。
- on(type: 'message')：订阅 WebSocket 的接收到服务器消息事件。
- off(type: 'message')：取消订阅 WebSocket 的接收到服务器消息事件。
- on(type: 'close')：订阅 WebSocket 的关闭事件。
- off(type: 'close')：取消订阅 WebSocket 的关闭事件
- on(type: 'error')：订阅 WebSocket 的 Erro 事件。
- off(type: 'error')：取消订阅 WebSocket 的 Erro 事件。

实例 9-2，使用 WebSocket 开发了一个在线聊天系统，包括聊天页面、消息发送和接收功能。用户通过顶部栏切换连接状态，通过自定义对话框输入服务 IP 地址实现连接。整体提供了简单而实用的聊天体验，支持与服务器进行实时通信。

实例 9-2：基于 WebSocket 客户端 / 服务端的聊天系统（源码路径 :codes\9\WebSocket）

（1）编写文件 src/main/ets/common/BindServiceIp.ets，定义一个 WebSocket 组件，构建一个简单的用户界面用于 WebSocket 服务的 IP 地址绑定。允许用户通过输入框输入服务 IP 地址，然后点击按钮执行绑定操作。当输入框发生变化时更新 IP 地址属性，点击按钮时会触发绑定操作，并调用预设回调函数。

扫码看视频

```
@Component
export default struct BindServiceIp {
  @Link ipAddress: string
  private onBind: () => void = () => {
  }

  build() {
    Column() {
      Text($r('app.string.welcome'))
        .fontSize(25)
```

```
            .margin({ top: 20 })
            .fontWeight(FontWeight.Bold)
        TextInput({ placeholder: $r('app.string.ip_placeholder') })
            .height(50)
            .fontSize(15)
            .width('70%')
            .margin({ top: 20 })
            .onChange((value: string) => {
              this.ipAddress = `ws://${value}/string`
            })
        Button() {
          Text($r('app.string.bind_ip'))
            .fontSize(20)
            .fontColor(Color.White)
        }
        .margin({ top: 20 })
        .width(200)
        .height(50)
        .type(ButtonType.Capsule)
        .onClick(() => {
          this.onBind()
        })
      }
      .width('100%')
    }
  }
```

（2）编写文件 src/main/ets/common/SendMessage.ets，构建一个简单的用户界面，用于发送文本消息。首先定义一个发送消息的组件，用户可以通过文本区域输入消息内容，然后点击按钮发送消息。当文本区域发生变化时触发更新消息属性，点击按钮时执行发送消息的操作。

```
@Component
export default struct SendMessage {
  @Link message: string
  private sendMessage: () => void = () => {
  }

  build() {
    Row() {
      TextArea({ placeholder: this.message, text: this.message })
        .height(50)
        .fontSize(25)
        .layoutWeight(3)
        .backgroundColor(Color.White)
        .margin({ left: 2, right: 2 })
        .onChange((value: string) => {
          this.message = value
        })

      Button() {
```

```
          Text($r('app.string.send_message'))
            .fontSize(23)
            .fontColor(Color.White)
        }
        .height(50)
        .layoutWeight(1)
        .borderRadius(10)
        .type(ButtonType.Normal)
        .backgroundColor('#ffadf58e')
        .margin({ left: 2, right: 2 })
        .onClick(() => {
          this.sendMessage()
        })
      }
      .height(70)
      .width('100%')
      .backgroundColor('#f5f5f5')
    }
  }
```

（3）编写文件 src/main/ets/common/ChatsPage.ets，构建一个基于 WebSocket 的聊天页面，支持显示服务器和用户之间的交互消息。首先定义了一个聊天页面的组件，用于展示 WebSocket 源的聊天信息。通过导入 ChatData 模型和 WebSocketSource 数据源，组件通过嵌套的方式构建了聊天消息的展示。每条消息包括发送者名称、消息内容和方向，并通过不同的背景颜色区分消息发送者。

```
import ChatData from '../model/ChatData'
import { WebSocketSource } from "../model/DataSource"

@Component
export default struct ChatsPage {
  @Link chats: WebSocketSource

  @Builder
  ChatsMessage(name: Resource, message: string, direction: Direction) {
    Row() {
      Text(name)
        .width(40)
        .height(40)
        .padding(5)
        .fontSize(30)
        .borderRadius(10)
        .margin({ right: 10 })
        .backgroundColor('#e5e5e5')
        .textAlign(TextAlign.Center)
      Text(message)
        .textOverflow({ overflow: TextOverflow.Clip })
        .padding(10)
        .maxLines(5)
        .fontSize(20)
        .borderRadius(10)
```

```
          .margin({ top: 20 })
          .alignSelf(ItemAlign.Start)
          .backgroundColor('#ff78dd4d')
      }
      .width('100%')
      .direction(direction)
      .margin({ top: 5, bottom: 10 })
  }

  build() {
    Column() {
      List() {
        LazyForEach(this.chats, (item: ChatData) => {
          ListItem() {
            if (item.isServer as boolean) {
              this.ChatsMessage($r('app.string.server'), item.message, Direction.Ltr)
            } else {
              this.ChatsMessage($r('app.string.me'), item.message, Direction.Rtl)
            }
          }
          .padding(10)
          .width('100%')
        }, (item: ChatData, index?: number) => item.message + index)
      }.width('100%').height('100%')
    }
    .width('100%')
    .layoutWeight(1)
    .backgroundColor(Color.White)
  }
}
```

（4）编写文件 src/main/ets/pages/Chats.ets，构建一个包含 WebSocket 通信的聊天页面。首先定义一个聊天页面的组件，该页面包括顶部栏、聊天内容展示、消息发送组件以及与 WebSocket 相关的连接和断开功能。通过导入相关模块和组件，以及 WebSocket 的库，实现了与服务器的连接、消息的发送和接收功能。页面还包含一个用于绑定服务 IP 地址的自定义对话框。用户可以通过顶部栏切换至连接状态，发送和接收聊天消息，以及通过对话框输入服务 IP 地址实现连接。

```
const TAG: string = '[Chats]'
let socket: webSocket.WebSocket = webSocket.createWebSocket()

@CustomDialog
struct BindCustomDialog {
  @State ipAddress: string = ''
  private controller?: CustomDialogController
  onBind: (ipAddress: string) => void = (ipAddress: string) => {
  }

  build() {
    Column() {
```

```
          BindServiceIP({ ipAddress: $ipAddress, onBind: () => {
            this.onBind(this.ipAddress)
          } })
        }
        .width('100%')
        .margin({ bottom: 20 })
    }
  }

  @Entry
  @Component
  struct Chats {
    @State numberOfPeople: number = 1
    @State message: string = ''
    @State chats: WebSocketSource = new WebSocketSource([])
    @State isConnect: boolean = false
    @State ipAddress: string = ''
    controller: CustomDialogController = new CustomDialogController({
      builder: BindCustomDialog({ onBind: (ipAddress: string): void => this.
onBind(ipAddress) }),
      autoCancel: false
    })

    aboutToAppear() {
      this.controller.open()
    }

    onBind(ipAddress: string) {
      this.ipAddress = ipAddress
      this.controller.close()
    }

    onConnect() {
      let promise = socket.connect(this.ipAddress)
      Logger.info(TAG, `ipAddress:${JSON.stringify(this.ipAddress)}`)
      promise.then(() => {
        Logger.info(TAG, `connect success`)
      }).catch((err: Error) => {
        Logger.info(TAG, `connect fail, error:${JSON.stringify(err)}`)
      })
      socket.on('open', () => {
        // 当收到on('open')事件时,可以通过send()方法与服务器进行通信
        promptAction.showToast({ message: '连接成功,可以聊天了！', duration: 1500 })
      })
      socket.on('message', (err: Error, value: Object) => {
        Logger.info(TAG, `on message, value = ${value}`)
        let receiveMessage = new ChatData(JSON.stringify(value), true)
        this.chats.pushData(receiveMessage)
      })
    }

    disConnect() {
```

```
    socket.off('open', (err, value) => {
      let val: Record<string, Object> = value as Record<string, Object>;
      Logger.info(TAG, `on open, status:${val['status']}, message:${val['message']}`);
    })
    socket.off('message')
    promptAction.showToast({ message: ' 连接已断开! ', duration: 1500 })
    socket.close()
  }

  sendMessage() {
    let sendMessage = new ChatData(this.message, false)
    this.chats.pushData(sendMessage)
    let sendResult = socket.send(this.message)
    sendResult.then(() => {
      Logger.info(TAG, `[send]send success:${this.message}`)
    }).catch((err: Error) => {
      Logger.info(TAG, `[send]send fail, err:${JSON.stringify(err)}`)
    })
    this.message = ''
  }

  build() {
    Column() {
      Text($r('app.string.EntryAbility_label'))
        .height(50)
        .fontSize(25)
        .width('100%')
        .padding({ left: 10 })
        .fontColor(Color.White)
        .textAlign(TextAlign.Start)
        .backgroundColor('#0D9FFB')
        .fontWeight(FontWeight.Bold)
      TopBar({ isConnect: $isConnect, connect: () => {
        this.isConnect = !this.isConnect
        if (this.isConnect) {
          this.onConnect()
        } else {
          this.disConnect()
        }
      } })
      ChatsPage({ chats: $chats })
      SendMessage({ message: $message, sendMessage: () => {
        this.sendMessage()
      } })
    }
    .width('100%')
    .height('100%')
  }
}
```

执行后，输入服务器的 IP 地址，点击绑定服务器 IP 地址按钮，即可绑定该 IP 并退出对话框，如果要解绑 IP，重启应用即可。点击顶部栏的连接按钮，当按钮颜色从灰色变为绿色后即可与服务器建立 WebSocket 连接。执行效果如图 9-2 所示。

9.1.3 Socket 连接

Socket 连接是指通过计算机网络中的套接字（Socket）实现的两台计算机之间的通信连接。这是一种双向的通信管道，允许数据在客户端和服务器之间双向流动。Socket 是一种抽象的通信端点，允许应用程序在网络上发送和接收数据。

在 HarmonyOS 系统中，通过 TCP/UDP Socket 进行数据传输，通过 TLS Socket 进行加密数据传输。其中 Socket 功能主要由 socket 模块实现，在 socket 模块中包含以下接口。

图 9-2　建立 WebSocket 连接

- constructUDPSocketInstance()：创建一个 UDPSocket 对象。
- constructTCPSocketInstance()：创建一个 TCPSocket 对象。
- bind()：绑定 IP 地址和端口。
- send()：发送数据。
- close()：关闭连接。
- getState()：获取 Socket 状态。
- connect()：连接到指定的 IP 地址和端口（仅 TCP 支持）。
- getRemoteAddress()：获取对端 Socket 地址（仅 TCP 支持，需要先调用 connect 方法）。
- on(type: 'message')：订阅 Socket 连接的接收消息事件。
- off(type: 'message')：取消订阅 Socket 连接的接收消息事件。
- on(type: 'close')：订阅 Socket 连接的关闭事件。
- off(type: 'close')：取消订阅 Socket 连接的关闭事件。
- on(type: 'error')：订阅 Socket 连接的 Error 事件。
- off(type: 'error')：取消订阅 Socket 连接的 Error 事件。
- on(type: 'listening')：订阅 UDPSocket 连接的数据包消息事件（仅 UDP 支持）。
- off(type: 'listening')：取消订阅 UDPSocket 连接的数据包消息事件（仅 UDP 支持）。
- on(type: 'connect')：订阅 TCPSocket 的连接事件（仅 TCP 支持）。
- off(type: 'connect')：取消订阅 TCPSocket 的连接事件（仅 TCP 支持）。

TLS Socket 连接功能主要由 tls_socket 模块实现，主要包含以下接口。

- constructTLSSocketInstance()：创建一个 TLSSocket 对象。
- bind()：绑定 IP 地址和端口号。
- close(type: 'error')：关闭连接。
- connect()：连接到指定的 IP 地址和端口。
- getCertificate()：返回表示本地证书的对象。
- getCipherSuite()：返回包含协商的密码套件信息的列表。
- getProtocol()：返回包含当前连接协商的 SSL/TLS 协议版本的字符串。
- getRemoteAddress()：获取 TLSSocket 连接的对端地址。
- getRemoteCertificate()：返回表示对等证书的对象。
- getSignatureAlgorithms()：在服务器和客户端之间共享的签名算法列表，按优先级降序排列。
- getState()：获取 TLSSocket 连接的状态。
- off(type:'close')：取消订阅 TLSSocket 连接的关闭事件。
- off(type:'error')：取消订阅 TLSSocket 连接的 Error 事件。
- off(type:'message')：取消订阅 TLSSocket 连接的接收消息事件。
- on(type:'close')：订阅 TLSSocket 连接的关闭事件。
- on(type:'error')：订阅 TLSSocket 连接的 Error 事件。
- on(type:'message')：订阅 TLSSocket 连接的接收消息事件。
- send()：发送数据。
- setExtraOptions()：设置 TLSSocket 连接的其他属性。

以下代码，展示了在 HarmonyOS 应用中实现 TCP 通信功能的过程。首先使用 socket 创建了一个 TCPSocket 连接，通过订阅事件实现了消息接收、连接建立和连接关闭的处理。绑定了本地 IP 地址和端口，连接到远程 IP 地址和端口，发送数据到服务器，并在 30 秒后主动关闭连接，同时取消相关事件的订阅。

```
import socket from '@ohos.net.socket';

// 创建一个TCPSocket连接，返回一个TCPSocket对象。
let tcp = socket.constructTCPSocketInstance();

// 订阅TCPSocket相关的订阅事件
tcp.on('message', value => {
  console.log("on message")
  let buffer = value.message
  let dataView = new DataView(buffer)
  let str = ""
  for (let i = 0; i < dataView.byteLength; ++i) {
    str += String.fromCharCode(dataView.getUint8(i))
  }
  console.log("on connect received:" + str)
});
tcp.on('connect', () => {
  console.log("on connect")
});
tcp.on('close', () => {
  console.log("on close")
});

// 绑定IP地址和端口。
let bindAddress = {
  address: '192.168.xx.xx',
  port: 1234, // 绑定端口，如1234
  family: 1
};
tcp.bind(bindAddress, err => {
  if (err) {
    console.log('bind fail');
```

```
      return;
    }
    console.log('bind success');
    // 连接到指定的IP地址和端口。
    let connectAddress = {
      address: '192.168.xx.xx',
      port: 5678, // 连接端口，如5678
      family: 1
    };
    tcp.connect({
      address: connectAddress, timeout: 6000
    }, err => {
      if (err) {
        console.log('connect fail');
        return;
      }
      console.log('connect success');
      // 发送数据
      tcp.send({
        data: 'Hello, server!'
      }, err => {
        if (err) {
          console.log('send fail');
          return;
        }
        console.log('send success');
      })
    });
  });
  // 连接使用完毕后，主动关闭。取消相关事件的订阅。
  setTimeout(() => {
    tcp.close((err) => {
      console.log('close socket.')
    });
    tcp.off('message');
    tcp.off('connect');
    tcp.off('close');
  }, 30 * 1000);
```

对上述代码的具体说明如下。

- 引入Socket库：使用@ohos.net.socket库引入Socket相关功能。
- 创建TCPSocket对象：通过socket.constructTCPSocketInstance()创建一个TCPSocket对象，用于建立TCP连接。
- 订阅Socket事件：使用.on()方法订阅了消息到达（message）、连接建立（connect）和连接关闭（close）等事件，以便在发生时执行相应的回调函数。
- 绑定IP地址和端口：使用tcp.bind()方法绑定本地IP地址和端口。如果绑定失败，输出相应的日志。
- 连接到指定的IP地址和端口：使用tcp.connect()方法连接到指定的远程IP地址和端口。如果连接失败，输出相应的日志。

- 发送数据：通过 tcp.send() 方法发送数据到服务器。如果发送失败，输出相应的日志。
- 定时关闭连接：使用 setTimeout 定时器，在 30 秒后调用 tcp.close() 关闭连接，并取消对相关事件的订阅（tcp.off()）。

9.2 IPC 与 RPC 通信

在鸿蒙系统中，进程间通信（Inter-Process Communication，IPC）和远程过程调用（Remote Procedure Call，RPC）都是实现进程间通信的关键技术。这两种通信方式都被广泛应用于构建分布式系统、多模态设备间的协同工作，以及实现应用的模块化和解耦。IPC 通常用于同一设备上的组件之间，而 RPC 则用于不同设备或进程之间，使鸿蒙系统更好地适应各种复杂的联网场景。

9.2.1 IPC 与 RPC 的基本概念

在鸿蒙系统中，IPC（进程间通信）和 RPC（远程过程调用）有不同的应用场景和实现方式。

1. IPC
- IPC 用于在同一设备上的不同进程之间进行通信。
- 鸿蒙系统提供了一系列的 IPC 机制，如 Binder、管道、消息队列等。
- 进程间通信可以帮助不同组件、服务或应用之间共享数据、传递消息，以实现协同工作。

2. RPC
- RPC 用于在网络上的不同设备或进程之间进行通信，允许调用远程服务器上的函数。
- 鸿蒙系统支持基于分布式技术的 RPC 通信，使设备间的协同工作更加灵活和高效。
- RPC 可以让应用程序远程调用另一设备上的服务，实现分布式系统中的协同处理。

IPC 和 RPC 通常采用"客户端/服务器"（Client-Server）模型，在使用时，请求服务的（Client）一端进程可获取服务提供端（Server）所在进程的代理（Proxy），并通过此代理读写数据来实现进程间的数据通信。更具体来说，首先请求服务的一端会建立一个服务提供端的代理对象，这个代理对象具备和服务提供端一样的功能。若想访问服务提供端中的某一个方法，只需访问代理对象中对应的方法即可，代理对象会将请求发送给服务提供端。然后服务提供端处理接受到的请求，处理完之后通过驱动返回处理结果给代理对象。最后代理对象将请求结果进一步返回给请求服务的一端。通常，Server 会先注册系统能力（System Ability）到系统能力管理者（System Ability Manager，SAMgr）中，SAMgr 负责管理这些 SA 并向 Client 提供相关的接口。Client 要和某个具体的 SA 通信，必须先从 SAMgr 中获取该 SA 的代理，然后使用代理和 SA 通信。

在鸿蒙系统中，实现 IPC 与 RPC 通信的基本步骤如下。

（1）实现接口类：需继承 IRemoteBroker，定义消息码，可声明不在此类实现的方法。

（2）实现服务提供端：需继承 IRemoteStub 或 RemoteObject，重写 AsObject 方法及 OnRemoteRequest 方法。

（3）实现服务请求端：需继承 IRemoteProxy 或 RemoteProxy，重写 AsObject 方法，封装所需方法调用 SendRequest。

（4）注册 SA：申请 SA 的唯一 ID，向 SAMgr 注册 SA。

(5）获取 SA：通过 SA 的 ID 和设备 ID 获取 Proxy，使用 Proxy 与远端通信。

9.2.2 开发 IPC 与 RPC 通信程序

在 HarmonyOS 系统中，使用 JavaScript 或其衍生语言（如 TypeScript）开发 IPC 与 RPC 通信程序的步骤如下。

（1）添加依赖：在代码中引入了 HarmonyOS 系统中两个关键的模块。

```
import rpc from "@ohos.rpc"
import featureAbility from "@ohos.ability.featureAbility"
```

- rpc 模块：在 HarmonyOS 系统中用于支持 RPC 的模块。远程过程调用允许一个进程（模块）调用另一个进程（模块）中的函数或方法，即实现了 IPC。通过 RPC，HarmonyOS 应用可以在不同模块或设备之间进行高效的远程调用，实现分布式系统中的协同工作。
- featureAbility 模块：在 HarmonyOS 中用于支持能力分发和启动的模块。能力分发是指在 HarmonyOS 中通过定义能力（Ability）来组织和管理应用的功能，而 featureAbility 提供了与这些能力进行交互的接口。通过 featureAbility，应用可以启动其他模块的能力，实现模块间的交互和通信。

这两个模块结合使用，可以实现在 HarmonyOS 应用中的模块间远程调用和能力分发。这对于构建分布式、模块化的 HarmonyOS 应用是非常重要的。

（2）绑定 Ability：首先，构造变量 want，指定要绑定的 Ability 所在应用的包名、组件名。如果是跨设备的场景，还需要绑定目标设备 NetworkId。其次，构造变量 connect，指定绑定成功、绑定失败、断开连接时的回调函数。最后，使用 featureAbility 提供的接口绑定 Ability。

```
import rpc from "@ohos.rpc"
import featureAbility from "@ohos.ability.featureAbility"

let proxy = null
let connectId = null

// 单个设备绑定 Ability
let want = {
    // 包名和组件名写实际的值
    "bundleName": "ohos.rpc.test.server",
    "abilityName": "ohos.rpc.test.server.ServiceAbility",
}
let connect = {
    onConnect:function(elementName, remote) {
        proxy = remote
    },
    onDisconnect:function(elementName) {
    },
    onFailed:function() {
        proxy = null
    }
}
```

```
    }
    connectId = featureAbility.connectAbility(want, connect)

    // 如果是跨设备绑定,可以使用 deviceManager 获取目标设备 NetworkId
    import deviceManager from '@ohos.distributedHardware.deviceManager'
    function deviceManagerCallback(deviceManager) {
        let deviceList = deviceManager.getTrustedDeviceListSync()
        let networkId = deviceList[0].networkId
        let want = {
            "bundleName": "ohos.rpc.test.server",
            "abilityName": "ohos.rpc.test.service.ServiceAbility",
            "networkId": networkId,
            "flags": 256
        }
        connectId = featureAbility.connectAbility(want, connect)
    }
    // 第一个参数是本应用的包名,第二个参数是接收 deviceManager 的回调函数
    deviceManager.createDeviceManager("ohos.rpc.test", deviceManagerCallback)
```

(3)服务端处理客户端请求:服务端被绑定的 Ability 在 onConnect 方法里返回继承自 rpc.RemoteObject 的对象,该对象需要实现 onRemoteMessageRequest 方法,处理客户端的请求。下面的代码演示了在 HarmonyOS 中实现远程调用(RPC)的用法,这是通过使用 @ohos.rpc 和 @ohos.ability.featureAbility 模块实现的。通过 featureAbility.connectAbility 方法,实现了在本地设备和跨设备两种情况下的 Ability 绑定,通过回调函数获得连接状态和远程代理。在跨设备情况下,使用 deviceManager 获取目标设备的 NetworkId,并通过创建设备管理器的方式实现跨设备的 Ability 绑定。

```
onConnect(want: Want) {
    var robj:rpc.RemoteObject = new Stub("rpcTestAbility")
    return robj
}
class Stub extends rpc.RemoteObject {
    constructor(descriptor) {
        super(descriptor)
    }
    onRemoteMessageRequest(code, data, reply, option) {
        // 根据 code 处理客户端的请求
        return true
    }
}
```

(4)客户端处理服务端响应:客户端在 onConnect 回调中接收到代理对象,调用 sendRequestAsync 方法发起请求,在期约(JavaScript 中的 Promise:用于表示一个异步操作的最终完成或失败及其结果值)或者回调函数中接收结果。以下代码展示了在 HarmonyOS 中使用 RPC 进行远程过程调用的两种方式:一种是通过 Promise 发送异步请求,另一种是使用回调函数处理同步请求。通过创建消息选项和消息包裹,设置参数后,实现了向远程服务发送请求,并处理请求成功、异常及释放资源的逻辑。

```
    // 使用期约
```

```
let option = new rpc.MessageOption()
let data = rpc.MessageParcel.create()
let reply = rpc.MessageParcel.create()
// 往 data 里写入参数
proxy.sendRequestAsync(1, data, reply, option)
    .then(function(result) {
        if (result.errCode != 0) {
            console.error("send request failed, errCode: " + result.errCode)
            return
        }
        // 从 result.reply 里读取结果
    })
    .catch(function(e) {
        console.error("send request got exception: " + e)
    }
    .finally(() => {
        data.reclaim()
        reply.reclaim()
    })

// 使用回调函数
function sendRequestCallback(result) {
    try {
        if (result.errCode != 0) {
            console.error("send request failed, errCode: " + result.errCode)
            return
        }
        // 从 result.reply 里读取结果
    } finally {
        result.data.reclaim()
        result.reply.reclaim()
    }
}
let option = new rpc.MessageOption()
let data = rpc.MessageParcel.create()
let reply = rpc.MessageParcel.create()
// 往 data 里写入参数
proxy.sendRequest(1, data, reply, option, sendRequestCallback)
```

（5）断开连接：在 IPC 通信工作结束后，使用 featureAbility 的接口断开连接。

```
import rpc from "@ohos.rpc"
import featureAbility from "@ohos.ability.featureAbility"
function disconnectCallback() {
    console.info("disconnect ability done")
}
featureAbility.disconnectAbility(connectId, disconnectCallback)
```

第 10 章 数据管理

在 HarmonyOS 系统中,数据管理至关重要,因为它为应用程序和服务提供了高效、安全的存储和检索机制。它支持设备间的数据共享与同步,为用户提供了一致且流畅的跨设备体验。数据管理在系统架构中扮演着关键角色,促进了各种设备的协同工作,提升了系统的整体性能,并优化了用户体验。本章将详细讲解如何在 HarmonyOS 应用程序中实现数据管理的知识。

10.1 HarmonyOS 数据管理介绍

数据管理为开发者提供了数据存储和管理能力。比如，联系人应用数据可以保存到数据库中，并享有数据库的安全、可靠的管理机制。在移动设备中，需要对数据进行存储和管理。

- 数据存储：提供通用数据持久化能力，根据数据特点，分为用户首选项、键值型数据库和关系型数据库。
- 数据管理：提供高效的数据管理能力，包括权限管理、数据备份恢复、数据共享框架等。

在移动智能系统中，应用程序创建的数据库都保存在应用程序的沙盒中。当这个应用程序被卸载时，也会自动删除这个数据库。

HarmonyOS 系统的数据管理模块包括用户首选项（Preferences）、键值型数据管理（KV-Store）、关系型数据管理（RelationalStore）、分布式数据对象（DataObject）和跨应用数据管理（DataShare），如图 10-1 所示。Interface 接口层提供了标准的 JS API 接口，定义了这些组件接口的描述，供开发者参考。Frameworks&System service 层负责实现组件数据存储功能，以及一些 SQLite 和其他子系统的依赖。

图 10-1　HarmonyOS 系统的数据管理模块

- 用户首选项：提供了轻量级配置数据的持久化能力，并支持订阅数据变化的通知能力。并不支持分布式同步，常用于保存应用配置信息、用户偏好设置等。
- 键值型数据管理：提供了键值型数据库的读写、加密、手动备份能力，目前暂不支持分布式功能。
- 关系型数据管理：提供了关系型数据库的增删改查、加密、手动备份能力，目前也暂不支持分布式功能。
- 分布式数据对象：独立提供对象型结构数据的分布式能力，但目前暂不支持分布式功能。
- 跨应用数据管理：提供了向其他应用共享及管理其数据的方法。这仅适用于系统应用，非系统应用无须关注，因此本书不作具体介绍。

10.2 应用数据持久化

在 HarmonyOS 系统中，应用数据持久化指的是应用程序将数据保存在设备上，确保数据在应用关闭或设备重启后仍然保持存在和可访问的状态。这可以通过使用数据库、文件系统或其他持久化存储技术来实现。应用数据持久化，允许用户在不同的应用生命周期和设备重启之间保留其数据，从而提供更加连贯和无缝的用户体验。这对支持跨设备的 HarmonyOS 生态系统尤为重要，因为它使数据在多个设备之间的同步和共享成为可能。

目前，HarmonyOS 系统支持以下三种数据持久化技术。
- 用户首选项：通常用于保存应用的配置信息。数据以文本形式保存在设备中，应用在使用过程中会将文本中的数据全量加载到内存中，因此访问速度快、效率高。但这种方式不适合需要存储大量数据的场景。
- 键值型数据库：一种非关系型数据库，数据以键值对的形式进行组织、索引和存储，其中"键"作为唯一标识符。它适合数据关系和业务关系较少的业务数据存储。由于在分布式场景中降低了解决数据库版本兼容问题的复杂度，以及数据同步过程中冲突解决的复杂度，因此被广泛使用。与关系型数据库相比，键值型数据库更容易实现跨设备和跨版本兼容。
- 关系型数据库：一种关系型数据库，数据以行和列的形式存储，它广泛用于应用中的关系型数据的处理，包括增、删、改、查等操作接口。开发者也可以运行自定义的 SQL 语句来满足复杂业务场景的需要。

本节将详细讲解使用以上三种数据持久化技术的知识。

10.2.1 使用用户首选项存储数据

在 HarmonyOS 系统中，用户首选项是指使用"Key-Value"键值对的方式保存数据，支持对数据进行修改和查询操作。当用户希望有一个全局且唯一的存储数据的地方时，建议采用用户首选项方式存储数据。用户首选项会将该数据缓存在内存中，当用户读取时，能够快速从内存中获取数据。然而，用户首选项随着存放的数据量增加，会导致应用占用的内存增大，因此，用户首选项不适合存放大量数据，适用场景一般为应用保存用户的个性化设置（如字体大小，是否开启夜间模式等）。

HarmonyOS 系统对用户首选项的约束限制如下。
- Key 键为 string 类型，要求非空且长度不超过 80 个字节。
- 如果 Value 值为 string 类型，可以为空；不为空时，长度不超过 8192 个字节。
- 内存占用量会随着存储数据量的增大而增大，所以存储的数据量应是轻量级的。建议存储的数据不超过一万条，否则可能会在内存方面产生较大的开销。

HarmonyOS 提供了多个实现用户首选项功能的接口，大多数为异步接口。这些异步接口均有 callback 和 Promise 两种返回形式。以下是一些常用的 Preferences 接口，以 callback 形式为例。

- getPreferences(context: Context, name: string, callback: AsyncCallback<Preferences>)：获取 Preferences 实例。
- put(key: string, value: ValueType, callback: AsyncCallback<void>)：将数据写入 Preferences 实例，可以通过 flush 将 Preferences 实例持久化。
- has(key: string, callback: AsyncCallback<boolean>)：检查 Preferences 实例是否包含名为给定 Key 的存储键值对。给定的 Key 值不能为空。
- get(key: string, defValue: ValueType, callback: AsyncCallback<ValueType>)：获取键对应的值，如果值为 null 或者非默认值类型，返回默认数据 defValue。
- delete(key: string, callback: AsyncCallback<void>)：从 Preferences 实例中删除名为给定 Key 的存储键值对。
- flush(callback: AsyncCallback<void>)：将当前 Preferences 实例的数据异步存储到用户首选项持久化文件中。
- on(type:'change', callback: Callback<{ key : string }>)：订阅数据变更。订阅的 Key 的值发生变更后，在执行 flush 方法后，触发 callback 回调。
- off(type:'change', callback?: Callback<{ key : string }>)：取消订阅数据变更。
- deletePreferences(context: Context, name: string, callback: AsyncCallback<void>)：从内存中移除指定的 Preferences 实例。若 Preferences 实例有对应的持久化文件，则同时删除其持久化文件。

实例 10-1 展示了使用用户首选项切换手机主题的过程。为应用程序提供了 key-value 键值型的数据处理能力，支持应用持久化轻量级数据，并对其修改和查询。数据的存储形式为键值对，键的类型为字符串型，值的存储数据类型包括数字型、字符型、布尔型以及这 3 种类型的数组类型。

实例 10-1：切换手机主题（源码路径 :codes\10\Preferences）

（1）编写文件 src/main/ets/common/ThemeDesktop.ets，定义一个名为 ThemeDesktop 的组件，用于展示手机桌面主题的图标和名称。首先，通过导入 MyDataSource 获取主题数据。其次，通过类 arrayType 定义图标和名称的数据结构。最后，利用 UI 框架的 Grid 布局，展示主题数据中的图标和名称，确保界面布局美观且灵活。

```
import { MyDataSource } from '../util/DataSource'

class arrayType {
  image: Resource | null = null;
  name: string = '';
}

@Component
export default struct ThemeDesktop {
  @Link themeDatas: Array<arrayType>
```

```
@State default: Array<arrayType> = [
  { image: $r('app.media.dialer'), name: '电话' },
  { image: $r('app.media.shopping'), name: '商城' },
  { image: $r('app.media.notes'), name: '备忘录' },
  { image: $r('app.media.settings'), name: '设置' },
  { image: $r('app.media.camera'), name: '相机' },
  { image: $r('app.media.gallery'), name: '相册' },
  { image: $r('app.media.music'), name: '音乐' },
  { image: $r('app.media.video'), name: '视频' },
]

build() {
  Grid() {
    ForEach(this.themeDatas, (item: arrayType) => {
      GridItem() {
        Column() {
          Image(item.image!)
            .width(70)
            .height(70)
            .objectFit(ImageFit.Fill)
          Text(item.name).fontSize(15)
        }
        .width(90)
        .height(90)
      }
    })
  }
  .rowsGap(10)
  .width('100%')
  .columnsGap(10)
  .layoutWeight(1)
  .padding({ top: 20 })
  .backgroundColor('#e5e5e5')
  .columnsTemplate('1fr 1fr 1fr 1fr')
  }
}
```

（2）编写文件 src/main/ets/util/DataSource.ets，定义一个基本的数据源类 BasicDataSource，实现了 IDataSource 接口，用于管理数据变化的监听器。

```
class BasicDataSource implements IDataSource {
  private listeners: DataChangeListener[] = []

  public totalCount(): number {
    return 0
  }

  public getData(index: number): undefined | Object {
    return undefined
  }

  registerDataChangeListener(listener: DataChangeListener) {
```

```
      if (this.listeners.indexOf(listener) < 0) {
        this.listeners.push(listener)
      }
    }

    unregisterDataChangeListener(listener: DataChangeListener) {
      const pos = this.listeners.indexOf(listener)
      if (pos >= 0) {
        this.listeners.splice(pos, 1)
      }
    }

    notifyDataReload() {
      this.listeners.forEach((listener: DataChangeListener) => {
        listener.onDataReloaded()
      })
    }

    notifyDataAdd(index: number) {
      this.listeners.forEach((listener: DataChangeListener) => {
        listener.onDataAdd(index)
      })
    }

    notifyDataChange(index: number) {
      this.listeners.forEach((listener: DataChangeListener) => {
        listener.onDataChange(index)
      })
    }

    notifyDataDelete(index: number) {
      this.listeners.forEach((listener: DataChangeListener) => {
        listener.onDataDelete(index)
      })
    }

    notifyDataMove(from: number, to: number) {
      this.listeners.forEach((listener: DataChangeListener) => {
        listener.onDataMove(from, to)
      })
    }
}

export class MyDataSource extends BasicDataSource {
  private dataArray: Array<number> = []

  constructor(dataArray: Array<number>) {
    super()
    this.dataArray = dataArray
  }

  public totalCount(): number {
```

```
    return this.dataArray.length
  }

  public getData(index: number): Object {
    return this.dataArray[index]
  }

  public addData(index: number) {
    this.dataArray.splice(index, 0)
    this.notifyDataAdd(index)
  }

  public pushData(index: number) {
    this.dataArray.push(index)
    this.notifyDataAdd(this.dataArray.length - 1)
  }

  public replaceData(result: Array<number>) {
    this.dataArray = result
    this.notifyDataReload()
  }
}
```

对上述代码的具体说明如下。

- 首先，通过 registerDataChangeListener 方法和 unregisterDataChangeListener 方法，可以注册和注销数据变化监听器。
- 其次，通过 notifyDataReload、notifyDataAdd、notifyDataChange、notifyDataDelete 和 notifyDataMove 方法，通知注册的监听器数据的重新加载、添加、修改、删除和移动等变化。
- 再次，定义了一个具体的数据源类 MyDataSource，并初始化了一个存储数字的数组。通过重写 totalCount 方法和 getData 方法，实现了获取数据总数和具体数据的功能。
- 最后，通过自定义的方法如 addData、pushData 和 replaceData，分别实现了对数据的添加、推送和替换操作，并通过相应的通知方法通知监听器数据的变化。确保在数据发生变化时，可以及时通知相关监听器进行相应的处理。

（3）编写文件 src/main/ets/pages/Index.ets，定义一个名为 Index 的页面组件，实现轻量级存储和主题切换功能。首先，通过 preferences 模块实现了对主题偏好的存储和获取，以及通过改变主题实时更新界面。其次，通过 ThemeDesktop 组件展示了不同主题下的图标和名称。最后，页面顶部包含应用入口标题，同时提供了切换主题的选项，用户可通过点击切换图标选择不同的主题，从而改变整体应用的外观。

```
import Logger from '../model/Logger'
import preferences from '@ohos.data.preferences'
import ThemeDesktop from '../common/ThemeDesktop'

const THEMES: Record<string, string | Resource>[][] = [
  [
```

```
      { 'image': $r('app.media.dialer'), 'name': '电话' },
      { 'image': $r('app.media.shopping'), 'name': '商城' },
      { 'image': $r('app.media.notes'), 'name': '备忘录' },
      { 'image': $r('app.media.settings'), 'name': '设置' },
      { 'image': $r('app.media.camera'), 'name': '相机' },
      { 'image': $r('app.media.gallery'), 'name': '相册' },
      { 'image': $r('app.media.music'), 'name': '音乐' },
      { 'image': $r('app.media.video'), 'name': '视频' },
  ],
  [
      { 'image': $r('app.media.simplicityCall'), 'name': '电话' },
      { 'image': $r('app.media.simplicityShop'), 'name': '商城' },
      { 'image': $r('app.media.simplicityNotes'), 'name': '备忘录' },
      { 'image': $r('app.media.simplicitySetting'), 'name': '设置' },
      { 'image': $r('app.media.simplicityCamera'), 'name': '相机' },
      { 'image': $r('app.media.simplicityPhotos'), 'name': '相册' },
      { 'image': $r('app.media.simplicityMusic'), 'name': '音乐' },
      { 'image': $r('app.media.simplicityVideo'), 'name': '视频' },
  ],
  [
      { 'image': $r('app.media.pwcall'), 'name': '电话' },
      { 'image': $r('app.media.pwshop'), 'name': '商城' },
      { 'image': $r('app.media.pwnotes'), 'name': '备忘录' },
      { 'image': $r('app.media.pwsetting'), 'name': '设置' },
      { 'image': $r('app.media.pwcamera'), 'name': '相机' },
      { 'image': $r('app.media.pwphotos'), 'name': '相册' },
      { 'image': $r('app.media.pwmusic'), 'name': '音乐' },
      { 'image': $r('app.media.pwvideo'), 'name': '视频' },
  ]
]
const TAG: string = '[Index]'
const PREFERENCES_NAME = 'theme.db'
const THEME_NAMES: string[] = ['default', 'simplicity', 'pomeloWhtie']
let preferenceTheme: preferences.Preferences | null = null


@Entry
@Component
struct Index {
  @State nowTheme: string = ''
  @State themeDatas: Record<string, string | Resource>[] = []

  async aboutToAppear() {
    // 从内存中获取轻量级存储 db 文件
    await this.getPreferencesFromStorage()
    // 从轻量级存储 db 文件中获取键名为 theme 的键值
    this.nowTheme = await this.getPreference()
    let index = THEME_NAMES.indexOf(this.nowTheme)
    this.themeDatas = THEMES[index]
  }

  async getPreferencesFromStorage() {
```

```
      let context = getContext(this) as Context
      preferenceTheme = await preferences.getPreferences(context, PREFERENCES_
NAME)
  }

  async putPreference(data: string) {
    Logger.info(TAG, `Put begin`)
    if (preferenceTheme !== null) {
      await preferenceTheme.put('theme', data)
      await preferenceTheme.flush()
    }
  }

  async getPreference(): Promise<string> {
    Logger.info(TAG, `Get begin`)
    let theme: string = ''
    if (preferenceTheme !== null) {
      theme = await preferenceTheme.get('theme', 'default') as string
    }
    return theme
  }

  changeTheme(themeNum: number) {
    this.themeDatas = THEMES[themeNum]
    this.putPreference(THEME_NAMES[themeNum])
  }

  build() {
    Column() {
      Row() {
        Text($r('app.string.EntryAbility_label'))
          .fontSize(25)
          .layoutWeight(5)
          .padding({ left: 10 })
          .fontColor(Color.White)
          .fontWeight(FontWeight.Bold)
        Image($r('app.media.change'))
          .height(30)
          .layoutWeight(1)
          .objectFit(ImageFit.ScaleDown)
          .bindMenu([
            {
              value: THEME_NAMES[0],
              action: () => {
                this.changeTheme(0)
              }
            },
            {
              value: THEME_NAMES[1],
              action: () => {
                this.changeTheme(1)
              }
```

```
          },
          {
            value: THEME_NAMES[2],
            action: () => {
              this.changeTheme(2)
            }
          }
        ])
      }
      .width('100%')
      .height(50)
      .backgroundColor('#0D9FFB')

      ThemeDesktop({ themeDatas: $themeDatas })
    }
    .width('100%')
    .height('100%')
  }
}
```

在上述 build 方法中，通过 UI 组件构建了页面布局，包括页面顶部的标题和切换主题的图标。通过点击不同主题的切换图标，触发 changeTheme 方法切换主题，同时更新界面显示。整体代码实现了通过轻量级存储管理主题偏好，以及通过 UI 组件展示不同主题下的图标和名称，使用户能够灵活切换应用的整体外观。执行后点击顶部右侧的切换按钮后弹出选择主题菜单，选择任意主题则切换至相应的主题界面。执行效果如图 10-2 所示。

图 10-2　切换手机主题界面

10.2.2　使用键值型数据库存储数据

键值型数据库是指将信息存储为键值对的形式，当需要存储的数据没有复杂的关系模型，例如，存储商品名称及对应价格、员工工号及当日是否已出勤等，数据复杂度低，更易兼容不同数据库版本和设备类型，因此推荐使用键值型数据库持久化此类数据。

在 HarmonyOS 系统中，对键值型数据库的限制约束如下。

- 设备协同数据库，针对每条记录，Key 的长度为 ≤ 896 Byte，Value 的长度为 <4 MB。
- 单版本数据库，针对每条记录，Key 的长度为 ≤ 1 KB，Value 的长度为 <4 MB。
- 每个应用程序最多支持同时打开 16 个键值型分布式数据库。
- 键值型数据库事件回调方法中不允许进行阻塞操作，如修改 UI 组件。

HarmonyOS 提供了多个实现键值型数据库存储的相关接口，其中大部分为异步接口。异步接口均有 callback 和 Promise 两种返回形式，在以下列出的常用接口均以 callback 形式为例。

- createKVManager(config: KVManagerConfig)：创建一个 KVManager 对象实例，用于管理数据库对象。

- getKVStore<T>(storeId: string, options: Options, callback: AsyncCallback<T>)：指定 Options 和 storeId，创建并获取指定类型的 KVStore 数据库。
- put(key: string, value: Uint8Array|string|number|boolean, callback: AsyncCallback<void>)：添加指定类型的键值对到数据库。
- get(key: string, callback: AsyncCallback<Uint8Array|string|boolean|number>)：获取指定键的值。
- delete(key: string, callback: AsyncCallback<void>)：从数据库中删除指定键值的数据。

在 HarmonyOS 系统中，基于 Stage 模型使用键值型数据库的步骤如下。

（1）如果要使用键值型数据库，首先要获取一个 KVManager 实例，用于管理数据库对象。例如，以下代码实现了在页面能力创建时初始化分布式键值存储的管理器（KVManager），为后续的数据库操作提供了基础支持。通过引入 distributedKVStore 模块和 UIAbility 模块，创建了 KVManager 实例，使页面能力得以利用分布式键值存储进行数据管理。

```
// 导入模块
import distributedKVStore from '@ohos.data.distributedKVStore';

// Stage 模型
import UIAbility from '@ohos.app.ability.UIAbility';

let kvManager;

export default class EntryAbility extends UIAbility {
  onCreate() {
    let context = this.context;
    const kvManagerConfig = {
      context: context,
      bundleName: 'com.example.datamanagertest'
    };
    try {
      // 创建 KVManager 实例
      kvManager = distributedKVStore.createKVManager(kvManagerConfig);
      console.info('Succeeded in creating KVManager.');
      // 继续创建获取数据库
    } catch (e) {
      console.error(`Failed to create KVManager. Code:${e.code},message:${e.message}`);
    }
  }
}
```

（2）创建并获取键值数据库：初始化和获取分布式键值数据库实例，并提供相应的配置选项。例如，以下代码段通过分布式键值存储的管理器（kvManager）获取一个具有指定配置选项的键值数据库实例。通过 kvManager.getKVStore 方法，传递数据库标识符（storeId）和一组配置选项，如是否创建、是否加密、是否备份等。在获取数据库实例的回调函数中，如果成功获取，则输出成功信息到控制台，接着可以进行相关的数据操作。如果在获取过程中出现错误，通过异常捕获（try-catch 块）输出错误代码和错误消息到控制台，提示获取 KVStore 失败或发生意外错误。

```
try {
  const options = {
    createIfMissing: true, // 当数据库文件不存在时是否创建数据库，默认创建
    encrypt: false, // 设置数据库文件是否加密，默认不加密
    backup: false, // 设置数据库文件是否备份，默认备份
    kvStoreType: distributedKVStore.KVStoreType.SINGLE_VERSION, // 设置要
创建的数据库类型，默认为多设备协同数据库
    securityLevel: distributedKVStore.SecurityLevel.S2 // 设置数据库安全级别
  };
  // storeId 为数据库唯一标识符
  kvManager.getKVStore('storeId', options, (err, kvStore) => {
    if (err) {
      console.error(`Failed to get KVStore. Code:${err.code},message: ${err.message}`);
      return;
    }
    console.info('Succeeded in getting KVStore.');
    // 进行相关数据操作
  });
} catch (e) {
  console.error(`An unexpected error occurred. Code:${e.code}, message: ${e.message}`);
}
```

（3）调用 put() 方法向键值数据库中插入数据：例如，在以下代码中，通过 kvStore.put 方法将测试键值对数据存储到分布式键值数据库中。在存储完成的回调函数中，根据操作结果输出相应的成功或失败信息到控制台。通过定义测试键和值，以及异常捕获机制，代码实现了向分布式键值数据库中存储数据并处理可能的错误情况。

```
const KEY_TEST_STRING_ELEMENT = 'key_test_string';
const VALUE_TEST_STRING_ELEMENT = 'value_test_string';
try {
  kvStore.put(KEY_TEST_STRING_ELEMENT, VALUE_TEST_STRING_ELEMENT, (err) => {
    if (err !== undefined) {
      console.error(`Failed to put data. Code:${err.code},message:${err.message}`);
      return;
    }
    console.info('Succeeded in putting data.');
  });
} catch (e) {
  console.error(`An unexpected error occurred. Code:${e.code},message:${e.message}`);
}
```

注意：当 Key 值存在时，put() 方法会修改其值，否则将新增一条数据记录。

（4）调用 get() 方法获取指定键的值：例如，以下代码首先尝试将测试键值对数据存储到分布式键值数据库（kvStore）中，然后通过 kvStore.get 方法获取相同的键值对数据。在存储和获取数据的回调函数中，通过条件判断输出相应的成功或失败信息到控制台，并在成功获取数据时输出具体的数据内容。在异常捕获块中，处理存储或获取数据过程中的意外错误，输出相应的错误代码和错误消息到控制台。

```
const KEY_TEST_STRING_ELEMENT = 'key_test_string';
const VALUE_TEST_STRING_ELEMENT = 'value_test_string';
try {
  kvStore.put(KEY_TEST_STRING_ELEMENT, VALUE_TEST_STRING_ELEMENT, (err) => {
    if (err !== undefined) {
      console.error(`Failed to put data. Code:${err.code},message:${err.message}`);
      return;
    }
    console.info('Succeeded in putting data.');
    kvStore.get(KEY_TEST_STRING_ELEMENT, (err, data) => {
      if (err !== undefined) {
        console.error(`Failed to get data. Code:${err.code},message:${err.message}`);
        return;
      }
      console.info(`Succeeded in getting data. data:${data}`);
    });
  });
} catch (e) {
  console.error(`Failed to get data. Code:${e.code},message:${e.message}`);
}
```

（5）调用 delete() 方法删除指定键值的数据：例如，在以下代码中，首先将测试键值对数据存储到分布式键值数据库（kvStore）中，然后通过 kvStore.delete 方法删除相同的键值对数据。在存储和删除数据的回调函数中，通过条件判断输出相应的成功或失败信息到控制台。在成功存储数据后尝试删除该数据，如果删除失败，则输出相应的错误信息。在异常捕获块中，处理存储或删除数据过程中的意外错误，输出相应的错误代码和错误消息到控制台。

```
const KEY_TEST_STRING_ELEMENT = 'key_test_string';
const VALUE_TEST_STRING_ELEMENT = 'value_test_string';
try {
  kvStore.put(KEY_TEST_STRING_ELEMENT, VALUE_TEST_STRING_ELEMENT, (err) => {
    if (err !== undefined) {
      console.error(`Failed to put data. Code:${err.code},message:${err.message}`);
      return;
    }
    console.info('Succeeded in putting data.');
    kvStore.delete(KEY_TEST_STRING_ELEMENT, (err) => {
      if (err !== undefined) {
        console.error(`Failed to delete data. Code:${err.code},message:${err.message}`);
        return;
      }
      console.info('Succeeded in deleting data.');
    });
  });
} catch (e) {
  console.error(`An unexpected error occurred. Code:${e.code},message:${e.message}`);
}
```

10.2.3 使用关系型数据库存储数据

关系型数据库是一种以表格（表）为基础的数据库，它使用结构化查询语言（SQL）来定义和操作数据。在关系型数据库中，数据被组织成一个或多个表，每个表包含行和列。每一行代表一个记录，而每一列代表记录中的一个属性。关系型数据库的核心概念是表与表之间的关系，这种关系通过共享数据中的键（键值）来建立。典型的关系型数据库管理系统（RDBMS）包括 MySQL、Oracle Database、Microsoft SQL Server、PostgreSQL 和 SQLite 等。

在开发 HarmonyOS 应用程序的过程中，开发者可以选择使用合适的数据库引擎来满足其应用的需求。HarmonyOS 支持多种数据库系统，其中最常用的关系型数据库是 SQLite。SQLite 是一种轻量级的嵌入式数据库引擎，适用于嵌入式系统和移动设备等资源受限的环境。

在 HarmonyOS 系统中，关系型数据库对应用提供通用的操作接口，底层使用 SQLite 作为持久化存储引擎，支持 SQLite 具有的数据库特性，包括但不限于事务、索引、视图、触发器、外键、参数化查询和预编译 SQL 语句。HarmonyOS 关系型数据库的运作机制，如图 10-3 所示。

图 10-3　HarmonyOS 关系型数据库的运作机制

HarmonyOS 系统为关系型数据库提供了多个相关接口，其中大部分为异步接口。

- getRdbStore(context: Context, config: StoreConfig, callback: AsyncCallback<RdbStore>)：获得一个相关的 RdbStore，操作关系型数据库，用户可以根据自己的需求配置 RdbStore 的参数，然后通过 RdbStore 调用相关接口可以执行相关的数据操作。
- executeSql(sql: string, bindArgs: Array<ValueType>, callback: AsyncCallback<void>)：执行包含指定参数但不返回值的 SQL 语句。
- insert(table: string, values: ValuesBucket, callback: AsyncCallback<number>)：向目标表中插入一行数据。
- update(values: ValuesBucket, predicates: RdbPredicates, callback: AsyncCallback<number>)：根据 RdbPredicates 的指定实例对象更新数据库中的数据。
- delete(predicates: RdbPredicates, callback: AsyncCallback<number>)：根据 RdbPredicates 的指定实例对象从数据库中删除数据。
- query(predicates: RdbPredicates, columns: Array<string>, callback: AsyncCallback<ResultSet>)：根据指定条件查询数据库中的数据。
- deleteRdbStore(context: Context, name: string, callback: AsyncCallback<void>)：删除数据库。

实例 10-2 实现了一个基于关系型数据库的记账系统。这个例子实现了账目的展示和管理功能，包括搜索、编辑、删除和新增功能。通过自定义对话框组件（DialogComponent）进行账目插入和更新，支持用户交互，并展示相应的图标和金额信息。整个项目基于关系模型来管理账目数据，提供了增、删、改、查等接口，也可以运行输入的 SQL 语句满足复杂场景的需要。

实例 10-2：基于关系型数据库的记账系统（源码路径 :codes\10\ORM）

（1）编写文件 src/main/ets/common/constants/CommonConstants.ets，这段代码是 HarmonyOS 常量定义文件，使用了鸿蒙的关系型数据库模块 @ohos.data.relationalStore，定义了一系列常量，包括数据库配置、账目表配置、搜索文本、提示文本、组件尺寸和位置以及日志标签等。其中，STORE_CONFIG 定义了数据库配置，ACCOUNT_TABLE 定义了账目表配置，而其他常量则包括搜索文本、提示文本、组件尺寸和位置，以及日志标签等。这些常量将在整个项目中用于保持统一的数值和文本标识，提高代码的可维护性和可读性。

```
import relationalStore from '@ohos.data.relationalStore';
import { AccountTable } from '../../viewmodel/ConstantsInterface';

export default class CommonConstants {
  /**
   * Rdb database config.
   */
  static readonly STORE_CONFIG: relationalStore.StoreConfig = {
    name: 'database.db',
    securityLevel: relationalStore.SecurityLevel.S1
  };
  /**
   * Account table config.
   */
  static readonly ACCOUNT_TABLE: AccountTable = {
    tableName: 'accountTable',
    sqlCreate: 'CREATE TABLE IF NOT EXISTS accountTable(id INTEGER PRIMARY KEY AUTOINCREMENT, accountType INTEGER, ' +
      'typeText TEXT, amount INTEGER)',
    columns: ['id', 'accountType', 'typeText', 'amount']
  };
  /**
   * Search text of Search component.
   */
  static readonly SEARCH_TEXT = '搜索';
  /**
   * toast text of prompt component.
   */
  static readonly TOAST_TEXT_1 = '账目类型不能为空';
  static readonly TOAST_TEXT_2 = '账目金额不为正整数';
  /**
   * Component size.
   */
  static readonly FULL_WIDTH = '100%';
  static readonly FULL_HEIGHT = '100%';
  static readonly DIALOG_HEIGHT = '55%';
  static readonly TABS_HEIGHT = '45%';
  static readonly MINIMUM_SIZE = 0;
  static readonly FULL_SIZE = 1;
  static readonly PROMPT_BOTTOM = '70vp';
```

```
  /**
   * Component location.
   */
  static readonly EDIT_POSITION_X = '80%';
  static readonly EDIT_POSITION_Y = '90%';
  static readonly DELETE_POSITION_X = '50%';
  static readonly DELETE_POSITION_Y = '90%';
  /**
   * Log tag.
   */
  static readonly RDB_TAG = '[Debug.Rdb]';
  static readonly TABLE_TAG = '[Debug.AccountTable]';
  static readonly INDEX_TAG = '[Debug.Index]';
}
```

（2）编写文件 src/main/ets/common/database/Rdb.ets，定义一个名为 Rdb 的类，作为鸿蒙项目中与关系型数据库（Rdb）交互的模块。该类封装了数据库表的创建、数据插入、删除、更新和查询等操作，通过鸿蒙的关系型数据库模块 @ohos.data.relationalStore 实现数据的持久化存储。具体功能包括配置数据库表信息、获取数据库实例、插入、删除、更新和查询数据，并通过回调函数处理异步操作。

```
export default class Rdb {
  private rdbStore: relationalStore.RdbStore | null = null;
  private tableName: string;
  private sqlCreateTable: string;
  private columns: Array<string>;

  constructor(tableName: string, sqlCreateTable: string, columns: Array<string>) {
    this.tableName = tableName;
    this.sqlCreateTable = sqlCreateTable;
    this.columns = columns;
  }

  getRdbStore(callback: Function = () => {
  }) {
    if (!callback || typeof callback === 'undefined' || callback === undefined) {
      Logger.info(CommonConstants.RDB_TAG, 'getRdbStore() has no callback!');
      return;
    }
    if (this.rdbStore !== null) {
      Logger.info(CommonConstants.RDB_TAG, 'The rdbStore exists.');
      callback();
      return
    }
    let context: Context = getContext(this) as Context;
    relationalStore.getRdbStore(context, CommonConstants.STORE_CONFIG, (err, rdb) => {
      if (err) {
        Logger.error(CommonConstants.RDB_TAG, `gerRdbStore() failed, err: ${err}`);
        return;
      }
      this.rdbStore = rdb;
```

```
      this.rdbStore.executeSql(this.sqlCreateTable);
      Logger.info(CommonConstants.RDB_TAG, 'getRdbStore() finished.');
      callback();
    });
  }

  insertData(data: relationalStore.ValuesBucket, callback: Function = () => {
  }) {
    if (!callback || typeof callback === 'undefined' || callback === undefined) {
      Logger.info(CommonConstants.RDB_TAG, 'insertData() has no callback!');
      return;
    }
    let resFlag: boolean = false;
    const valueBucket: relationalStore.ValuesBucket = data;
    if (this.rdbStore) {
      this.rdbStore.insert(this.tableName, valueBucket, (err, ret) => {
        if (err) {
          Logger.error(CommonConstants.RDB_TAG, `insertData() failed, err: ${err}`);
          callback(resFlag);
          return;
        }
        Logger.info(CommonConstants.RDB_TAG, `insertData() finished: ${ret}`);
        callback(ret);
      });
    }
  }

  deleteData(predicates: relationalStore.RdbPredicates, callback: Function = () => {
  }) {
    if (!callback || typeof callback === 'undefined' || callback === undefined) {
      Logger.info(CommonConstants.RDB_TAG, 'deleteData() has no callback!');
      return;
    }
    let resFlag: boolean = false;
    if (this.rdbStore) {
      this.rdbStore.delete(predicates, (err, ret) => {
        if (err) {
          Logger.error(CommonConstants.RDB_TAG, `deleteData() failed, err: ${err}`);
          callback(resFlag);
          return;
        }
        Logger.info(CommonConstants.RDB_TAG, `deleteData() finished: ${ret}`);
        callback(!resFlag);
      });
    }
  }

  updateData(predicates: relationalStore.RdbPredicates, data: relationalStore.ValuesBucket, callback: Function = () => {
  }) {
    if (!callback || typeof callback === 'undefined' || callback === undefined) {
      Logger.info(CommonConstants.RDB_TAG, 'updateDate() has no callback!');
```

```
        return;
      }
      let resFlag: boolean = false;
      const valueBucket: relationalStore.ValuesBucket = data;
      if (this.rdbStore) {
        this.rdbStore.update(valueBucket, predicates, (err, ret) => {
          if (err) {
            Logger.error(CommonConstants.RDB_TAG, `updateData() failed, err: ${err}`);
            callback(resFlag);
            return;
          }
          Logger.info(CommonConstants.RDB_TAG, `updateData() finished: ${ret}`);
          callback(!resFlag);
        });
      }
    }

    query(predicates: relationalStore.RdbPredicates, callback: Function = () => {
    }) {
      if (!callback || typeof callback === 'undefined' || callback === undefined) {
        Logger.info(CommonConstants.RDB_TAG, 'query() has no callback!');
        return;
      }
      if (this.rdbStore) {
        this.rdbStore.query(predicates, this.columns, (err, resultSet) => {
          if (err) {
            Logger.error(CommonConstants.RDB_TAG, `query() failed, err: ${err}`);
            return;
          }
          Logger.info(CommonConstants.RDB_TAG, 'query() finished.');
          callback(resultSet);
          resultSet.close();
        });
      }
    }
  }
```

（3）编写文件 src/main/ets/view/DialogComponent.ets，定义一个名为 DialogComponent 的结构体，用于创建自定义对话框界面。该对话框包含用于选择账目类型、输入金额和确认操作的交互组件。这个文件主要包含以下函数。

- Constructor（构造函数）：定义对话框组件的结构和初始状态。
- aboutToAppear()：在对话框即将显示时执行的操作，用于初始化输入金额、当前账目类型索引和类型文本。
- selectAccount(item: AccountItem)：选择账目类型的回调函数，更新当前账目类型和类型文本。参数 item 表示账目项的信息。
- build()：构建对话框组件的界面和交互逻辑，返回对话框组件的 UI 结构。
- TabBuilder(index: number)：构建对话框中的选项卡，参数 index 表示选项卡的索引，返回选项卡的 UI 结构。
- onChange((index) => {...})：当选项卡切换时的回调函数，用于更新当前选中的账目类型索引。参数 index 表示选中的选项卡索引。
- onClick(() => {...})：按钮单击事件的回调函数，用于确认输入的金额和账目类型。
- TextInput({...})：创建用于输入金额的文本输入框，返回文本输入框的 UI 结构。
- Button() {...}：创建确认按钮，返回确认按钮的 UI 结构。

（4）编写文件 src/main/ets/pages/MainPage.ets，定义一个名为 MainPage 的 HarmonyOS 应用主页面组件，实现了展示和管理账目信息的功能。它包括搜索栏、账目列表，支持编辑、删除和新增操作，通过自定义对话框组件实现账目的插入和更新，同时展示相应的图标和金额信息。具体实现流程如下。

（1）定义 MainPage 结构体：用于构建主页面的组件，使用 @Entry 和 @Component 注解表示这是一个入口组件。

```
@Entry
@Component
struct MainPage {
  @State accounts: Array<AccountData> = [];
  @State searchText: string = '';
  @State isEdit: boolean = false;
  @State isInsert: boolean = false;
  @State newAccount: AccountData = { id: 0, accountType: 0, typeText: '', amount: 0 };
  @State index: number = -1;
  private AccountTable = new AccountTable(() => {});
  private deleteList: Array<AccountData> = [];
  searchController: SearchController = new SearchController();
  dialogController: CustomDialogController = new CustomDialogController({
    builder: DialogComponent({
      isInsert: $isInsert,
      newAccount: $newAccount,
      confirm: (isInsert: boolean, newAccount: AccountData) => this.accept(isInsert, newAccount)
    }),
    customStyle: true,
    alignment: DialogAlignment.Bottom
  });
```

（2）定义 accept(isInsert: boolean, newAccount: AccountData) 函数，用于处理确认对话框操作的回调函数，根据操作类型执行插入或更新账目的操作。

```
accept(isInsert: boolean, newAccount: AccountData): void {
  if (isInsert) {
    Logger.info(`${CommonConstants.INDEX_TAG}`, `The account inserted is: ${JSON.stringify(newAccount)}`);
    this.AccountTable.insertData(newAccount, (id: number) => {
      newAccount.id = id;
      this.accounts.push(newAccount);
    });
```

```
    } else {
      this.AccountTable.updateData(newAccount, () => {
      });
      let list = this.accounts;
      this.accounts = [];
      list[this.index] = newAccount;
      this.accounts = list;
      this.index = -1;
    }
  }
```

（3）定义 aboutToAppear() 函数，功能是在页面即将显示时执行的操作，包括获取数据库存储、查询账目数据，并更新页面显示。

```
  aboutToAppear() {
    this.AccountTable.getRdbStore(() => {
      this.AccountTable.query(0, (result: AccountData[]) => {
        this.accounts = result;
      }, true);
    });
  }
```

（4）定义 selectListItem(item: AccountData) 函数，功能是选择列表项的回调函数，用于更新编辑状态、保存选中的账目索引和初始化新账目数据。

```
  selectListItem(item: AccountData) {
    this.isInsert = false;
    this.index = this.accounts.indexOf(item);
    this.newAccount = {
      id: item.id,
      accountType: item.accountType,
      typeText: item.typeText,
      amount: item.amount
    };
  }
```

（5）定义 deleteListItem() 函数，功能是删除列表项的函数，用于删除选中的账目并更新页面显示。

```
  deleteListItem() {
    for (let i = 0; i < this.deleteList.length; i++) {
      let index = this.accounts.indexOf(this.deleteList[i]);
      this.accounts.splice(index, 1);
      this.AccountTable.deleteData(this.deleteList[i], () => {
      });
    }
    this.deleteList = [];
    this.isEdit = false;
  }
```

（6）定义 build() 函数，功能是构建页面的 UI 结构，包括搜索栏、账目列表、编辑按钮、新增按钮和删除按钮等。使用 Stack、Column、Row、Text、Image、Search、List、

ListItem、Button 等组件构建页面布局。根据编辑状态显示不同的按钮，点击按钮时执行相应的操作，如切换编辑模式、打开对话框等。

```
build() {
  Stack() {
    Column() {
      Row() {
        Text($r('app.string.MainAbility_label'))
          .height($r('app.float.component_size_SP'))
          .fontSize($r('app.float.font_size_L'))
          .margin({ left: $r('app.float.font_size_L') })

        Image($rawfile('ic_public_edit.svg'))
          .width($r('app.float.component_size_S'))
          .aspectRatio(CommonConstants.FULL_SIZE)
          .margin({ right: $r('app.float.font_size_L') })
          .onClick(() => {
            this.isEdit = true;
          })
      }
      .width(CommonConstants.FULL_WIDTH)
      .justifyContent(FlexAlign.SpaceBetween)
      .margin({ top: $r('app.float.edge_size_M'), bottom: $r('app.float.edge_size_MM') })

      Row() {
        Search({
          value: this.searchText,
          placeholder: CommonConstants.SEARCH_TEXT,
          controller: this.searchController
        })
          .width(CommonConstants.FULL_WIDTH)
          .borderRadius($r('app.float.radius_size_M'))
          .borderWidth($r('app.float.border_size_S'))
          .borderColor($r('app.color.border_color'))
          .placeholderFont({ size: $r('app.float.font_size_M') })
          .textFont({ size: $r('app.float.font_size_M') })
          .backgroundColor(Color.White)
          .onChange((searchValue: string) => {
            this.searchText = searchValue;
          })
          .onSubmit((searchValue: string) => {
            if (searchValue === '') {
              this.AccountTable.query(0, (result: AccountData[]) => {
                this.accounts = result;
              }, true);
            } else {
              this.AccountTable.query(Number(searchValue), (result: AccountData[]) => {
                this.accounts = result;
              }, false);
            }
```

```
            })
          }
          .width(CommonConstants.FULL_WIDTH)
          .padding({ left: $r('app.float.edge_size_M'), right: $r('app.float.edge_size_M') })
          .margin({ top: $r('app.float.edge_size_S'), bottom: $r('app.float.edge_size_S') })

          Row() {
            List({ space: CommonConstants.FULL_SIZE }) {
              ForEach(this.accounts, (item: AccountData) => {
                ListItem() {
                  Row() {
                    Image(ImageList[item.typeText])
                      .width($r('app.float.component_size_M'))
                      .aspectRatio(CommonConstants.FULL_SIZE)
                      .margin({ right: $r('app.float.edge_size_MP') })

                    Text(item.typeText)
                      .height($r('app.float.component_size_SM'))
                      .fontSize($r('app.float.font_size_M'))

                    Blank()
                      .layoutWeight(1)

                    if (!this.isEdit) {
                      Text(item.accountType === 0 ? '-' + item.amount.toString() : '+' + item.amount.toString())
                        .fontSize($r('app.float.font_size_M'))
                        .fontColor(item.accountType === 0 ? $r('app.color.pay_color') : $r('app.color.main_color'))
                        .align(Alignment.End)
                        .flexGrow(CommonConstants.FULL_SIZE)
                    } else {
                      Row() {
                        Toggle({ type: ToggleType.Checkbox })
                          .onChange((isOn) => {
                            if (isOn) {
                              this.deleteList.push(item);
                            } else {
                              let index = this.deleteList.indexOf(item);
                              this.deleteList.splice(index, 1);
                            }
                          })
                      }
                      .align(Alignment.End)
                      .flexGrow(CommonConstants.FULL_SIZE)
                      .justifyContent(FlexAlign.End)
                    }

                  }
                  .width(CommonConstants.FULL_WIDTH)
```

```
                    .padding({ left: $r('app.float.edge_size_M'), right: $r('app.
float.edge_size_M') })
              }
              .width(CommonConstants.FULL_WIDTH)
              .height($r('app.float.component_size_LM'))
              .onClick(() => {
                this.selectListItem(item);
                this.dialogController.open();
              })
            })
          }
          .width(CommonConstants.FULL_WIDTH)
          .borderRadius($r('app.float.radius_size_L'))
          .backgroundColor(Color.White)
        }
        .width(CommonConstants.FULL_WIDTH)
        .padding({ left: $r('app.float.edge_size_M'), right: $r('app.float.
edge_size_M') })
        .margin({ top: $r('app.float.edge_size_SM') })

      }
      .width(CommonConstants.FULL_WIDTH)
      .height(CommonConstants.FULL_HEIGHT)

      if (!this.isEdit) {
        Button() {
          Image($rawfile('add.png'))
        }
        .width($r('app.float.component_size_MP'))
        .height($r('app.float.component_size_MP'))
        .position({ x: CommonConstants.EDIT_POSITION_X, y: CommonConstants.
EDIT_POSITION_Y })
        .onClick(() => {
          this.isInsert = true;
          this.newAccount = { id: 0, accountType: 0, typeText: '', amount:
0 };
          this.dialogController.open();
        })
      }

      if (this.isEdit) {
        Button() {
          Image($rawfile('delete.png'))
        }
        .width($r('app.float.component_size_MP'))
        .height($r('app.float.component_size_MP'))
        .backgroundColor($r('app.color.background_color'))
        .markAnchor({ x: $r('app.float.mark_anchor'), y: CommonConstants.
MINIMUM_SIZE })
        .position({ x: CommonConstants.DELETE_POSITION_X, y: CommonConstants.
DELETE_POSITION_Y })
        .onClick(() => {
```

```
        this.deleteListItem();
      })
    }
  }
  .width(CommonConstants.FULL_WIDTH)
  .height(CommonConstants.FULL_HEIGHT)
  .backgroundColor($r('app.color.background_color'))
  }
}
```

执行后来到应用程序首页,可以进行如下操作。
- 点击应用程序首页右下角的"添加"图标,在弹出的窗口中选择账目类型并填写金额,点击"确定"按钮,即可添加一条账目。
- 点击应用程序首页右上角的"编辑"图标,选中想要删除的账目,然后点击下方的"删除"图标,即可删除所选账目。
- 在应用程序首页点击想要编辑的账目,在弹出的窗口中更改账目类型或金额,点击"确定"按钮,即可修改一条账目。
- 在应用首页点击搜索栏,填写想要查找的账目金额,点击"搜索"图标后,将仅显示与输入金额匹配的账目;搜索栏为空时则显示全部账目。

执行效果如图 10-4 所示。

添加收入金额

成功添加收入金额

图 10-4 关系型数据库的记账系统

第 11 章
电话和短信服务

在 HarmonyOS 系统中，电话和短信服务的开发至关重要，尤其对那些需要与通信功能整合的应用程序。这包括通信类应用、社交媒体平台以及需要实现身份验证的应用。通过充分利用 HarmonyOS 的通信接口，开发者能够提供更加综合和创新的用户体验，使应用在系统生态中更为完善和互联。本章将详细讲解在 HarmonyOS 系统中开发电话和短信服务程序的知识。

11.1 电话服务开发概述

HarmonyOS 系统的电话服务开发主要涉及与通信功能的集成，以实现与电话相关的功能。HarmonyOS 电话服务系统提供了一系列的 API，用于拨打电话、获取无线蜂窝网络和 SIM 卡相关信息。在开发应用程序时，可以通过调用对应的 API 来获取当前注册网络名称、网络服务状态、信号强度及 SIM 卡的相关信息。

开发者开发电话服务程序的步骤如下。

- 权限获取：确保应用已经获取了必要的权限，如访问通话记录、拨打电话和读取联系人等权限。在 HarmonyOS 系统中，直接拨打电话需要系统权限 ohos.permission.PLACE_CALL。建议使用 makeCall()，跳转到拨号界面，并显示拨号的号码，具体可参考跳转拨号界面开发指导。
- 调用 Telephony API：利用 HarmonyOS 系统提供的 Telephony API，开发者可以实现电话功能的调用，包括拨打电话、接听来电、挂断通话等操作。
- 监听电话状态：使用电话状态监听器，应用可以实时获取电话的状态变化，如来电、去电、接听、挂断等，以便根据不同状态执行相应的逻辑。
- 通话管理：实现通话管理功能，包括多通道通话、静音、扬声器切换等，以提供更全面的通话体验。
- 集成联系人：将应用与系统联系人集成，使用户能够方便地在通话过程中访问和管理联系人信息。

需要注意的是，HarmonyOS 系统电话服务的开发需要结合系统提供的 Telephony API，以实现各种与电话通信相关的功能，为用户提供一体化的通信体验。

11.2 跳转拨号界面

在 HarmonyOS 应用程序中，当拨打电话时，需要先跳转到拨号界面。拨号功能调用 makeCall 接口，设备会自动跳转到拨号界面。用户可以选择音频或视频呼叫，可选择卡 1 或卡 2 拨出。

11.2.1 拨号接口

在 HarmonyOS 系统中，call 模块为开发者提供了呼叫管理功能，observer 模块为开发者提供了通话业务状态订阅和取消订阅功能。各个接口的具体说明如表 11-1 所示。

表 11-1 拨号接口说明

功能分类	接口名	描述	所需权限
能力获取	call.hasVoiceCapability()	是否具有语音功能	无

续表

功能分类	接口名	描述	所需权限
跳转拨号界面	call.makeCall()	跳转到拨号界面，并显示要拨的号码	无
订阅通话业务状态变化	observer.on('callStateChange')	订阅通话业务状态变化	ohos.permission.READ_CALL_LOG（获取通话号码需要该权限）
取消订阅通话业务状态变化	observer.off('callStateChange')	取消订阅通话业务状态变化	无

开发 HarmonyOS 拨号应用程序的基本步骤如下。

（1）使用 import 导入需要的模块。

（2）调用 hasVoiceCapability() 接口获取当前设备呼叫能力，如果支持就继续下一步操作；如果不支持则无法发起呼叫。

（3）跳转到拨号界面，并显示要拨的号码。

（4）订阅通话业务状态变化，这一步骤是可选的。

以下代码，演示了创建拨号程序的过程。首先，检查设备是否支持语音通话功能。其次，如果支持，发起一次号码为"13xxxx"的呼叫，并在成功或失败时分别输出相应的消息。最后，可选择性地订阅通话状态变化，以实时监测通话状态的更新。

```
// import 需要的模块
import call from '@ohos.telephony.call';
import observer from '@ohos.telephony.observer';

// 调用查询能力接口
let isSupport = call.hasVoiceCapability();
if (!isSupport) {
    console.log("not support voice capability, return.");
    return;
}
// 如果设备支持呼叫能力，则继续跳转到拨号界面，并显示拨号的号码
call.makeCall("13xxxx", (err)=> {
    if (!err) {
        console.log("make call success.");
    } else {
        console.log("make call fail, err is:" + JSON.stringify(err));
    }
});
// 订阅通话业务状态变化（可选）
observer.on("callStateChange", (data) => {
    console.log("call state change, data is:" + JSON.stringify(data));
});
```

11.2.2 开发一个拨号程序

实例 11-1 演示了使用 ohos.telephony.call 实现一个拨号程序的过程。

实例 11-1：一个拨号程序（源码路径 :codes\11\harmonyos_call）

（1）编写文件 src/main/module.json5，这是一个 HarmonyOS 应用程序的模块配置文件，

扫描看视频

用于描述应用的基本信息和结构。

```json
{
  "module": {
    "name": "entry",
    "type": "entry",
    "description": "$string:module_desc",
    "mainElement": "EntryAbility",
    "deviceTypes": [
      "phone",
      "tablet"
    ],
    "deliveryWithInstall": true,
    "installationFree": false,
    "pages": "$profile:main_pages",
    "abilities": [
      {
        "name": "EntryAbility",
        "srcEntry": "./ets/entryability/EntryAbility.ts",
        "description": "$string:EntryAbility_desc",
        "icon": "$media:icon",
        "label": "$string:EntryAbility_label",
        "startWindowIcon": "$media:icon",
        "startWindowBackground": "$color:start_window_background",
        "exported": true,
        "skills": [
          {
            "entities": [
              "entity.system.home"
            ],
            "actions": [
              "action.system.home"
            ]
          }
        ]
      }
    ]
  }
}
```

（2）编写文件 src/main/ets/entryability/EntryAbility.ts，实现了一个 HarmonyOS 应用程序的入口能力（EntryAbility），继承自 UIAbility 类。当能力被创建时进行初始化，在销毁时清理资源。在窗口舞台创建时加载名为"pages/Index"的内容，并在销毁时释放与 UI 相关的资源。此外，能力在进入前台和后台时分别触发相应的事件，通过日志输出展示生命周期各阶段。

```typescript
// 引入所需的模块和类
import UIAbility from '@ohos.app.ability.UIAbility';
import hilog from '@ohos.hilog';
import window from '@ohos.window';

// 定义 EntryAbility 类，继承自 UIAbility
```

```
export default class EntryAbility extends UIAbility {
  // 当能力被创建时调用
  onCreate(want, launchParam) {
    hilog.info(0x0000, 'testTag', '%{public}s', 'Ability onCreate');
  }

  // 当能力被销毁时调用
  onDestroy() {
    hilog.info(0x0000, 'testTag', '%{public}s', 'Ability onDestroy');
  }

  // 当窗口舞台被创建时调用
  onWindowStageCreate(windowStage: window.WindowStage) {
    // 主窗口被创建,为该能力设置主页面
    hilog.info(0x0000, 'testTag', '%{public}s', 'Ability onWindowStageCreate');

    // 加载名为 'pages/Index' 的内容
    windowStage.loadContent('pages/Index', (err, data) => {
      if (err.code) {
        hilog.error(0x0000, 'testTag', 'Failed to load the content. Cause: %{public}s', JSON.stringify(err) ?? '');
        return;
      }
        hilog.info(0x0000, 'testTag', 'Succeeded in loading the content. Data: %{public}s', JSON.stringify(data) ?? '');
    });
  }

  // 当窗口舞台被销毁时调用
  onWindowStageDestroy() {
    // 主窗口被销毁,释放与 UI 相关的资源
    hilog.info(0x0000, 'testTag', '%{public}s', 'Ability onWindowStageDestroy');
  }

  // 当能力进入前台时调用
  onForeground() {
    // 能力进入前台
    hilog.info(0x0000, 'testTag', '%{public}s', 'Ability onForeground');
  }

  // 当能力进入后台时调用
  onBackground() {
    // 能力进入后台
    hilog.info(0x0000, 'testTag', '%{public}s', 'Ability onBackground');
  }
}
```

对上述代码的具体说明如下。
- 类 EntryAbility 继承自 UIAbility,这是鸿蒙框架提供的 UI 能力基类。
- onCreate 方法在能力被创建时调用,可以用于执行初始化操作。
- onDestroy 方法在能力被销毁时调用,可以用于清理资源。

- onWindowStageCreate 方法在窗口舞台被创建时调用，这里加载了名为"pages/Index"的内容。
- onWindowStageDestroy 方法在窗口舞台被销毁时调用，用于释放与 UI 相关的资源。
- onForeground 方法和 onBackground 方法分别在能力进入前台和后台时被调用，可以执行相应的处理逻辑。

整体上，类 EntryAbility 实现了鸿蒙应用程序的入口能力，并定义了一些生命周期方法，用于处理能力的创建、销毁及窗口舞台的创建和销毁等事件。

（3）编写文件 src/main/ets/pages/Index.ets，定义一个 HarmonyOS 应用程序的页面组件（Index），包含一个按钮，点击按钮时会调用电话拨号功能。

```
import call from '@ohos.telephony.call';

function Make_Call() {
  call.makeCall("18900000000", (err, data) => {
    console.log(`makeCall callback: err->${JSON.stringify(err)}`);
    console.log(`makeCall callback: err->${JSON.stringify(data)}`);
  });
}

@Entry
@Component
struct Index {
  @State message: string = 'Make Call'

  build() {
    Row() {
      Column({ space:10}) {
        Text(' 点击按钮唤起手机拨号键盘 ')
        Text(' 并自动填入指定手机号 ').margin({bottom:20})
        Button(this.message)
          .fontSize(22)
          .padding({left:35, right:35, top:10, bottom:10})
          .onClick(()=>{
            Make_Call()
          })
      }
      .width('100%')
      .margin({top:40})
    }
    .height('100%')
    .alignItems(VerticalAlign.Top)
  }
}
```

上述代码具体说明如下。
- 导入了"@ohos.telephony.call"模块，用于处理电话拨号功能。
- 定义了一个名为"Make_Call"的函数，其中使用 call.makeCall 方法实现拨打指定手机号的功能，并在回调中输出相关信息。
- 使用 @Entry 和 @Component 装饰器，定义了名为"Index"的页面组件结构。

- 在界面中包含一个按钮，点击按钮时触发 Make_Call 函数，实现点击按钮唤起手机拨号键盘并自动填入指定手机号的功能。页面上还包含一些文本信息，说明按钮的作用。

执行效果如图 11-1 所示。

初始界面　　拨号界面
图 11-1　拨号程序

11.3 获取当前蜂窝网络信号信息

在日常应用中，经常需要获取用户所在蜂窝网络下的信号信息，以便了解当前电话信号的质量。在 HarmonyOS 系统中，可以使用 radio 模块获取当前网络信号信息，使用 observer 模块为开发者提供蜂窝网络状态的订阅和取消订阅功能。这两个模块包含的相关接口信息，如表 11-2 所示。

表 11-2　接口信息

功能分类	接口名	描　　述	所需权限
信号强度信息	radio.getSignalInformation()	获取当前注册蜂窝网络信号强度信息	无
订阅蜂窝网络信号变化	observer.on('signalInfoChange')	订阅蜂窝网络信号变化	无
取消订阅蜂窝网络信号变化	observer.off('signalInfoChange')	取消订阅蜂窝网络信号变化	无

在 HarmonyOS 系统中，获取当前蜂窝网络信号信息的步骤如下。

（1）使用 import 语句导入需要的模块 ohos.telephony.radio 和 ohos.telephony.observer。
（2）调用 getSignalInformation() 方法，返回所有的 SignalInformation 列表。
（3）遍历 SignalInformation 数组，并根据不同的 signalType 获取不同制式的信号强度。
（4）订阅蜂窝网络信号变化，这一步是可选的。

以下代码，演示了获取当前蜂窝网络信号信息的过程。使用了 HarmonyOS 的 Telephony 模块，通过 radio 对象获取卡槽 1 的信号强度信息，并订阅了蜂窝网络信号变化。

```
import radio from '@ohos.telephony.radio'
import observer from '@ohos.telephony.observer';

// 以获取卡 1 的信号强度为例
let slotId = 0;
radio.getSignalInformation(slotId, (err, data) => {
    if (!err) {
        console.log("get signal information success.");
        // 遍历数组，输出不同网络制式下的信号强度
        for (let j = 0; j < data.length; j++) {
            console.log("type:" + data[j].signalType + ", level:" + data[j].signalLevel);
        }
    } else {
        console.log("get signal information fail, err is:" + JSON.stringify(err));
    }
});
```

```
// 订阅蜂窝网络信号变化（可选）
observer.on("signalInfoChange", (data) => {
    console.log("signal info change, data is:" + JSON.stringify(data));
});
```

上述代码的主要功能是监控蜂窝网络信号的变化，获取卡槽 1 的信号强度信息，并在控制台输出相应的信息。同时，通过订阅事件，在信号发生变化时实时获取并处理相关信息。

11.4 短信服务

短信作为一种简便而广泛应用的通信方式，为用户提供了快捷、稳定的信息传递途径。鸿蒙系统通过集成短信服务，不仅能够实现即时通信功能，还为应用程序提供了便捷的身份验证和信息推送手段，提升了用户体验。此外，鸿蒙系统的短信服务还有助于推动开发者社区的创新，为构建更强大、多功能的应用生态系统奠定了基础。

11.4.1 sms 模块介绍

在 HarmonyOS 系统中，ohos.telephony.sms 模块提供了短信服务功能。它提供了管理短信的一些基础能力，包括创建、发送短信，获取、设置发送短信的默认 SIM 卡槽 ID，获取、设置短信服务中心（SMSC）地址，以及检查当前设备是否具备短信发送和接收能力等。在使用 ohos.telephony.sms 模块前，先导入这个模块：

```
import sms from '@ohos.telephony.sms';
```

在 ohos.telephony.sms 模块中包含以下成员。

（1）sms.createMessage：根据协议数据单元（PDU）和指定的短信协议创建短信实例，使用 callback 异步回调。原型是：

```
createMessage(pdu: Array<number>, specification: string, callback: AsyncCallback<ShortMessage>): void
```

各个参数的具体说明如下。
- pdu：协议数据单元，从收到的信息中获取。
- specification：短信协议类型。3gpp 表示 GSM/UMTS/LTE SMS，3gpp2 表示 CDMA SMS
- callback：回调函数。

（2）sms.createMessage：根据协议数据单元（PDU）和指定的短信协议创建短信实例。使用 Promise 异步回调，原型是：

```
createMessage(pdu: Array<number>, specification: string): Promise<ShortMessage>
```

各个参数的具体说明如下。
- pdu：协议数据单元，从收到的信息中获取。
- specification：短信协议类型。3gpp 表示 GSM/UMTS/LTE SMS，3gpp2 表示 CDMA SMS。

（3）sms.sendMessage：发送短信，原型是：

```
sendMessage(options: SendMessageOptions): void
```

需要权限：

```
ohos.permission.SEND_MESSAGES
```

参数 options 是发送短信的参数和回调，具体说明如表 11-3 所示。

表 11-3　参数 options 的取值选项

名　　称	说　　明
hasReplyPath	收到的短信是否包含"TP-Reply-Path"，默认为 false； "TP-Reply-Path"：设备根据发送 SMS 消息的短消息中心进行回复
isReplaceMessage	收到的短信是否为"替换短信"，默认为 false； "替换短信"有关详细信息，参见"3GPP TS 23.040 9.2.3.9"
isSmsStatusReportMessage	当前消息是否为"短信状态报告"，默认为 false； "短信状态报告"是一种特定格式的短信，被用来从 Service Center 到 Mobile Station 传送状态报告
messageClass	短信类型
pdu	SMS 消息中的协议数据单元（PDU）
protocolId	发送短信时使用的协议标识
scAddress	短消息服务中心（SMSC）地址
scTimestamp	SMSC 时间戳
status	SMS-STATUS-REPORT 消息中的短信状态指示短信服务中心（SMSC）发送的短信状态
visibleMessageBody	短信正文
visibleRawAddress	发送者地址

（4）sms.getDefaultSmsSlotId：获取发送短信的默认 SIM 卡槽 ID，使用 callback 异步回调。原型是：

```
getDefaultSmsSlotId(callback: AsyncCallback<number>): void
```

参数 callback 表示回调函数，- 0 表示卡槽 1，- 1 表示卡槽 2。

（5）sms.getDefaultSmsSlotId：获取发送短信的默认 SIM 卡槽 ID，使用 Promise 异步回调。原型是：

```
getDefaultSmsSlotId(): Promise<number>
```

参数 Promise<number> 表示以 Promise 形式返回发送短信的默认 SIM 卡：- 0 表示卡槽 1，- 1 表示卡槽 2。

（6）sms.hasSmsCapability：检查当前设备是否具备短信发送和接收能力，该方法是同步方法。原型是：

```
hasSmsCapability(): boolean
```

返回值是 boolean，如果是 true 表示设备具备短信发送和接收能力，如果是 false 表示设备不具备短信发送和接收能力。

(7) ShortMessage：ShortMessage 实例包含了一系列属性和系统能力，用于处理短信相关的功能。以下是其中一些重要的属性和能力。

- hasReplyPath（是否包含回复路径）：用于标识收到的短信是否包含"TP-Reply-Path"，以便设备可以根据发送短信到短消息中心进行回复。
- isReplaceMessage（是否为替换短信）：用于标识收到的短信是否为"替换短信"，相关详细信息参见 3GPP TS 23.040 9.2.3.9。
- isSmsStatusReportMessage（是否为短信状态报告）：用于标识当前消息是否为"短信状态报告"，这种报告是一种特定格式的短信，用于传送状态报告。
- messageClass（短信类型）：指定短信的类型，包括 UNKNOWN（未知类型）、INSTANT_MESSAGE（即时消息）、OPTIONAL_MESSAGE（存储在设备或 SIM 卡上的短信）等。
- pdu（协议数据单元）：包含在 SMS 消息中的协议数据单元。
- protocolId（协议标识）：发送短信时使用的协议标识。
- scAddress（短消息服务中心地址）：指定短信服务中心（SMSC）的地址。
- scTimestamp（SMSC 时间戳）：指定 SMSC 的时间戳。
- status（短信状态）：包含在 SMS-STATUS-REPORT 消息中的短信状态，指示短信服务中心发送的短信状态。
- visibleMessageBody（短信正文）：包含短信的正文内容。
- visibleRawAddress（发送者地址）：包含发送者的地址。

(8) ISendShortMessageCallback：是一个回调实例，用于返回短信发送的结果、存储已发送短信的 URI 以及指示是否为长短信的最后一部分。以下是其中一些关键属性。

- isLastPart（是否为长短信的最后一部分）：用于指定是否为长短信的最后一部分。如果为 true，则表示这是长短信的最后一部分；如果为 false，则表示不是。默认值为 false。
- result（短信发送结果）：返回有关短信发送操作的结果。可能包括成功、失败等不同状态。
- url（存储发送短信的 URI）：包含已发送短信的存储 URI，可以用于后续的检索或管理操作。

这个回调实例的信息可以帮助开发者更好地处理短信发送操作的结果，包括了发送是否成功、存储位置等重要信息。在使用这个回调时，开发者可以根据 isLastPart 属性判断是否为长短信的最后一部分，以及根据 result 属性获取短信发送的具体结果状态。而 url 属性则提供了一个指向已发送短信存储位置的引用。这些信息为开发者提供了在短信发送操作中更精细控制和处理的能力，以确保用户获得良好的短信发送体验。

(9) IDeliveryShortMessageCallback：是一个回调实例，用于返回短信送达报告。以下是其中的关键属性。

- pdu（短信送达报告）：包含短信送达报告的协议数据单元（PDU）。短信送达报告是一种特殊的短信，用于指示先前发送的短信是否已经成功地被接收。

这个回调实例的信息提供了有关短信送达报告的详细数据，其中 PDU 属性包含了报告的协议数据单元。开发者可以利用这些信息来确定特定短信的送达状态以及其他相关的信息。

(10) SendSmsResult：是一个用于表示短信发送结果的枚举类型，该类型的不同取值和对应的说明如下。
- SEND_SMS_SUCCESS（发送短信成功，值为 0）：表示短信成功发送。
- SEND_SMS_FAILURE_UNKNOWN（发送短信失败，原因未知，值为 1）：表示短信发送失败，但失败的具体原因未知。
- SEND_SMS_FAILURE_RADIO_OFF（发送短信失败，原因为调制解调器关机，值为 2）：表示短信发送失败，因为调制解调器（通常指手机上的通信模块）处于关闭状态。
- SEND_SMS_FAILURE_SERVICE_UNAVAILABLE（发送短信失败，原因为网络不可用、不支持发送或接收短信，值为 3）：表示短信发送失败，原因可能包括网络不可用、设备不支持发送或接收短信等。

这些枚举值提供了开发者在处理短信发送结果时的详细信息，使其能够根据具体的失败原因采取适当的处理措施。例如，可以根据失败原因为用户提供友好的错误提示，或者尝试重新发送短信等操作。

11.4.2 sms 实战：发送指定内容的短信

实例 11-2 演示了使用 sms 模块发送短信的过程。在本例中，将短信发送逻辑封装在独立的模块 smsModule.ets 中，使在主页面 Index.ets 中实现了点击按钮即可触发发送短信的功能。在模块中定义了发送短信的参数和回调，并提供了一个简单的 sendSms 方法。主页面通过导入该模块并在按钮单击事件中调用该方法，实现了页面与短信功能的分离，提高了代码的模块化和可维护性。

实例 11-2：发送指定内容的短信（源码路径 :codes\11\SMS）

（1）编写文件 src/main/ets/modules/smsModule.ets，定义一个名为 SmsModule 的模块，其中封装了发送短信的功能。通过设置短信的各项参数，包括 SIM 卡槽 ID、短信内容、接收者地址等，以及定义短信发送结果和送达结果的回调函数。最终，通过调用 sms.sendMessage(options) 实现了发送短信的功能。这使在其他模块或页面中可以轻松调用 SmsModule.sendSms() 实现短信发送。

扫码看视频

```
import sms from '@ohos.telephony.sms';

export default class SmsModule {
  static sendSms() {
    // 定义发送短信的参数
    let slotId = 0; // SIM 卡槽 ID, 0 为卡槽 1
    let content = 'Hello, this is a test message.'; // 短信内容
    let destinationHost = '+861234567890'; // 接收者地址
    let serviceCenter = '+8611123456789'; // 短信中心地址
    let destinationPort = 1000; // 如果发送数据消息,destinationPort 是必需的。否则是可选的。

    // 定义短信发送结果回调函数
    let sendCallback = function (err, data) {
        console.log(`sendCallback: err->${JSON.stringify(err)}, data->${JSON.stringify(data)}`);
    }
```

```
        // 定义短信送达结果回调函数
        let deliveryCallback = function (err, data) {
            console.log(`deliveryCallback: err->${JSON.stringify(err)}, data-
>${JSON.stringify(data)}`);
        }

        // 构建发送短信的参数对象
        let options = { slotId, content, destinationHost, serviceCenter, destinationPort,
sendCallback, deliveryCallback };

        // 调用 sendMessage 方法发送短信
        sms.sendMessage(options);
    }
}
```

（2）编写文件 src/main/ets/pages/Index.ets，定义一个 HarmonyOS 页面（Index.ets），通过导入短信模块（SmsModule）实现按钮点击发送短信的功能。页面简单显示"Hello World"文本和一个"Send SMS"按钮。点击按钮即可调用短信模块完成短信发送，实现了代码逻辑的模块化和可维护性。

```
import SmsModule from '../modules/smsModule';

@Entry
@Component
struct Index {
  @State message: string = 'Hello World'

  build() {
    Row() {
      Column() {
        Text(this.message)
          .fontSize(50)
          .fontWeight(FontWeight.Bold)
        Button('Send SMS').onClick(this.handleSendSms)
      }
      .width('100%')
    }
    .height('100%')
  }

  private handleSendSms() {
    // 在按钮单击事件中调用发送短信的逻辑
    SmsModule.sendSms();
  }
}
```

执行效果如图 11-2 所示。

图 11-2　发送指定内容的短信执行效果

第 12 章
设备管理

在 HarmonyOS 系统应用开发中，设备管理是至关重要的一环，它涵盖了 USB 连接、位置服务、传感器利用及震动反馈等方面。这些功能不仅为应用提供了多样化的用户体验，还为开发者提供了丰富的硬件交互可能性，提高了应用的实用性和创新性。因此，有效的设备管理在鸿蒙应用的开发过程中具有重要意义，能够充分发挥鸿蒙系统的跨设备协同特性。本章将详细讲解在 HarmonyOS 系统开发中设备管理应用程序的知识。

12.1 USB 开发

在 HarmonyOS 系统中，USB 服务是应用访问底层设备的一种抽象概念。开发者可以根据提供的 USB API，获取设备列表、控制设备访问权限、与连接的设备进行数据传输及控制命令传输等。

12.1.1 HarmonyOS USB API 介绍

HarmonyOS 系统中的 USB 服务系统包含 USB API、USB Service、USB HAL，具体运作机制，如图 12-1 所示。

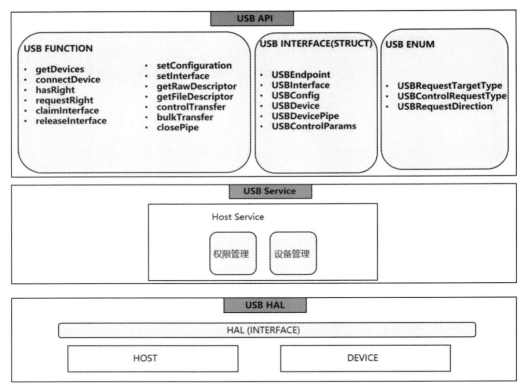

图 12-1 USB 服务运作机制

- USB API：提供 USB 的基础 API，主要包含查询 USB 设备列表、批量数据传输、控制命令传输、权限控制等。
- USB Service：主要实现 HAL 层数据的接收、解析、分发，前后台的策略管控及对设备的管理等。
- USB HAL：提供给用户可直接调用的驱动能力接口。

在 HOST 模式下，可以获取已经连接的设备列表，并根据需要打开和关闭设备、控制权限、进行数据传输等。在鸿蒙系统中，HOST 模式是 USB 主机模式（USB Host Mode）。USB 主机模式是指设备充当 USB 总线上的主机，能够控制和管理连接到其上的其他 USB 设备，如键盘、鼠标、打印机等。在 HOST 模式下，设备能够主动发起和管理 USB 连接，与其他设备进行通信和数据交换。

在鸿蒙系统中，通过 @ohos.usb 模块实现 USB 管理功能，分别实现了查询 USB 设备列表、批量数据传输、控制命令传输、权限控制等功能。在使用 @ohos.usb 模块之前，需要先通过以下代码导入：

```
import usb from "@ohos.usb";
```

在 @ohos.usb 模块中包含以下成员。

（1）usb.getDevices()：一个用于获取 USB 设备列表的方法，属于鸿蒙系统的 USB 管理功能，可返回一个包含 USB 设备信息的只读数组。每个数组元素都是一个 USBDevice 对象，其中包含有关连接的 USB 设备的信息。调用这个方法可以帮助开发者获取当前连接到系统的 USB 设备列表，从而进行后续的管理和交互操作。开发者可以遍历返回的数组，了解每个连接设备的详细信息，如设备类型、厂商信息等。

（2）connectDevice(device: USBDevice): Readonly<USBDevicePipe>：打开 USB 设备，建立与设备的传输通道，以便进行数据传输和通信。参数 device（类型：USBDevice，必填）表示 USB 设备的信息，需通过 usb.getDevices 获取。使用步骤如下。

- 调用 usb.getDevices 获取 USB 设备信息。
- 使用获取的设备信息调用 usb.connectDevice 打开 USB 设备。
- 在调用前，通过 usb.requestRight 获取设备请求权限。

返回一个只读的 USBDevicePipe 对象，表示与 USB 设备建立的传输通道，可用于后续的数据传输和通信。

（3）hasRight(deviceName: string): boolean：用于判断是否具有访问指定设备的权限。
- 参数 deviceName（类型：string，必填）：设备名称，用于判断是否有权访问该设备。
- 返回值：类型 boolean，返回 true 表示具有访问设备的权限，返回 false 表示没有访问设备的权限。

（4）requestRight(deviceName: string): Promise<boolean>：用于请求软件包的临时权限，以便访问指定设备。
- 参数 deviceName（类型：string，必填）：设备名称，用于请求访问权限。
- 返回值：类型 Promise<boolean>，返回一个 Promise 对象，获取到 true 表示软件包的临时权限已访问成功，获取到 false 表示软件包的临时权限已访问失败。

（5）claimInterface(pipe: USBDevicePipe, iface: USBInterface, force?: boolean): number：用于注册通信接口，以确保可以与设备的指定接口进行通信。参数说明如下。
- pipe（类型：USBDevicePipe，必填）：用于确定总线号和设备地址。
- iface（类型：USBInterface，必填）：用于确定需要获取接口的索引。
- force（类型：boolean，可选）：是否强制获取，默认值为 false，表示不强制获取。

使用步骤如下。
- 调用 usb.getDevices 获取 USB 设备信息及接口信息。

- 调用 usb.requestRight 获取设备请求权限。
- 调用 usb.connectDevice 接口得到 devicepipe 作为参数。
- 使用获得的 devicepipe、接口信息及可选的 force 参数调用 usb.claimInterface 注册通信接口。

返回值类型：number，如果注册通信接口成功返回 0，如果注册通信接口失败返回其他错误码。

（6）releaseInterface(pipe: USBDevicePipe, iface: USBInterface): number：用于释放注册过的通信接口，确保在不再需要与设备的特定接口进行通信时释放相应资源。参数说明如下。

- pipe（类型：USBDevicePipe，必填）：用于确定总线号和设备地址。
- iface（类型：USBInterface，必填）：用于确定需要释放的接口的索引。

使用步骤是：在不需要与设备的特定接口进行通信时，使用获得的 devicepipe 和接口信息调用 usb.releaseInterface 释放注册过的通信接口。

返回值类型 number，如果释放接口成功返回 0，如果释放接口失败返回其他错误码。

（7）setConfiguration(pipe: USBDevicePipe, config: USBConfig): number：用于设置 USB 设备的配置，以确保设备按照指定的配置进行操作。参数说明如下。

- pipe（类型：USBDevicePipe，必填）：用于确定总线号和设备地址。
- config（类型：USBConfig，必填）：用于确定需要设置的配置。

使用步骤如下。

- 调用 usb.getDevices() 获取 USB 设备信息以及配置信息。
- 调用 usb.requestRight() 获取设备请求权限。
- 调用 usb.connectDevice() 得到 devicepipe 作为参数。
- 使用获得的 devicepipe 和配置信息调用 usb.setConfiguration（）设置设备配置。

返回值类型 number，如果设置设备配置成功返回 0，设置设备配置失败返回其他错误码。

（8）setInterface(pipe: USBDevicePipe, iface: USBInterface): number：用于设置 USB 设备的接口，以确保设备按照指定的接口进行操作。参数说明如下。

- pipe（类型：USBDevicePipe，必填）：用于确定总线号和设备地址。
- iface（类型：USBInterface，必填）：用于确定需要设置的接口。

使用步骤如下。

- 调用 usb.getDevices() 获取 USB 设备列表及接口信息。
- 调用 usb.requestRight() 获取设备请求权限。
- 调用 usb.connectDevice() 得到 devicepipe 作为参数。
- 调用 usb.claimInterface() 注册通信接口。
- 使用获得的 devicepipe 和接口信息调用 usb.setInterface（）设置设备接口。

返回值类型 number，如果设置设备接口成功返回 0，设置设备接口失败返回其他错误码。

12.1.2 开发 HarmonyOS USB 程序

在 HarmonyOS 系统中，USB 设备可作为 Host 设备连接 Device 设备进行数据传输。开发 HarmonyOS USB 程序的基本步骤如下。

1）获取设备列表

以下代码调用 usb.getDevices() 方法获取了当前连接到系统的 USB 设备列表，返回一个包含 USB 设备信息的数组。每个设备信息包含名称、制造商、产品等详细信息，便于开发者了解和管理连接的 USB 设备，为后续交互和配置提供基础。

```
let devicesList = usb.getDevices();
console.log(`devicesList = ${devicesList}`);
//devicesList   返回的数据结构
// 此处提供一个简单的示例，如下
[
  {
    name: "1-1",
    serial: "",
    manufacturerName: "",
    productName: "",
    version: "",
    vendorId: 7531,
    productId: 2,
    clazz: 9,
    subClass: 0,
    protocol: 1,
    devAddress: 1,
    busNum: 1,
    configs: [
      {
        id: 1,
        attributes: 224,
        isRemoteWakeup: true,
        isSelfPowered: true,
        maxPower: 0,
        name: "1-1",
        interfaces: [
          {
            id: 0,
            protocol: 0,
            clazz: 9,
            subClass: 0,
            alternateSetting: 0,
            name: "1-1",
            endpoints: [
              {
                address: 129,
                attributes: 3,
                interval: 12,
                maxPacketSize: 4,
                direction: 128,
                number: 1,
                type: 3,
                interfaceId: 0,
              },
            ],
```

```
              },
            ],
          },
        ],
      },
    ]
```

2）获取设备操作权限

以下代码，主要功能是通过 USB 管理接口申请对指定 USB 设备的操作权限，并根据申请结果输出相应的信息。

```
let deviceName = deviceList[0].name;
// 申请操作指定的 device 的操作权限。
usb.requestRight(deviceName).then(hasRight => {
  console.info("usb device request right result: " + hasRight);
}).catch(error => {
  console.info("usb device request right failed : " + error);
});
```

在上述代码中，首先从 deviceList 中获取第一个设备的名称 deviceName，然后使用 usb.requestRight(deviceName) 来申请操作指定设备的权限。在这个过程中，返回一个 Promise 对象，通过 .then() 和 .catch() 处理异步操作的结果。如果权限请求成功，hasRight 为 true，打印相应的成功信息；如果失败，捕获错误并打印失败信息。

3）打开 Device 设备

以下代码主要功能是通过 USB 管理接口打开设备，获取数据传输通道，并选择一个接口进行打开，以确保可以与该接口进行通信。

```
// 打开设备，获取数据传输通道。
let pipe = usb.connectDevice(deviceList[0]);
let interface1 = deviceList[0].configs[0].interfaces[0];
/*
 打开对应接口，在设备信息（deviceList）中选取对应的 interface。
interface1 为设备配置中的一个接口。
*/
usb.claimInterface(pipe , interface1, true);
```

在上述代码中，首先，使用 usb.connectDevice(deviceList[0]) 打开设备，获取数据传输通道 pipe。其次，从设备信息 deviceList 中选择需要打开的接口 interface1，通常是从设备的配置中选择一个接口。最后，通过 usb.claimInterface(pipe, interface1, true) 来打开对应的接口，确保可以与该接口进行通信。

4）数据传输

以下代码，主要功能是通过 USB 管理接口，使用 usb.bulkTransfer 方法进行数据的读取和发送操作。在读取数据时，选择对应的输入端点，将数据存储到 dataUint8Array 中，并将 Uint8Array 数据转换为字符串进行输出。在发送数据时，选择对应的输出端点，将数据从 dataUint8Array 发送到设备。

```
/*
 读取数据，在 device 信息中选取对应数据接收的 endpoint 来做数据传输
```

```
  (endpoint.direction == 0x80);dataUint8Array 是要读取的数据,类型为 Uint8Array。
  */
  let inEndpoint = interface1.endpoints[2];
  let outEndpoint = interface1.endpoints[1];
  let dataUint8Array = new Uint8Array(1024);
  usb.bulkTransfer(pipe, inEndpoint, dataUint8Array, 15000).then(dataLength => {
  if (dataLength >= 0) {
    console.info("usb readData result Length : " + dataLength);
    let resultStr = this.ab2str(dataUint8Array); // uint8 数据转 string。
    console.info("usb readData buffer : " + resultStr);
  } else {
    console.info("usb readData failed : " + dataLength);
  }
  }).catch(error => {
  console.info("usb readData error : " + JSON.stringify(error));
  });
  // 发送数据,在 device 信息中选取对应数据发送的 endpoint 来做数据传输。(endpoint.
direction == 0)
  usb.bulkTransfer(pipe, outEndpoint, dataUint8Array, 15000).then
(dataLength => {
    if (dataLength >= 0) {
      console.info("usb writeData result write length : " + dataLength);
    } else {
      console.info("writeData failed");
    }
  }).catch(error => {
    console.info("usb writeData error : " + JSON.stringify(error));
  });
```

在上述代码中,首先从 interface1 中选择用于数据接收的输入端点(inEndpoint)和用于数据发送的输出端点(outEndpoint)。然后,通过 usb.bulkTransfer 方法进行数据传输,包括读取数据和发送数据的过程。

5)释放接口,关闭设备

```
  usb.releaseInterface(pipe, interface1);
  usb.closePipe(pipe);
```

12.2 位置服务

移动终端设备已经深入人们日常生活的方方面面,如查看所在城市的天气、新闻逸事、出行打车、旅行导航、运动记录等。这些习以为常的活动,都离不开定位用户终端设备的位置。当用户处于这些丰富的使用场景中时,系统的位置能力可以提供实时准确的位置数据。对于开发者,设计基于位置体验的服务,也可以使应用的使用体验更贴近每个用户。当应用在实现基于设备位置的功能时,如驾车导航,记录运动轨迹等,可以调用该模块的 API 接口,完成位置信息的获取。

12.2.1 位置开发概述

位置能力用于确定用户设备在哪里,系统使用位置坐标标示设备的位置,并用多种定位技术提供服务，如 GNSS 定位、基站定位、WLAN/蓝牙定位（基站定位、WLAN/蓝牙定位以下统称"网络定位技术"）。通过这些定位技术，无论用户设备在室内或是室外，都可以准确地确定设备位置。

1. 基本概念

和位置开发相关的几个关键概念如下。

1）坐标

系统以 1984 年世界大地坐标系统为参考，使用经度、纬度数据描述地球上的一个位置。

2）GNSS 定位

基于全球导航卫星系统，包括：GPS、GLONASS、北斗、Galileo 等，通过导航卫星、设备芯片提供的定位算法，来确定设备准确位置。定位过程具体使用哪些定位系统，取决于用户设备的硬件能力。

3）基站定位

根据设备当前驻网基站和相邻基站的位置，估算设备当前位置。此定位方式的定位结果精度相对较低，并且需要设备可以访问蜂窝网络。

4）WLAN、蓝牙定位

根据设备可搜索到的周围 WLAN、蓝牙设备位置，估算设备当前位置。此定位方式的定位结果精度依赖设备周围可见的固定 WLAN、蓝牙设备的分布，密度较高时，精度也相对于基站定位方式更高，同时也需要设备可以访问网络。

2. 运作机制

位置能力作为系统为应用提供的一种基础服务，需要应用在所使用的业务场景，向系统主动发起请求，并在业务场景结束时，主动结束此请求，在此过程中系统会将实时的定位结果上报给应用。

3. 约束与限制

使用设备的位置能力，需要用户进行确认并主动开启位置开关。如果位置开关没有开启，系统不会向任何应用提供位置服务。因为设备位置信息属于用户敏感数据，所以即使用户已经开启位置开关，应用在获取设备位置前仍需向用户申请位置访问权限。在用户确认允许后，系统才会向应用提供位置服务。

12.2.2 获取设备的位置信息

HarmonyOS 系统中，开发者可以调用位置相关接口获取设备实时位置，或者最近的历史位置。对于位置敏感的应用业务，建议获取设备实时的位置信息。如果不需要设备实时位置信息，并且希望尽可能地节省耗电，开发者可以考虑获取最近的历史位置。

如果想获取设备的位置信息，可以使用表 12-1 中的 API 接口。

表 12-1 获取位置信息的 API 接口

接口名	功能描述
on(type: 'locationChange', request: LocationRequest, callback: Callback<Location>) : void	开启位置变化订阅，并发起定位请求
off(type: 'locationChange', callback?: Callback<Location>) : void	关闭位置变化订阅，并删除对应的定位请求

续表

接口名	功能描述
on(type: 'locationServiceState', callback: Callback<boolean>) : void	订阅位置服务状态变化
off(type: 'locationServiceState', callback: Callback<boolean>) : void	取消订阅位置服务状态变化
on(type: 'cachedGnssLocationsReporting', request: CachedGnssLoactionsRequest, callback: Callback<Array<Location>>) : void;	订阅缓存 GNSS 位置上报
off(type: 'cachedGnssLocationsReporting', callback?: Callback<Array<Location>>) : void;	取消订阅缓存 GNSS 位置上报
on(type: 'gnssStatusChange', callback: Callback<SatelliteStatusInfo>) : void;	订阅卫星状态信息更新事件
off(type: 'gnssStatusChange', callback?: Callback<SatelliteStatusInfo>) : void;	取消订阅卫星状态信息更新事件
on(type: 'nmeaMessageChange', callback: Callback<string>) : void;	订阅 GNSS NMEA 信息上报
off(type: 'nmeaMessageChange', callback?: Callback<string>) : void;	取消订阅 GNSS NMEA 信息上报
on(type: 'fenceStatusChange', request: GeofenceRequest, want: WantAgent) : void;	添加围栏，并订阅该围栏事件上报
off(type: 'fenceStatusChange', request: GeofenceRequest, want: WantAgent) : void;	删除围栏，并取消订阅该围栏事件
getCurrentLocation(request: CurrentLocationRequest, callback: AsyncCallback<Location>) : void	获取当前位置，使用 callback 回调异步返回结果
getCurrentLocation(request?: CurrentLocationRequest) : Promise<Location>	获取当前位置，使用 Promise 方式异步返回结果
getLastLocation(callback: AsyncCallback<Location>) : void	获取上一次位置，使用 callback 回调异步返回结果
getLastLocation() : Promise<Location>	获取上一次位置，使用 Promise 方式异步返回结果
isLocationEnabled(callback: AsyncCallback<boolean>) : void	判断位置服务是否已经打开，使用 callback 回调异步返回结果
isLocationEnabled() : Promise<boolean>	判断位置服务是否已经开启，使用 Promise 方式异步返回结果
requestEnableLocation(callback: AsyncCallback<boolean>) : void	请求打开位置服务，使用 callback 回调异步返回结果
requestEnableLocation() : Promise<boolean>	请求打开位置服务，使用 Promise 方式异步返回结果
getCachedGnssLocationsSize(callback: AsyncCallback<number>) : void;	获取缓存 GNSS 位置的个数，使用 callback 回调异步返回结果
getCachedGnssLocationsSize() : Promise<number>;	获取缓存 GNSS 位置的个数，使用 Promise 方式异步返回结果
flushCachedGnssLocations(callback: AsyncCallback<boolean>) : void;	获取所有的 GNSS 缓存位置，并清空 GNSS 缓存队列，使用 callback 回调异步返回结果

续表

接口名	功能描述
flushCachedGnssLocations() : Promise<boolean>;	获取所有的 GNSS 缓存位置，并清空 GNSS 缓存队列，使用 Promise 方式异步返回结果
sendCommand(command: LocationCommand, callback: AsyncCallback<boolean>) : void;	给位置服务子系统发送扩展命令，使用 callback 回调异步返回结果
sendCommand(command: LocationCommand) : Promise<boolean>;	给位置服务子系统发送扩展命令，使用 Promise 方式异步返回结果

在下面的内容中，将详细讲解在 HarmonyOS 系统中获取设备位置信息的方法。

（1）在使用位置功能前，需要检查是否获取用户授权访问设备位置信息。如果未获得授权，可以向用户申请以下相关的位置权限：

- ohos.permission.LOCATION
- ohos.permission.LOCATION_IN_BACKGROUND

要访问设备的位置信息，必须申请 ohos.permission.LOCATION 权限，并且获得用户授权。如果应用程序在后台运行时也需要访问设备位置，除需要将应用声明为允许后台运行外，还必须申请 ohos.permission.LOCATION_IN_BACKGROUND 权限，这样应用在切入后台之后，系统可以继续上报位置信息。

开发者可以在应用程序的 config.json 文件中声明所需要的权限，例如，下面的代码请求了位置权限，用于某个指定的场景（LocationAbility）在应用处于活动状态时使用。

```
{
    "module": {
        "reqPermissions": [
          {
            "name": "ohos.permission.LOCATION",
            "reason": "$string:reason_description",
            "usedScene": {
                "ability": ["com.myapplication.LocationAbility"],
                "when": "inuse"
            }
          }
        ]
    }
}
```

（2）导入 geolocation 模块，所有与基础定位能力相关的功能 API，都是通过该模块提供的。

```
import geolocation from '@ohos.geolocation';
```

（3）实例化 LocationRequest 对象，用于告知系统应该向应用提供何种类型的位置服务，以及位置结果上报的频率。

- 方式一：例如，以下代码描述了一种 API 设计方式，通过系统定义一组常见的位置能力使用场景的枚举类型 LocationRequestScenario，其中包括导航、轨迹跟踪、打车服务、日常生活服务等场景。开发者可以直接选择适当的场景，简化了位置能力的集成和使用，提高了开发效率。

```
export enum LocationRequestScenario {
    UNSET = 0x300,
    NAVIGATION,
    TRAJECTORY_TRACKING,
    CAR_HAILING,
    DAILY_LIFE_SERVICE,
    NO_POWER,
}
```

在 HarmonyOS 系统中，定位场景类型的说明，如表 12-2 所示。

表 12-2 定位场景类型的说明

场景名称	常量定义	说　　明
导航场景	NAVIGATION	适用于户外定位设备实时位置的场景，如车载、步行导航。在此种场景下，为保证系统提供的位置结果精度最优，主要使用 GNSS 定位技术提供定位服务。结合场景特点，在导航启动之初，用户很可能在室内、车库等遮蔽环境下，GNSS 技术可能难以提供位置服务。为解决此问题，我们会在 GNSS 提供稳定位置结果之前，使用系统的网络定位技术，向应用提供位置服务，以在导航初始阶段提升用户体验； 此场景默认以最小 1 秒间隔上报定位结果，使用此场景的应用必须申请 ohos.permission.LOCATION 权限，同时获得用户授权
轨迹跟踪场景	TRAJECTORY_TRACKING	适用于记录用户位置轨迹的场景，如运动类应用记录轨迹功能，主要使用 GNSS 定位技术提供定位服务； 此场景默认以最小 1 秒间隔上报定位结果，并且应用必须申请 ohos.permission.LOCATION 权限，同时获得用户授权
出行约车场景	CAR_HAILING	适用于用户出行打车时定位当前位置的场景，如网约车类应用； 此场景默认以最小 1 秒间隔上报定位结果，并且应用必须申请 ohos.permission.LOCATION 权限，同时获得用户授权
生活服务场景	DAILY_LIFE_SERVICE	生活服务场景，适用于不需要定位用户精确位置的使用场景，如新闻资讯、网购、点餐类应用，在进行推荐、推送时定位用户大致位置即可。 此场景默认以最小 1 秒间隔上报定位结果，并且应用至少需要申请 ohos.permission.LOCATION 权限，同时获得用户授权
无功耗场景	NO_POWER	无功耗场景，适用于不需要主动启动定位业务的情况，系统在响应其他应用启动定位业务并上报位置结果时，会同时向请求此场景的应用程序上报定位结果，此时的应用程序不产生定位功耗。 此场景默认以最小 1 秒间隔上报定位结果，并且应用需要申请 ohos.permission.LOCATION 权限，同时获得用户授权

以导航场景为例，实例化方式如下：

```
var requestInfo = {'scenario': 0x301, 'timeInterval': 0,
'distanceInterval': 0, 'maxAccuracy': 0};
```

在上述代码中定义了一个变量 requestInfo，其中包含了位置请求的相关信息。具体的请求信息包括使用场景（scenario）、时间间隔（timeInterval）、距离间隔（distanceInterval）和最大精度（maxAccuracy）。在这个例子中，scenario 的值为 0x301，而其他参数都被设置为 0。这可能用于发起一个位置服务请求，但具体请求的详细配置需要根据实际需求进行调整。

- 方式二：如果定义的现有场景类型不能满足所需的开发场景，系统提供了基本的定位优先级策略类型。例如，以下代码描述了第二种 API 设计方式，当预定义的位置场景无法满足开发需求时，系统提供了基本的定位优先级策略类型，通过枚举类型 LocationRequestPriority 包括未设置、高精度、低功耗和首次定位等几种优先级，使开发者能够更灵活地定义定位的优先级，以适应不同的开发场景。

```
export enum LocationRequestPriority {
    UNSET = 0x200,
    ACCURACY,
    LOW_POWER,
    FIRST_FIX,
}
```

在 HarmonyOS 系统中，定位优先级策略类型的说明，如表 12-3 所示。

表 12-3　定位优先级策略类型说明

策略类型	常量定义	说　　明
定位精度优先策略	ACCURACY	定位精度优先策略主要以 GNSS 定位技术为主，在开阔场景下可以提供米级的定位精度。具体性能指标依赖于用户设备的定位硬件能力。但在室内等强遮蔽定位场景下，可能无法提供准确的位置服务。 应用必须申请 ohos.permission.LOCATION 权限，同时获得用户授权
快速定位优先策略	FAST_FIRST_FIX	快速定位优先策略会同时使用 GNSS 定位、基站定位和 WLAN、蓝牙定位技术，以便在室内和户外场景下都能通过此策略获得位置结果。当各种定位技术都提供位置结果时，系统会选择其中精度较好的结果返回给应用。由于这种策略同时使用多种定位技术，对设备的硬件资源消耗较高，功耗也较大。 应用必须申请 ohos.permission.LOCATION 权限，同时获得用户授权
低功耗定位优先策略	LOW_POWER	低功耗定位优先策略主要使用基站定位和 WLAN、蓝牙定位技术，也可以同时提供室内和户外场景下的位置服务。由于其依赖于周边基站、可见 WLAN、蓝牙设备的分布情况，定位结果的精度波动范围较大。如果对定位结果的精度要求不高，或者使用场景多在有基站、可见 WLAN、蓝牙设备高密度分布的情况下，推荐使用此策略，因为它可以有效节省设备功耗。 应用至少需要申请 ohos.permission.LOCATION 权限，同时获得用户授权

以定位精度优先策略为例，实例化方式如下：

```
var requestInfo = {'priority': 0x201, 'timeInterval': 0, 'distanceInterval': 0, 'maxAccuracy': 0};
```

在上述代码中定义了一个变量 requestInfo，其中包含了位置请求的相关信息。具体的请求信息包括定位优先级（priority）、时间间隔（timeInterval）、距离间隔（distanceInterval）和最大精度（maxAccuracy）。在这个例子中，priority 的值为 0x201，而其他参数都被设置为 0。上述代码用于发起一个位置服务请求，其中自定义的优先级为 0x201，其他配置参数暂时为默认值 0。开发者可以根据实际需求调整这些参数以满足特定的定位要求。

（4）实例化 Callback 对象，用于向系统提供位置上报的途径。

应用程序需要自行实现系统定义好的回调接口，并将其实例化。系统在定位成功确定设备的实时位置结果时，会通过该接口上报给应用。以下代码定义了一个名为 locationChange 的函数表达式（箭头函数），用于处理位置变化事件。当位置发生变化时，该函数将接收到的位置信息转换为 JSON 字符串并输出到控制台。该函数可作为位置变化事件的回调函数，用于处理定位信息的更新或其他相关操作。

```
var locationChange = (location) => {
    console.log('locationChanger: data: ' + JSON.stringify(location));
};
```

（5）启动定位：下面代码使用 geolocation.on 方法，通过监听 locationChange 事件，将之前定义的 requestInfo 和 locationChange 函数关联起来。当定位信息发生变化时，系统将会触发 locationChange 函数，并将最新的位置信息作为参数传递给该函数。这种方式允许开发者在位置发生变化时执行自定义的处理逻辑，以满足应用程序的需求。

```
geolocation.on('locationChange', requestInfo, locationChange);
```

（6）结束定位：下面代码使用 geolocation.off 方法，通过取消监听 locationChange 事件，解除了之前与 locationChange 函数的关联。这意味着当定位信息发生变化时，将不再触发之前定义的 locationChange 函数。

```
geolocation.off('locationChange', locationChange);
```

如果应用使用场景不需要实时的设备位置，可以获取系统缓存的最近一次历史定位结果。以下代码使用 geolocation.getLastLocation 方法，通过回调函数获取最后一次记录的位置信息。当调用该方法时，系统将返回最近一次获取到的位置信息，并通过传入的回调函数（箭头函数）将该信息转换为 JSON 字符串并输出到控制台。这可用于获取设备当前的位置信息，而不需要等待实时的位置变化事件。

```
geolocation.getLastLocation((data) => {
    console.log('getLastLocation: data: ' + JSON.stringify(data));
});
```

此接口的使用需要向用户申请 ohos.permission.LOCATION 权限。

12.2.3 地理编码转化

在智能设备中，使用坐标描述一个位置，非常准确，但是并不直观，面向用户表达并不友好。HarmonyOS 系统向开发者提供了地理编码转化能力（将地理描述转化为具体坐标），以及逆地理编码转化能力（将坐标转化为地理描述）。其中地理编码包含多个属性来描述位置，包括国家、行政区划、街道、门牌号、地址描述等，这样的信息更便于用户理解。

在 HarmonyOS 系统中提供了专用 API 接口进行坐标和地理编码信息的相互转化，这些接口的具体说明如下。

- isGeoServiceAvailable(callback: AsyncCallback<boolean>)：void：判断（逆）地理编码服务状态，使用 callback 回调异步返回结果。
- isGeoServiceAvailable()：Promise<boolean>：判断（逆）地理编码服务状态，使用 Promise 方式异步返回结果。
- getAddressesFromLocation(request: ReverseGeoCodeRequest, callback: AsyncCallback<Array<GeoAddress>>)：void：调用逆地理编码服务，将坐标转换为地理描述，使用 callback 回调异步返回结果。
- getAddressesFromLocation(request: ReverseGeoCodeRequest)：Promise<Array<GeoAddress>>：调用逆地理编码服务，将坐标转换为地理描述，使用 Promise 方式异步返回结果。
- getAddressesFromLocationName(request: GeoCodeRequest, callback: AsyncCallback<Array<GeoAddress>>)：void：调用地理编码服务，将地理描述转换为具体坐标，使用 callback 回调异步返回结果。
- getAddressesFromLocationName(request: GeoCodeRequest)：Promise<Array<GeoAddress>>：调用地理编码服务，将地理描述转换为具体坐标，使用 Promise 方式异步返回结果。

以下代码，演示了在 HarmonyOS 系统中开发地理编码转化应用程序的流程。

```
// 导入geolocation模块,所有与（逆）地理编码转化能力相关的功能API都是通过该模块提供的。
import geolocation from '@ohos.geolocation';

// 获取逆地理编码结果。
const reverseGeocodeRequest = {"latitude": 31.12, "longitude": 121.11, "maxItems": 1};
geolocation.getAddressesFromLocation(reverseGeocodeRequest, (data) => {
    console.log('逆地理编码结果：' + JSON.stringify(data));
});
// 参考接口API说明，应用可以获得与此坐标匹配的GeoAddress列表，应用可以根据实际使用需求，读取相应的参数数据。

// 获取正地理编码结果。
const geocodeRequest = {"description": "上海市浦东新区xx路xx号", "maxItems": 1};
geolocation.getAddressesFromLocationName(geocodeRequest, (data) => {
    console.log('正地理编码结果：' + JSON.stringify(data));
});
// 参考接口API说明，应用可以获得与位置描述相匹配的GeoAddress列表，其中包含对应的坐标数据，请参考API使用。

// 如果需要查询的位置描述可能出现多地重名的请求,可以设置GeoCodeRequest,通过设置一个经纬度范围,以高效地获取期望的准确结果。
```

在上述代码中，使用 geolocation 模块进行逆地理编码和正地理编码的操作。逆地理编码将坐标转换为地理位置信息，而正地理编码将位置描述转换为坐标。通过相应的请求参数，开发者可以获取匹配的地理位置信息或坐标，并根据实际需求读取相应的参数数据。

12.3 传感器

传感器是一种能够感知和测量特定物理量或环境参数的设备或装置。这些物理量包括光线、温度、声音、湿度、压力、加速度、方向等。传感器通过将这些物理量转换成电信号或数字信号，使计算机或其他电子系统能够理解和处理这些信息。传感器在各个领域都有广泛的应用，包括工业控制、医疗设备、汽车技术、环境监测、智能手机和可穿戴设备等。

12.3.1 HarmonyOS 系统传感器介绍

在 HarmonyOS 系统中，传感器是应用访问底层硬件传感器的一种设备抽象概念。开发者根据传感器提供的 Sensor API，可以查询设备上的传感器，订阅传感器数据，并根据传感器数据定制相应的算法开发各类应用，如指南针、运动健康、游戏等。

在移动设备中，传感器用于侦测环境中所发生的事件或变化，并将此消息发送至其他电子设备（如中央处理器）。传感器通常由敏感组件和转换组件组成。传感器是实现物联网智能化的重要基石，为实现全场景智慧化战略，支撑"1+8+N"产品需求，需要构筑统一的传感器管理框架，达到为各产品/业务提供低时延、低功耗的感知数据的目的。

HarmonyOS 系统支持的传感器信息，如表 12-4 所示。

表 12-4 HarmonyOS 系统支持的传感器信息

传感器类型	描述	说明	主要用途
SENSOR_TYPE_ACCELEROMETER	加速度传感器	测量设备在三个物理轴（x、y 和 z）上所受到的加速度（包括重力加速度），单位：m/s^2	检测运动状态
SENSOR_TYPE_ACCELEROMETER_UNCALIBRATED	未校准加速度传感器	测量设备在三个物理轴（x、y 和 z）上所受到的未校准的加速度（包括重力加速度），单位：m/s^2	检测加速度偏差估值
SENSOR_TYPE_LINEAR_ACCELERATION	线性加速度传感器	测量设备在三个物理轴（x、y 和 z）上所受到的线性加速度（不包括重力加速度），单位：m/s^2	检测每个单轴方向上的线性加速度
SENSOR_TYPE_GRAVITY	重力传感器	测量设备在三个物理轴（x、y 和 z）上所受到的重力加速度，单位：m/s^2	测量重力大小
SENSOR_TYPE_GYROSCOPE	陀螺仪传感器	测量设备在三个物理轴（x、y 和 z）上所受到的旋转角速度，单位：rad/s	测量旋转的角速度
SENSOR_TYPE_GYROSCOPE_UNCALIBRATED	未校准陀螺仪传感器	测量设备在三个物理轴（x、y 和 z）上所受到的未校准旋转角速度，单位：rad/s	测量旋转的角速度及偏差估值
SENSOR_TYPE_SIGNIFICANT_MOTION	大幅度动作传感器	测量设备在三个物理轴（x、y 和 z）上是否存在大幅度运动；如果取值为 1 则代表存在大幅度运动，取值为 0 则代表没有大幅度运动	用于检测设备是否存在大幅度运动
SENSOR_TYPE_PEDOMETER_DETECTION	计步器检测传感器	检测用户的计步动作；如果取值为 1 则代表用户产生了计步行走的动作；取值为 0 则代表用户没有发生运动	用于检测用户是否有计步的动作

续表

传感器类型	描 述	说 明	主要用途
SENSOR_TYPE_PEDOMETER	计步器传感器	统计用户的行走步数	用于提供用户行走的步数数据
SENSOR_TYPE_AMBIENT_TEMPERATURE	环境温度传感器	测量环境温度，单位：摄氏度(°C)	测量环境温度
SENSOR_TYPE_MAGNETIC_FIELD	磁场传感器	测量设备在三个物理轴（x、y 和 z）上所受到的境地磁场，单位：μT	创建指南针
SENSOR_TYPE_MAGNETIC_FIELD_UNCALIBRATED	未校准磁场传感器	测量设备在三个物理轴（x、y 和 z）上所受到的未校准环境地磁场，单位：μT	测量地磁偏差估值
SENSOR_TYPE_HUMIDITY	湿度传感器	测量环境的相对湿度，以百分比(%)表示	监测露点、绝对湿度和相对湿度
SENSOR_TYPE_BAROMETER	气压计传感器	测量环境气压，单位：hPa 或 mbar	测量环境气压
SENSOR_TYPE_ORIENTATION	方向传感器	测量设备在三个物理轴（x、y 和 z）上所受到的旋转角度值，单位：rad	用于提供屏幕旋转的3个角度值
SENSOR_TYPE_ROTATION_VECTOR	旋转矢量传感器	测量设备旋转矢量，复合传感器：由加速度传感器、磁场传感器、陀螺仪传感器合成	检测设备相对于东北天坐标系的方向
SENSOR_TYPE_PROXIMITY	接近光传感器	测量可见物体相对于设备显示屏的接近或远离状态	通话中设备相对人的位置
SENSOR_TYPE_AMBIENT_LIGHT	环境光传感器	测量设备周围光线强度，单位：lux	自动调节屏幕亮度，检测屏幕上方是否有遮挡
SENSOR_TYPE_HEART_RATE	心率传感器	测量用户的心率数值	用于提供用户的心率健康数据
SENSOR_TYPE_WEAR_DETECTION	佩戴检测传感器	检测用户是否佩戴	用于检测用户是否佩戴智能穿戴
SENSOR_TYPE_HALL	霍尔传感器	测量设备周围是否存在磁力吸引	设备的皮套模式

HarmonyOS 系统中的传感器包含 4 个模块：Sensor API、Sensor Framework、Sensor Service 和 HDF，具体结构如图 12-2 所示。各个模块的具体说明如下。

- Sensor API：提供传感器的基础 API，主要包含查询传感器列表、订阅/取消订阅传感器数据、执行控制命令等，简化应用开发。
- Sensor Framework：主要负责实现传感器的订阅管理，包括数据通道的创建与销毁、订阅与取消订阅操作，并负责与 SensorService 的通信。
- Sensor Service：主要负责 HD_IDL 层的数据接收、解析和分发，同时进行前后台的策略管控，管理设备上的 Sensor 以及 Sensor 权限控制等。
- HDF 层：负责对不同的 FIFO（先进先出队列）和频率进行策略选择，以及适配不同设备的需求。

在 HarmonyOS 系统中开发传感器应用程序时，需要先获取对应的权限。常用传感器及其对应的权限，如表 12-5 所示。

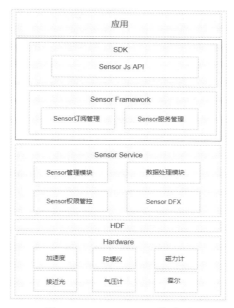

图 12-2 HarmonyOS 传感器

表 12-5 常用传感器对应的权限

传感器	权限名	敏感级别	权限描述
加速度传感器，加速度未校准传感器，线性加速度传感器	ohos.permission.ACCELEROMETER	system_grant	允许订阅 Motion 组对应的加速度传感器的数据
陀螺仪传感器，陀螺仪未校准传感器	ohos.permission.GYROSCOPE	system_grant	允许订阅 Motion 组对应的陀螺仪传感器的数据
计步器	ohos.permission.ACTIVITY_MOTION	user_grant	允许订阅运动状态
心率计	ohos.permission.READ_HEALTH_DATA	user_grant	允许读取健康数据

12.3.2 开发传感器应用程序

在 HarmonyOS 系统中，常用的传感器接口，如表 12-6 所示。

表 12-6 HarmonyOS 系统中常用的传感器接口

模块	接口名	描述
ohos.sensor	sensor.on(sensorType, callback:AsyncCallback<Response>): void	持续监听传感器数据变化
ohos.sensor	sensor.once(sensorType, callback:AsyncCallback<Response>): void	获取一次传感器数据变化
ohos.sensor	sensor.off(sensorType, callback:AsyncCallback<void>): void	注销传感器数据的监听

在开发 HarmonyOS 应用程序时，可以使用以下各类传感器实现不同的功能。

- 指南针传感器数据：感知用户设备当前的朝向，为用户指明方向。
- 接近光传感器数据：感知距离遮挡物的距离，使设备能够自动亮屏、灭屏，防止误触。
- 气压计传感器数据：准确判断设备当前所处的海拔。

- 环境光传感器数据：使设备能够实现背光自动调节。
- 霍尔传感器数据：使设备实现皮套功能等。
- 心率传感器数据：感知用户当前的心率。
- 计步传感器数据：统计用户的步数。
- 佩戴检测传感器：检测用户是否佩戴。

具体来说，开发传感器应用程序的基本步骤如下。

（1）获取设备上传感器的数据，需要在文件 config.json 里面配置请求权限。

```
"reqPermissions":[
  {
    "name":"ohos.permission.ACCELEROMETER",
    "reason":"",
    "usedScene":{
     "ability": ["sensor.index.MainAbility",".MainAbility"],
     "when":"inuse"
    }
  },
  {
    "name":"ohos.permission.GYROSCOPE",
    "reason":"",
    "usedScene":{
     "ability": ["sensor.index.MainAbility",".MainAbility"],
     "when":"inuse"
    }
  },
  {
    "name":"ohos.permission.ACTIVITY_MOTION",
    "reason":"ACTIVITY_MOTION_TEST",
    "usedScene":{
     "ability": ["sensor.index.MainAbility",".MainAbility"],
     "when":"inuse"
    }
  },
  {
    "name":"ohos.permission.READ_HEALTH_DATA",
    "reason":"HEALTH_DATA_TEST",
    "usedScene":{
     "ability": ["sensor.index.MainAbility",".MainAbility"],
     "when":"inuse"
    }
  },
]
```

（2）持续监听传感器数据变化：例如，以下代码使用 @ohos.sensor 模块中的 sensor 对象，订阅了加速度传感器（SENSOR_TYPE_ACCELEROMETER）的数据。

```
import sensor from "@ohos.sensor"
sensor.on(sensor.sensorType.SENSOR_TYPE_ACCELEROMETER,function(data){
        console.info("Subscription succeeded. data = " + data);//调用成功，
打印对应传感器的数据
```

```
    }
);
```

下面将 SensorType=SENSOR_TYPE_ID_ACCELEROMETER 为例展示运行结果，持续监听传感器接口的结果，如图 12-3 所示。

```
11-23 16:09:21.757 10498 10542 D 03B00/JSApp: app Log: SensorType.SENSOR_TYPE_ID_ACCELEROMETER , data.x:0.06667372584342957
11-23 16:09:21.757 10498 10542 D 03B00/JSApp: app Log: SensorType.SENSOR_TYPE_ID_ACCELEROMETER , data.y:0.6836786270141602
11-23 16:09:21.757 10498 10542 D 03B00/JSApp: app Log: SensorType.SENSOR_TYPE_ID_ACCELEROMETER , data.z:9.861286163330078
11-23 16:09:21.757 10498 10542 D 03B00/JSApp: app Log: sensor.on-SensorType.SENSOR_TYPE_ID_ACCELEROMETER - success
11-23 16:09:21.758 10498 10542 D 03B00/JSApp: app Log: SensorType.SENSOR_TYPE_ID_ACCELEROMETER , data.x:0.06613742560148239
11-23 16:09:21.758 10498 10542 D 03B00/JSApp: app Log: SensorType.SENSOR_TYPE_ID_ACCELEROMETER , data.y:0.6781432628631592
11-23 16:09:21.758 10498 10542 D 03B00/JSApp: app Log: SensorType.SENSOR_TYPE_ID_ACCELEROMETER , data.z:9.861123085021973
11-23 16:09:21.759 10498 10542 D 03B00/JSApp: app Log: sensor.on-SensorType.SENSOR_TYPE_ID_ACCELEROMETER - success
11-23 16:09:21.760 10498 10542 D 03B00/JSApp: app Log: SensorType.SENSOR_TYPE_ID_ACCELEROMETER , data.x:0.07091625034809113
11-23 16:09:21.760 10498 10542 D 03B00/JSApp: app Log: SensorType.SENSOR_TYPE_ID_ACCELEROMETER , data.y:0.6775399446487427
11-23 16:09:21.760 10498 10542 D 03B00/JSApp: app Log: SensorType.SENSOR_TYPE_ID_ACCELEROMETER , data.z:9.863909721374512
11-23 16:09:21.760 10498 10542 D 03B00/JSApp: app Log: sensor.on-SensorType.SENSOR_TYPE_ID_ACCELEROMETER - success
```

图 12-3 持续监听传感器接口的结果

（3）注销传感器数据监听：例如，以下代码使用 @ohos.sensor 模块中的 sensor 对象，取消订阅了加速度传感器（SENSOR_TYPE_ACCELEROMETER）的数据。

```
    import sensor from "@ohos.sensor"
    sensor.off(sensor.sensorType.SENSOR_TYPE_ACCELEROMETER,function() {
        console.info("Succeeded in unsubscribing from acceleration sensor data.");
// 注销成功，返回打印结果
    }
);
```

以 SensorType=SENSOR_TYPE_ID_ACCELEROMETER 为例展示运行结果，注销传感器成功结果，如图 12-4 所示。

```
11-23 16:15:19.116 11357 11407 D 03B00/JSApp: app Log: SensorType.SENSOR_TYPE_ID_ACCELEROMETER , data.x:0.06945100426673889
11-23 16:15:19.116 11357 11407 D 03B00/JSApp: app Log: SensorType.SENSOR_TYPE_ID_ACCELEROMETER , data.y:0.6964541077613831
11-23 16:15:19.116 11357 11407 D 03B00/JSApp: app Log: SensorType.SENSOR_TYPE_ID_ACCELEROMETER , data.z:9.890427589416504
11-23 16:15:19.116 11357 11407 D 03B00/JSApp: app Log: sensor.on-SensorType.SENSOR_TYPE_ID_ACCELEROMETER - success
11-23 16:15:19.126 11357 11407 D 03B00/JSApp: app Log: sensor.off-SensorType.SENSOR_TYPE_ID_ACCELEROMETER - success
```

图 12-4 注销传感器成功结果

（4）获取一次传感器数据变化：例如，以下代码使用 @ohos.sensor 模块中的 sensor 对象，通过 sensor.once 方法一次性订阅加速度传感器（SENSOR_TYPE_ACCELEROMETER）的数据。

```
    import sensor from "@ohos.sensor"
    sensor.once(sensor.sensorType.SENSOR_TYPE_ACCELEROMETER,function(data) {
            console.info("Data obtained successfully. data=" + data);// 获取
数据成功，打印对应传感器的数据
    }
);
```

以 SensorType=SENSOR_TYPE_ID_ACCELEROMETER 为例展示运行结果，获取数据成功的日志，如图 12-5 所示。

```
11-23 16:21:28.622 12636 12683 D 03B00/JSApp: app Log: SensorType.SENSOR_TYPE_ID_ACCELEROMETER , data.x:0.04590163379907608
11-23 16:21:28.622 12636 12683 D 03B00/JSApp: app Log: SensorType.SENSOR_TYPE_ID_ACCELEROMETER , data.y:0.6799160289764
11-23 16:21:28.622 12636 12683 D 03B00/JSApp: app Log: SensorType.SENSOR_TYPE_ID_ACCELEROMETER , data.z:9.883752822875977
11-23 16:21:28.622 12636 12683 D 03B00/JSApp: app Log: sensor.once-SensorType.SENSOR_TYPE_ID_ACCELEROMETER - success
```

图 12-5 获取数据成功日志

如果接口调用不成功,建议使用 try/catch 语句捕获代码中可能出现的错误信息。例如,下面代码使用了一个 try-catch 块,尝试一次性订阅加速度传感器(SENSOR_TYPE_ACCELEROMETER)的数据,同时捕获任何可能发生的异常。

```
try {
    sensor.once(sensor.sensorType.SENSOR_TYPE_ACCELEROMETER,function(data) {
        console.info("Data obtained successfully. data=" + data);// 获取数据成功,打印对应传感器的数据
    });
} catch (error) {
    console.error(error);
}
```

12.4 综合实战:健身计步器

实例 12-1 中,使用位置服务和传感器技术实现了一个健身计步器系统,分别实现了步数追踪和位置信息展示功能。利用 HarmonyOS 的 DSL 语法构建了清晰且美观的界面,包括任务完成状态、当前步数、起始位置和当前位置等信息展示。通过各模块间的协作,实现了步数目标的设定、实时步数展示、位置信息获取,并在任务完成状态界面提供了用户友好的交互体验。同时,项目中采用了常量定义、日志记录等工具类,提高了代码的可维护性和可读性。

实例 12-1:健身计步器系统(源码路径:codes\12\PedometerApp)

12.4.1 系统配置

扫码看视频

(1)编写文件 src/main/module.json5,这是一个应用程序入口模块,支持在手机和平板设备上运行,具备安装时交付和页面权限配置。主要包含一个名为"EntryAbility"的能力,定义了该能力的源文件、描述、图标、标签等信息,同时具备后台定位等背景模式。还申请了一系列权限,包括保持后台运行、活动动作和位置相关权限。

(2)编写文件 src/main/ets/common/constants/CommonConstants.ets,定义了一组常量,用于应用程序中的各种功能。这些常量涵盖了请求权限、文件存储、服务名称、通知标题、事件 ID、通知 ID、时间间隔等多个方面。它们还包括了与位置相关的常量,如位置优先级、场景、时间间隔、距离间隔等。此外,还定义了与界面显示和进度相关的常量,以及与步数和位置数据处理相关的键值和标签。这些常量的使用范围涉及应用程序的各个模块,提高了代码的可维护性和可读性。

(3)编写文件 src/main/ets/common/utils/LocationUtil.ets,定义类 LocationUtil,用于处理地理位置相关功能。首先导入了 @ohos.geoLocationManager 模块,定义了类 LocationUtil,通过 geoLocationManager 模块提供的接口实现了地理位置的监听和取消监听功能。geolocationOn 方法用于开启位置监听,配置监听参数并处理位置变化回调,而 geolocationOff 方法用于关闭位置监听。

(4)编写文件 src/main/ets/common/utils/StepsUtil.ets,这是一个工具类,通过整合 @ohos.app.ability.common 和 @ohos.data.preferences 等模块,提供了处理步数相关功能的方

法。它包含创建偏好设置、计算进度值、存储和获取偏好设置数据、清除步数相关数据等功能，便于在应用中管理和操作步数信息。最后，通过导出实例，使其在其他模块中易于引用。

```
import common from '@ohos.app.ability.common';
import data_preferences from '@ohos.data.preferences';
import { CommonConstants } from '../constants/CommonConstants';
import { GlobalContext } from './GlobalContext';
import Logger from './Logger';

const TAG: string = 'StepsUtil';
const PREFERENCES_NAME = 'myPreferences';

export class StepsUtil {
  createStepsPreferences(context: common.UIAbilityContext) {
    let preferences: Promise<data_preferences.Preferences> = data_preferences.getPreferences(context, PREFERENCES_NAME);
    GlobalContext.getContext().setObject('getStepsPreferences', preferences)
  }

  getProgressValue(setSteps: number, currentSteps: number): number {
    let progressValue: number = 0;
    if (setSteps > 0 && currentSteps > 0) {
      progressValue = Math.round((currentSteps / setSteps) * CommonConstants.ONE_HUNDRED);
    }
    return progressValue;
  }

  putStorageValue(key: string, value: string) {
      GlobalContext.getContext().getObject('getStepsPreferences')?.then((preferences: data_preferences.Preferences) => {
        preferences.put(key, value).then(() => {
          Logger.info(TAG, 'Storage put succeeded, key:' + key);
        }).catch((err: Error) => {
          Logger.error(TAG, 'Failed to put the value of startup with err:' + JSON.stringify(err));
        })
      }).catch((err: Error) => {
        Logger.error(TAG, 'Failed to get the storage with err:' + JSON.stringify(err));
      })
  }

  async getStorageValue(key: string): Promise<string> {
    let ret: data_preferences.ValueType = '';
    const preferences: data_preferences.Preferences | undefined = await GlobalContext.getContext().getObject('getStepsPreferences');
    if(preferences) {
      ret = await preferences?.get(key, ret);
    }
```

```
      return String(ret);
    }

    CleanStepsData(): void {
      this.putStorageValue(CommonConstants.OLD_STEPS, '');
      this.putStorageValue(CommonConstants.IS_START, CommonConstants.FALSE);
      this.putStorageValue(CommonConstants.START_POSITION, '');
      this.putStorageValue(CommonConstants.PROGRESS_VALUE_TAG, CommonConstants.
INITIALIZATION_VALUE);
    }

    checkStrIsEmpty(str: string): boolean {
      return str?.trim().length === 0;
    }
  }

  let stepsUtil = new StepsUtil();

  export default stepsUtil as StepsUtil;
```

对上述代码的具体说明如下。

- createStepsPreferences(context: common.UIAbilityContext): void：在应用上下文中创建步数相关的偏好设置，并将其保存在全局上下文中。
- getProgressValue(setSteps: number, currentSteps: number): number：计算并返回给定设定步数和当前步数的进度值。
- putStorageValue(key: string, value: string): void：将键值对存储到偏好设置中，使用全局上下文中保存的偏好设置实例。
- async getStorageValue(key: string): Promise<string>：异步地从偏好设置中获取指定键的值，返回一个 Promise 对象。
- CleanStepsData(): void：清除步数相关的存储数据，包括旧步数、启动状态、起始位置和进度值等。
- checkStrIsEmpty(str: string): boolean：检查给定字符串是否为空（包括只包含空格的情况），返回布尔值。

12.4.2 UI 视图

（1）编写文件 src/main/ets/view/CompletionStatus.ets，这是一个视图结构组件，依赖于 CommonConstants 常量、InputDialog 组件、Logger 和 StepsUtil 工具类。该组件通过 @Link 和 @Consume 注解建立了属性的数据关联。

```
import { CommonConstants } from '../common/constants/CommonConstants';
import InputDialog from '../view/InputDialog';
import Logger from '../common/utils/Logger';
import StepsUtil from '../common/utils/StepsUtil';

const TAG: string = 'CompletionStatus';
```

```
@Component
export struct CompletionStatus {
  @Link progressValue: number;
  @Consume stepGoal: string;

  inputDialogController: CustomDialogController = new CustomDialogController({
    builder: InputDialog({
      cancel: this.inputDialogCancel,
      confirm: this.inputDialogConfirm
    }),
    autoCancel: false,
    customStyle: true,
    alignment: DialogAlignment.Bottom,
    offset: { dx: CommonConstants.OFFSET_DX, dy: CommonConstants.OFFSET_DY }
  })

  inputDialogCancel() {
    Logger.info(TAG, 'Callback when the cancel button is clicked');
  }

  inputDialogConfirm() {
    if (StepsUtil.checkStrIsEmpty(this.stepGoal)) {
      return;
    }
    StepsUtil.putStorageValue(CommonConstants.STEP_GOAL, this.stepGoal);
  }

  build() {
    Stack({ alignContent: Alignment.TopStart }) {
      Column() {
        Progress({
          value: 0,
          total: CommonConstants.PROGRESS_TOTAL,
          type: ProgressType.Ring
        })
          .color(Color.White)
          .value(this.progressValue)
          .width($r('app.float.progress_width'))
          .style({
            strokeWidth: CommonConstants.PROGRESS_STROKE_WIDTH,
            scaleCount: CommonConstants.PROGRESS_SCALE_COUNT,
            scaleWidth: CommonConstants.PROGRESS_SCALE_WIDTH
          })
          .margin({ top: $r('app.float.progress_margin_top') })
          .borderRadius(CommonConstants.PROGRESS_BORDER_RADIUS)

        Button($r('app.string.set_target'))
          .width($r('app.float.target_button_width'))
          .height($r('app.float.target_button_height'))
          .borderRadius($r('app.float.target_button_radius'))
          .fontSize($r('app.float.target_button_font'))
          .fontColor($r('app.color.button_font'))
```

```
        .fontWeight(CommonConstants.SMALL_FONT_WEIGHT)
        .backgroundColor(Color.White)
        .margin({
          top: $r('app.float.target_margin_top'),
          bottom: $r('app.float.target_margin_bottom')
        })
        .onClick(() => {
          this.inputDialogController.open();
        })
    }
    .width(CommonConstants.FULL_WIDTH)
    .backgroundImage($r('app.media.ic_orange'))

    Row() {
      Text(this.progressValue.toString())
        .borderRadius($r('app.float.progress_text_radius'))
        .fontSize($r('app.float.progress_text_font'))
        .fontColor(Color.White)
        .fontWeight(CommonConstants.BIG_FONT_WEIGHT)
        .textAlign(TextAlign.Center)
        .margin({ top: $r('app.float.value_margin_top') })

      Text(CommonConstants.PERCENT_SIGN)
        .borderRadius($r('app.float.percent_text_radius'))
        .fontSize($r('app.float.percent_text_font'))
        .fontColor(Color.White)
        .fontWeight(CommonConstants.SMALL_FONT_WEIGHT)
        .textAlign(TextAlign.Center)
        .margin({ top: $r('app.float.percent_margin_top') })
    }
    .width(CommonConstants.FULL_WIDTH)
    .justifyContent(FlexAlign.Center)
  }
}
```

（2）编写文件 src/main/ets/view/CurrentSituation.ets，这是一个 HarmonyOS 视图组件，用于展示当前步数、起始位置和当前位置的信息。通过 DSL 语法构建了具有良好可读性和美观性的 UI 布局，其中包括描述性文本、显示文本和分隔线等元素。通过属性和样式的定义，实现了灵活的信息展示和界面样式设置，使用户能够直观地了解当前的步数和位置情况。

```
import { CommonConstants } from '../common/constants/CommonConstants';

@Component
export struct CurrentSituation {
  @Prop currentSteps: string = '';
  @Prop startPosition: string = '';
  @Prop currentLocation: string = '';

  @Styles descriptionTextStyle(){
```

```
      .width($r('app.float.description_text_width'))
      .height($r('app.float.description_text_height'))
  }
  @Styles displayTextStyle(){
    .width($r('app.float.display_text_width'))
    .height($r('app.float.display_text_height'))
  }

  build() {
    Row() {
      Column() {
        Text($r('app.string.walking_data'))
          .width(CommonConstants.FULL_WIDTH)
          .height($r('app.float.walling_height'))
          .fontSize($r('app.float.walling_text_font'))
          .fontColor($r('app.color.step_text_font'))
          .fontWeight(CommonConstants.SMALL_FONT_WEIGHT)
          .textAlign(TextAlign.Start)
          .margin({
            top: $r('app.float.walling_margin_top'),
            bottom: $r('app.float.walling_margin_bottom'),
            left: $r('app.float.walling_margin_left')
          })

        Row() {
          Text($r('app.string.current_steps'))
            .descriptionTextStyle()
            .fontSize($r('app.float.current_steps_font'))
            .fontColor($r('app.color.steps_text_font'))
            .fontWeight(CommonConstants.BIG_FONT_WEIGHT)
            .textAlign(TextAlign.Start)

          Text($r('app.string.step', this.currentSteps))
            .displayTextStyle()
            .fontSize($r('app.float.record_steps_font'))
            .fontColor($r('app.color.step_text_font'))
            .fontWeight(CommonConstants.BIG_FONT_WEIGHT)
            .textAlign(TextAlign.Start)
        }
        .width(CommonConstants.FULL_WIDTH)
        .height($r('app.float.current_row_height'))
        .margin({
          top: $r('app.float.current_margin_top'),
          bottom: $r('app.float.current_margin_bottom'),
          left: $r('app.float.current_margin_left')
        })

        Divider()
          .vertical(false)
          .color($r('app.color.divider'))
          .strokeWidth(CommonConstants.DIVIDER_STROKE_WIDTH)
          .margin({
```

```
      left: $r('app.float.divider_margin_left'),
      right: $r('app.float.divider_margin_right')
    })

  Row() {
    Text($r('app.string.start_position'))
      .descriptionTextStyle()
      .fontSize($r('app.float.start_position_font'))
      .fontColor($r('app.color.steps_text_font'))
      .fontWeight(CommonConstants.BIG_FONT_WEIGHT)
      .textAlign(TextAlign.Start)

    Text(this.startPosition)
      .displayTextStyle()
      .fontSize($r('app.float.position_font_size'))
      .fontColor($r('app.color.step_text_font'))
      .fontWeight(CommonConstants.BIG_FONT_WEIGHT)
      .textAlign(TextAlign.Start)
      .textOverflow({ overflow: TextOverflow.Ellipsis })
      .maxLines(CommonConstants.MAX_LINE)
  }
  .width(CommonConstants.FULL_WIDTH)
  .height($r('app.float.start_position_height'))
  .margin({
    top: $r('app.float.position_margin_top'),
    bottom: $r('app.float.position_margin_bottom'),
    left: $r('app.float.position_margin_left')
  })

  Divider()
    .vertical(false)
    .color($r('app.color.divider'))
    .strokeWidth(CommonConstants.DIVIDER_STROKE_WIDTH)
    .margin({
      left: $r('app.float.divider_margin_left'),
      right: $r('app.float.divider_margin_right')
    })

  Row() {
    Text($r('app.string.current_location'))
      .descriptionTextStyle()
      .fontSize($r('app.float.location_font_size'))
      .fontColor($r('app.color.steps_text_font'))
      .fontWeight(CommonConstants.BIG_FONT_WEIGHT)
      .textAlign(TextAlign.Start)

    Text(this.currentLocation)
      .displayTextStyle()
      .fontSize($r('app.float.current_font_size'))
      .fontColor($r('app.color.step_text_font'))
      .fontWeight(CommonConstants.BIG_FONT_WEIGHT)
      .textAlign(TextAlign.Start)
```

```
            .textOverflow({ overflow: TextOverflow.Ellipsis })
            .maxLines(CommonConstants.MAX_LINE)
        }
        .width(CommonConstants.FULL_WIDTH)
        .height($r('app.float.current_location_height'))
        .margin({
          top: $r('app.float.location_margin_top'),
          bottom: $r('app.float.location_margin_bottom'),
          left: $r('app.float.location_margin_left')
        })
      }
      .width(CommonConstants.SITUATION_WIDTH)
      .borderRadius($r('app.float.situation_border_radius'))
      .backgroundColor(Color.White)
    }
    .width(CommonConstants.FULL_WIDTH)
    .height($r('app.float.current_situation_height'))
    .margin({ top: $r('app.float.situation_margin_top') })
    .justifyContent(FlexAlign.Center)
  }
}
```

上述代码的主要功能如下。

- 属性定义：通过 @Prop 注解定义了三个属性：当前步数（currentSteps）、起始位置（startPosition）、当前位置（currentLocation）。
- 样式定义：使用 @Styles 注解定义了两个样式：descriptionTextStyle 和 displayTextStyle，分别设置了描述文本和显示文本的样式。
- UI 构建：使用 HarmonyOS 的 DSL 语法构建了一个 UI 布局，包括描述性文本、显示文本和分隔线等。三个信息（当前步数、起始位置、当前位置）分别以垂直排列的方式展示，每个信息包括一个描述性文本和一个显示文本。
- 文本样式设置：使用 Text 组件展示文本信息，设置了字体大小、颜色、粗细、对齐方式等样式。对于显示文本，通过 displayTextStyle 定义的样式设置了宽度和高度。
- 分隔线设置：使用 Divider 组件绘制分隔线，设置了颜色、宽度和边距。
- 整体样式设置：对整体布局设置了宽度、高度、边距、圆角和背景颜色等样式。

（3）编写文件 src/main/ets/view/InputDialog.ets，定义了一个名为 InputDialog 的自定义对话框组件，用于用户输入数据的场景。该组件包含一个输入框，用户可以在其中输入内容，并提供取消和确认按钮以执行相应操作。

```
import { CommonConstants } from '../common/constants/CommonConstants';

@CustomDialog
export default struct InputDialog {
  @Consume stepGoal: string;
  controller?: CustomDialogController;
  cancel?: () => void;
  confirm?: () => void;

  build() {
```

```
      Column() {
        Text($r('app.string.steps'))
          .width(CommonConstants.FULL_WIDTH)
          .height($r('app.float.input_text_height'))
          .fontSize($r('app.float.input_text_font_size'))
          .fontColor($r('app.color.step_text_font'))
          .fontWeight(CommonConstants.BIG_FONT_WEIGHT)
          .textAlign(TextAlign.Start)
          .margin({
            top: $r('app.float.input_margin_top'),
            bottom: $r('app.float.input_margin_bottom'),
            left: $r('app.float.input_margin_left')
          })

        TextInput({ placeholder: this.stepGoal === '' ? $r('app.string.placeholder') : this.stepGoal })
          .width(CommonConstants.FULL_WIDTH)
          .type(InputType.Number)
          .fontSize($r('app.float.input_font_size'))
          .alignSelf(ItemAlign.Start)
          .backgroundColor(Color.White)
          .margin({
            top: $r('app.float.text_margin_top'),
            bottom: $r('app.float.text_margin_bottom')
          })
          .onChange((value: string) => {
            this.stepGoal = value;
          })

        Divider()
          .width(CommonConstants.DIVIDER_WIDTH)
          .height($r('app.float.divider_height'))
          .vertical(false)
          .color($r('app.color.divider'))
          .strokeWidth(CommonConstants.DIVIDER_STROKE_WIDTH)

        Row() {
          Text($r('app.string.cancel'))
            .width($r('app.float.text_width'))
            .height($r('app.float.text_height'))
            .fontColor($r('app.color.input_text_font'))
            .fontWeight(CommonConstants.BIG_FONT_WEIGHT)
            .fontSize($r('app.float.text_font_size'))
            .textAlign(TextAlign.Center)
            .margin({ right: $r('app.float.text_margin_right') })
            .onClick(() => {
              if(this.controller) {
                this.controller.close();
              }
              if(this.cancel) {
                this.cancel();
              }
```

```
            })

          Divider()
            .height($r('app.float.divider_height'))
            .vertical(true)
            .color($r('app.color.divider'))
            .strokeWidth(CommonConstants.DIVIDER_STROKE_WIDTH)

          Text($r('app.string.confirm'))
            .width($r('app.float.text_width'))
            .height($r('app.float.text_height'))
            .fontColor($r('app.color.input_text_font'))
            .fontWeight(CommonConstants.BIG_FONT_WEIGHT)
            .fontSize($r('app.float.text_font_size'))
            .textAlign(TextAlign.Center)
            .margin({ left: $r('app.float.text_margin_left') })
            .fontColor($r('app.color.input_text_font'))
            .onClick(() => {
              if(this.controller) {
                this.controller.close();
              }
              if(this.confirm) {
                this.confirm();
              }
            })
```

12.4.3 项目主界面

编写文件 src/main/ets/pages/HomePage.ets，这是一个名为 HomePage 的 HarmonyOS 应用页面，主要功能包括获取计步和地理位置信息，提供开始/停止计步的按钮，并展示当前计步进度和位置信息。通过初始化、权限请求和事件处理，实现了与传感器、地理位置管理和其他模块的交互，为用户提供计步和位置监控的功能。

```
    const TAG: string = 'HomePage';

    @Entry
    @Component
    struct HomePage {
      @State currentSteps: string = CommonConstants.INITIALIZATION_VALUE;
      @Provide stepGoal: string = '';
      @State oldSteps: string = '';
      @State startPosition: string = '';
      @State currentLocation: string = '';
      @State locale: string = new Intl.Locale().language;
      @State latitude: number = 0;
      @State longitude: number = 0;
      @State progressValue: number = 0;
      @State isStart: boolean = false;
      private context: common.UIAbilityContext = getContext(this) as common.UIAbilityContext;
```

```typescript
  onPageShow() {
    this.init();
    this.requestPermissions();
  }

  onPageHide() {
    sensor.off(sensor.SensorId.PEDOMETER);
  }

  init() {
    StepsUtil.getStorageValue(CommonConstants.IS_START).then((res: string) => {
      if (res === CommonConstants.TRUE) {
        this.isStart = true;
        StepsUtil.getStorageValue(CommonConstants.CURRENT_STEPS).then((res: string) => {
          if (StepsUtil.checkStrIsEmpty(res)) {
            return;
          }
          this.currentSteps = res;
        });

        StepsUtil.getStorageValue(CommonConstants.PROGRESS_VALUE_TAG).then((res: string) => {
          if (StepsUtil.checkStrIsEmpty(res)) {
            return;
          }
          this.progressValue = NumberUtil._parseInt(res, 10);
        });

        StepsUtil.getStorageValue(CommonConstants.START_POSITION).then((res: string) => {
          if (StepsUtil.checkStrIsEmpty(res)) {
            return;
          }
          this.startPosition = res;
        });

        StepsUtil.getStorageValue(CommonConstants.OLD_STEPS).then((res: string) => {
          if (StepsUtil.checkStrIsEmpty(res)) {
            return;
          }
          this.oldSteps = res;
        });
      } else {
        this.isStart = false;
      }
    })

    StepsUtil.getStorageValue(CommonConstants.STEP_GOAL).then((res: string) => {
      if (StepsUtil.checkStrIsEmpty(res)) {
```

```
          return;
        }
        this.stepGoal = res;
      });
  }

  requestPermissions(): void {
    let atManager = abilityAccessCtrl.createAtManager();
    try {
      atManager.requestPermissionsFromUser(this.context, CommonConstants.
REQUEST_PERMISSIONS).then((data) => {
        if (data.authResults[0] !== 0 || data.authResults[1] !== 0) {
          return;
        }
        const that = this;
        try {
          sensor.on(sensor.SensorId.PEDOMETER, (data) => {
            try {
              if (that.isStart) {
                if (StepsUtil.checkStrIsEmpty(that.oldSteps)) {
                  that.oldSteps = data.steps.toString();
                  StepsUtil.putStorageValue(CommonConstants.OLD_STEPS, that.
oldSteps);
                } else {
                  that.currentSteps = (data.steps - NumberUtil._parseInt(that.
oldSteps, 10)).toString();
                }
              } else {
                that.currentSteps = data.steps.toString();
              }

              if (StepsUtil.checkStrIsEmpty(that.stepGoal) || !that.isStart) {
                return;
              }
              StepsUtil.putStorageValue(CommonConstants.CURRENT_STEPS, that.
currentSteps);
              that.progressValue = StepsUtil.getProgressValue(NumberUtil._
parseInt(that.stepGoal, 10),
                NumberUtil._parseInt(that.currentSteps, 10));
              StepsUtil.putStorageValue(CommonConstants.PROGRESS_VALUE_TAG,
String(that.progressValue));
            } catch (err) {
              Logger.error(TAG, 'Sensor on err' + JSON.stringify(err));
            }
          }, { interval: CommonConstants.SENSOR_INTERVAL });

        } catch (err) {
          console.error('On fail, errCode: ' + JSON.stringify(err));
        }

        LocationUtil.geolocationOn((location: geoLocationManager.Location)
```

```
      => {
                if (this.latitude === location.latitude && this.longitude === location.
longitude) {
                  return;
                }
                this.latitude = location.latitude;
                this.longitude = location.longitude;
                let reverseGeocodeRequest: geoLocationManager.ReverseGeoCodeRequest
= {
                  'locale': this.locale.toString().includes('zh') ? 'zh' : 'en',
                  'latitude': this.latitude,
                  'longitude': this.longitude
                };
                geoLocationManager.getAddressesFromLocation(reverseGeocodeRequ
est).then(data => {
                  if (data[0].placeName) {
                    this.currentLocation = data[0].placeName;
                  }
                }).catch((err: Error) => {
                  Logger.error(TAG, 'GetAddressesFromLocation err ' + JSON.stringify(err));
                });
              });
            }).catch((err: Error) => {
              Logger.error(TAG, 'requestPermissionsFromUser err' + JSON.stringify(err));
            })
      } catch (err) {
        Logger.error(TAG, 'requestPermissionsFromUser err' + JSON.stringify(err));
      }
    }

    build() {
      Stack({ alignContent: Alignment.TopStart }) {
        CompletionStatus({
          progressValue: $progressValue
        })

        CurrentSituation({
          currentSteps: this.currentSteps,
          startPosition: this.startPosition,
          currentLocation: this.currentLocation
        })

        Row() {
          Button(this.isStart ? $r('app.string.stop') : $r('app.string.start'))
            .width($r('app.float.start_button_width'))
            .height($r('app.float.start_button_height'))
            .borderRadius($r('app.float.start_button_radius'))
            .backgroundColor($r('app.color.button_background'))
            .fontSize($r('app.float.start_font_size'))
            .fontColor(Color.White)
            .fontWeight(CommonConstants.BIG_FONT_WEIGHT)
            .onClick(() => {
```

```
            if (this.isStart) {
              this.isStart = false;
              this.oldSteps = '';
              StepsUtil.CleanStepsData();
              BackgroundUtil.stopContinuousTask(this.context);
            } else {
              if (this.stepGoal === '' || this.currentLocation === '') {
                promptAction.showToast({ message: CommonConstants.WAIT });
              } else {
                this.isStart = true;
                this.startPosition = this.currentLocation;
                StepsUtil.putStorageValue(CommonConstants.START_POSITION,
this.startPosition);
                this.currentSteps = CommonConstants.INITIALIZATION_VALUE;
                this.progressValue = 0;
                BackgroundUtil.startContinuousTask(this.context);
              }
            }
            StepsUtil.putStorageValue(CommonConstants.IS_START, String(this.
isStart));
          })
      }
      .width(CommonConstants.FULL_WIDTH)
      .height($r('app.float.button_height'))
      .margin({ top: $r('app.float.button_margin_top') })
      .justifyContent(FlexAlign.Center)
    }
    .width(CommonConstants.FULL_WIDTH)
    .height(CommonConstants.FULL_HEIGHT)
    .backgroundColor(Color.White)
  }
}
```

上述代码的实现流程如下。

- 首先，定义了名为 HomePage 的 HarmonyOS 应用页面，其中声明了多个状态变量，用于存储当前计步、目标步数、旧步数、起始位置、当前位置等信息。
- 其次，在 onPageShow 函数中初始化页面，并调用 requestPermissions 请求用户权限，包括传感器和地理位置权限。在获得权限后，通过 sensor.on 注册计步传感器监听器，实时获取计步数据，同时通过 LocationUtil.geolocationOn 监听地理位置变化。
- 最后，通过 build 函数构建页面布局，包括展示计步进度和当前位置的组件，并提供开始 / 停止计步的按钮。按钮单击事件会触发相应的逻辑，包括启动 / 停止计步、清除步数数据等，并通过 BackgroundUtil 在后台执行相关任务，实现计步功能的持续监测。

执行效果，如图 12-6 所示。

图 12-6 项目主界面执行效果

第 13 章
综合实战：新闻客户端（Node.js 服务端 + HarmonyOS 客户端）

在 HarmonyOS 系统中，电话和短信服务的开发至关重要，尤其对需要与通信功能整合的应用程序。这包括通信类应用程序、社交媒体平台以及需要实现身份验证的应用程序。通过充分利用 HarmonyOS 的通信接口，开发者能够提供更加综合和创新的用户体验，使应用程序在系统生态中更加完善和互联。本章将详细讲解在 HarmonyOS 系统中开发电话和短信服务程序的知识。

13.1 背景介绍

在当今数字化时代，随时随地获取最新新闻信息已成为人们生活中不可或缺的一部分。移动设备的普及使用户对即时、个性化的新闻阅读体验有了更高的期望。为了满足这一需求，我们推出了移动新闻系统，旨在为用户提供流畅、直观、多元的新闻获取方式。移动新闻系统的愿景是打破传统新闻阅读的空间和时间限制，通过精心设计的用户界面和灵活的交互方式，让用户随时随地畅快阅读感兴趣的新闻。我们致力于创造一个信息丰富、个性化的新闻平台，使用户能够轻松快捷地捕捉到全球各类新闻资讯。

为了提供一款适用于鸿蒙操作系统的优质新闻阅读应用，我们创建了 HarmonyNews 项目。HarmonyNews 项目的创建旨在展示 HarmonyOS 应用开发的优越性，为开发者提供实践经验。同时，作为一款全面而强大的新闻阅读应用，HarmonyNews 将为用户提供便捷、高效的新闻阅读服务，助力鸿蒙生态系统的发展。

HarmonyNews 系统的目标是成为用户日常生活中的首选新闻阅读平台，为用户提供更加高效、愉悦的新闻获取体验。我们期待通过这一系统，为用户打造一个打破时空限制的数字新闻天地，为用户呈现丰富多彩、及时精准的新闻信息。

13.2 项目介绍

本项目是一个基于 HarmonyOS 和 Node.js 的新闻阅读系统，旨在为用户提供流畅的界面和丰富的新闻内容。通过 HarmonyOS 提供的 UI 构建语言（JSUI），结合 Node.js 后端，实现了前后端的协同工作。

13.2.1 主要特点

- HarmonyOS 前端：使用 HarmonyOS 提供的 View、Component 和其他组件构建用户界面，包括新闻列表、标签导航等。
- Node.js 后端：利用 Node.js 构建灵活的后端服务，处理前端请求，提供新闻数据和其他服务。
- 新闻类型切换：支持多种新闻类型，用户可以通过标签导航栏切换感兴趣的新闻类别。
- 下拉刷新与上拉加载：实现了下拉刷新和上拉加载更多的交互功能，提升用户体验。
- 自定义布局：通过自定义的刷新布局（CustomRefreshLoadLayout）和加载更多布局（LoadMoreLayout）增强用户界面的交互性。
- 数据交互：通过 Node.js 后端，实现了前后端之间的数据交互，获取新闻类型列表和新闻内容。

13.2.2 项目结构

- src/main/ets/pages/Index.ets：应用的入口页面。
- src/main/ets/view：前端的各种视图组件，包括新闻列表、刷新布局等。
- src/main/ets/viewmodel：前端的视图模型，负责处理业务逻辑和数据请求。
- HttpServerOfNews：Node.js 后端代码，处理前端请求，提供数据服务。

13.3 系统架构

本项目包括前端和后端两个部分，具体来说，本项目使用 Node.js 作为后端（服务端）技术，使用 HarmonyOS 作为前端（客户端）部分。

- 后端（Node.js）部分：使用 Node.js 作为后端技术，主要使用了 Express.js 框架。另外还包括后端路由、控制器和其他业务逻辑的实现。
- 前端（HarmonyOS）部分：使用 HarmonyOS 作为前端技术，采用了 View、Component、TabBar 等 HarmonyOS 组件。通过 HarmonyOS 提供的 UI 构建语言（如 JSUI）构建了前端界面。使用了自定义的 View Model（viewmodel 目录）、刷新布局（CustomRefreshLoadLayout.ets）、新闻列表（NewsList.ets）等组件和模块。
- 通信：前后端之间实现通信功能，例如，前端通过 HTTP 请求与后端进行数据交互，获取新闻类型列表和新闻数据。

整个项目的功能模块架构，如图 13-1 所示。

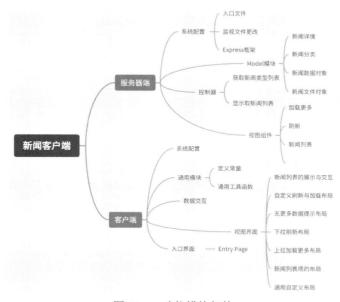

图 13-1 功能模块架构

13.4 服务器端

本项目的服务器端是通过 Node.js 实现的，具体而言，它使用了 Express 框架来搭建服务器。通过 Node.js 的能力来处理 HTTP 请求、定义路由、使用中间件以及提供静态文件服务等。这种选择使开发者能够使用 JavaScript（或者 TypeScript，如果项目中使用了该语言）来构建服务器端应用，同时利用 Node.js 的非阻塞 I/O 模型和事件驱动的特性，实现高效的异步编程。

13.4.1 系统配置

（1）编写文件 HttpServerOfNews/package.json，这是一个鸿蒙开源项目中的 Node.js 后端应用，名为"httpserverofnews"，用于构建新闻服务。通过 Express 框架，它实现了处理 HTTP 请求、文件上传、会话管理等核心功能，同时使用 cookie-parser 和 morgan 进行请求处理和日志记录。通过启动脚本，开发者可以使用 nodemon 在开发过程中实时监测代码变化，便于调试和快速开发。

```
{
  "name": "httpserverofnews",
  "version": "0.0.0",
  "private": true,
  "scripts": {
    "start": "node ./bin/www",
    "dev": "nodemon ./bin/www"
  },
  "dependencies": {
    "cookie-parser": "1.4.6",
    "debug": "2.6.9",
    "express": "4.16.1",
    "express-fileupload": "1.4.0",
    "express-session": "1.17.3",
    "http-errors": "1.6.3",
    "jade": "1.11.0",
    "morgan": "1.9.1"
  }
}
```

对上述代码的具体说明如下。

- "start"："node ./bin/www"：当运行 npm start 命令时，将执行"node ./bin/www"命令。./bin/www 是应用程序的入口文件，这个文件包含了启动应用程序的逻辑。
- "dev"："nodemon ./bin/www"：当在 Node.js 端运行 npm run dev 命令时，将执行"nodemon ./bin/www"命令。nodemon 是一个监视文件更改的工具，当代码发生变化时它会自动重启应用，使在开发过程中不需要手动重启服务器。这在开发阶段非常方便，因为可以实时看到对代码的修改效果。

总之，这两个脚本命令提供了两种不同的方式来启动应用程序：npm start 命令用于生产环境，而 npm run dev 命令则用于开发环境，其中后者利用了 nodemon 实现了自动重启。

（2）编写文件 HttpServerOfNews/app.js，这是一个 Node.js 应用程序的入口文件 app.js，用于创建一个基于 Express 框架的 HTTP 服务器。通过中间件和路由配置，实现了请求日志记录、静态文件服务、JSON 和 URL 编码数据解析，以及错误处理。服务器启动后监听端口 9588，同时提供 /news 路径的新闻服务和 /images 路径的静态图片资源。

```
  var app = express();
  app.set('views', path.join(__dirname, 'views'));
  app.set('view engine', 'jade');
  app.use('/images', express.static('images'));
  app.use('/news', newsRouter);
  app.use(function (req, res, next) {
    next(createError(404));
  });
  app.use(function (err, req, res, next) {
    // 设置本地变量,仅在开发环境提供错误信息
    res.locals.message = err.message;
    res.locals.error = req.app.get('env') === 'development' ? err : {};

    // 渲染错误页面
    res.status(err.status || 500);
    res.render('error');
  });
  app.listen(9588, () => {
    console.log('服务器启动成功! ');
  });
```

注意: Express 是一个基于 Node.js 的轻量级、灵活且功能强大的 Web 应用框架, 它提供了一组强大的特性和工具, 用于快速构建具有良好组织结构的 Web 应用和 API。Express 是 Node.js 生态系统中最受欢迎和广泛使用的 Web 框架之一, 被广泛应用于构建 Web 应用、API 和微服务。

13.4.2 Model 模块

在本项目的 "model" 目录中, 定义了与新闻相关的数据模型, 包括新闻内容和新闻文件, 用于存储和检索新闻数据。

(1) 编写文件 HttpServerOfNews/model/NewsDataModel.js, 定义名为 NewsDataModel 的模块, 其中包含了以下两个主要的数据结构。

- newsDataArray: 一个包含多个 NewsData 对象的数组, 每个对象表示一条新闻的详细信息, 包括唯一标识符、标题、内容、相关图片的 URL 和信息来源。
- newsType: 一个包含不同新闻类型的数组, 每个类型由 id 和 name 组成, 用于对新闻进行分类, 如国内、国际、娱乐等。

```
const NewsData = require('../model/NewsData');
const NewsFile = require('../model/NewsFile');

var newsDataArray = [
    new NewsData(0, '入春来, 百花香', '随着气温升高, 连日来, 某某县某某街道的各个角度盛开了各种各样的花朵, 让人眼花缭乱, 装点了春天的气息.', [new NewsFile(0, '/images/ic_news_1_1.png', 0, 0), new NewsFile(1, '/images/ic_news_1_2.png', 0, 0), new NewsFile(2, '/images/ic_news_1_3.png', 0, 0)], '2022年08月16日'),
    new NewsData(1, '第四届美食节正式启动', '8月5日, 某某市第四届美食节正式启动. 美食节将围绕各地特色美食展开, 推进各地美食文化交流.', [new NewsFile(0, '/images/ic_news_2_1.png', 0, 1), new NewsFile(1, '/images/ic_news_2_2.png', 0, 1)], '2022年8月5日'),
```

```
    new NewsData(2, '江南风景美如画', '南朝四百八十寺,多少楼台烟雨中。江南美,去
过风光绮丽的江南,便久久不能忘怀,江南的雨如沐春风。', [new NewsFile(0, '/images/ic_
news_3_1.png', 0, 2), new NewsFile(1, '/images/ic_news_3_2.png', 0, 2), new
NewsFile(1, '/images/ic_news_3_3.png', 0, 2)], '2022年7月26日'),
    new NewsData(3, '好物推荐', '精选好物,提升家居氛围,享受美好生活。', [new
NewsFile(0, '/images/ic_news_4_1.png', 0, 3), new NewsFile(1, '/images/ic_
news_4_2.png', 0, 3)], '2023年4月20日'),
    new NewsData(4, '某某大桥正式建成', '今日,历经了6年的某某大桥正式建成,自从
2018年起,这座大桥就汇聚了全世界的目光,它的建筑风格十分新颖。', [new NewsFile(0, '/
images/ic_yunnan.png', 0, 4) ], '2022年6月11日'),
    new NewsData(5, '黑白复古风复苏', '近几年,时尚风潮不断变化,复古风又开始风靡全球。
其中,黑白照片是提升复古质感的最佳利器,简洁优雅。', [new NewsFile(0, '/images/ic_
yiqing.png', 0, 5), new NewsFile(1, '/images/ic_student.png', 0, 5)], '2022
年6月10日'),
    new NewsData(6, '设计中的材质碰撞', '近几年,设计风格已不满足于传统的二维
图形,逐渐向三维立体图形靠近,但传统的三维图形操作复杂,未来方向仍值得深究。', [new
NewsFile(0, '/images/ic_student.png', 0, 6), new NewsFile(1, '/images/ic_
yunnan.png', 0, 6), new NewsFile(1, '/images/ic_news_3_3.png', 0, 6)], '2022
年5月19日'),
    new NewsData(7, '如何过好周末', '周末人们最喜爱做什么呢?有些人喜欢带着家人出
门旅行,有些人喜欢宅在家里玩游戏,有的人喜欢约会,你呢?', [new NewsFile(0, '/images/
ic_news_1_1.png', 0, 7), new NewsFile(1, '/images/ic_news_2_1.png', 0, 7)],
'2022年5月15日')
  ];

  var newsType = [
    { id: 0, name: '全部' },
    { id: 1, name: '国内' },
    { id: 2, name: '国际' },
    { id: 3, name: '娱乐' },
    { id: 4, name: '军事' },
    { id: 5, name: '体育' },
    { id: 6, name: '科技' },
    { id: 7, name: '财经' }
  ];

  module.exports = {
    newsDataArray,
    newsType
  }
```

通过 NewsDataModel 模块的导出功能,其他文件可以引用并使用这些预定义的新闻数据和类型。通常,这样的模块在整个应用程序中提供静态的、预定义的数据,方便其他部分的代码使用,如前端展示和后端逻辑。

(2) 编写文件 HttpServerOfNews/model/NewsData.js,定义了一个 JavaScript 类 NewsData,用于表示新闻数据对象。它包括新闻的唯一标识符、标题、内容、相关图片的 URL 以及信息来源。该类通过构造函数初始化这些属性,并通过 module.exports 导出,使其在其他文件中可引用,为鸿蒙项目提供了一种结构化和可重用的新闻数据模型。

```
    class NewsData {
```

```
  constructor(id, title, content, imagesUrl, source) {
    this.id = id;
    this.title = title;
    this.content = content;
    this.imagesUrl = imagesUrl;
    this.source = source;
  }
}

module.exports = NewsData;
```

（3）编写文件 HttpServerOfNews/model/NewsFile.js，定义了一个 JavaScript 类 NewsFile，用于表示新闻文件对象。

```
class NewsFile {
  constructor(id, url, type, newsId) {
    this.id = id;
    this.url = url;
    this.type = type;
    this.newsId = newsId;
  }
}

module.exports = NewsFile;
```

在上述代码中，类 NewsFile 具有以下属性。
- id：文件的唯一标识符。
- url：文件的 URL 地址。
- type：文件类型。
- newsId：与文件相关联的新闻的唯一标识符。

类 NewsFile 的构造函数通过参数初始化了这些属性，并通过 module.exports 语句将类 NewsFile 导出，以便其他文件可以引用和使用这个模型来表示新闻相关的文件信息。

13.4.3 控制器

编写文件 HttpServerOfNews/controllers/newsController.js，实现了项目中的新闻控制器功能。其中包含了获取新闻类型列表和根据当前页码及每页显示数量获取新闻列表的两个主要函数。这些函数通过引用预定义的新闻数据模型和文件模型，以 JSON 格式成功返回新闻数据，实现了对新闻信息请求的处理。

```
/* controllers/newsController.js */
const dbFile = require('../model/NewsDataModel');
const NewsData = require('../model/NewsData');
const NewsFile = require('../model/NewsFile');

const DEFAULT_PAGE_SIZE = 4; // 默认每页显示的新闻数量
const SUCCESS_CODE = 'success'; // 成功返回的状态码

/**
```

```
 * 获取新闻类型列表。
 *
 * @param req 请求对象
 * @param res 响应对象
 */
const getNewsType = (req, res) => {
  res.send({ code: SUCCESS_CODE, data: dbFile.newsType, msg: '' });
};

/**
 * 根据当前页码和每页显示数量获取新闻列表。
 *
 * @param req 请求对象
 * @param res 响应对象
 */
const getNewsList = (req, res) => {
  let { currentPage = 1, pageSize = DEFAULT_PAGE_SIZE } = req.query;
  let newsList = dbFile.newsDataArray.slice((currentPage - 1) * pageSize, currentPage * pageSize);
  res.send({ code: SUCCESS_CODE, data: newsList, msg: '' });
};

/**
 * 根据当前页码和每页显示数量获取额外的新闻列表。
 *
 * @param req 请求对象
 * @param res 响应对象
 */
const getExtraNewsList = (req, res) => {
  let { currentPage = 1, pageSize = DEFAULT_PAGE_SIZE } = req.query;
  let newsList = dbFile.newsExtraDataArray.slice((currentPage - 1) * pageSize, currentPage * pageSize);
  res.send({ code: SUCCESS_CODE, data: newsList, msg: '' });
};

module.exports = {
  getNewsType,
  getNewsList,
  getExtraNewsList
};
```

编写文件 HttpServerOfNews/routes/news.js，定义了一个 Express 路由模块 news.js。这个路由模块定义了三个路由，分别处理获取新闻类型列表、获取普通新闻列表和获取额外新闻列表的 HTTP GET 请求。这些路由将请求映射到相应的新闻控制器中的处理函数，实现了对不同类型新闻数据的请求路由。

```
var express = require('express');
var router = express.Router();
const newsController = require('../controllers/newsController');

router.get('/getNewsType', newsController.getNewsType);
```

```
router.get('/getNewsList', newsController.getNewsList);
router.get('/getExtraNewsList', newsController.getExtraNewsList);
module.exports = router;
```

13.4.4 视图组件

在本项目的"views"目录中，包含了用于构建前端界面的各种组件，包括加载更多、刷新、新闻列表等。这些组件共同构成了新闻系统的前端界面。

（1）编写文件 HttpServerOfNews/views/index.jade，这是一个使用 Jade（现在更名为 Pug）模板引擎的视图文件 index.jade，它继承了一个名为 layout 的布局模板。这个模板的目的是显示错误信息，其中 message 是一个变量，error 是一个错误对象，显示了错误的状态和堆栈信息。

```
extends layout

block content
  h1= message
  h2= error.status
  pre #{error.stack}
```

（2）编写文件 HttpServerOfNews/views/layout.jade，这是一个使用 Jade（现在更名为 Pug）模板引擎的布局文件 layout.jade，定义了一个基本的 HTML 结构。它包含了文档类型声明、HTML 标签、头部元素和主体元素。头部包括页面标题和链接到样式表的链接。主体元素的内容将由子视图文件填充，通过 block content 定义了一个内容块。这样的布局文件提供了一个基本的 HTML 结构，以确保一致性和可维护性。

```
doctype html
html
  head
    title= title
    link(rel='stylesheet', href='/stylesheets/style.css')
  body
    block content
```

13.5 客户端

本项目的 HarmonyOS 前端实现了一个功能丰富的新闻系统界面，包括自定义刷新和加载布局、新闻列表展示、加载更多、异常处理等特性。通过视图组件协调呈现新闻数据，为用户提供流畅的浏览体验。

13.5.1 系统配置

编写文件 src/main/module.json5，这是一个 HarmonyOS 模块配置文件，通常位于 src/

main/module.json5 路径下，用于定义应用程序的入口模块，名为"entry"，适用于手机设备。其中包括入口能力的信息，如源代码入口、描述和图标，同时请求了网络权限。这个配置文件用于指导 HarmonyOS 应用程序的模块行为，包括安装、交付、页面和能力定义及权限请求等。

```
{
  "module": {
    "name": "entry",
    "type": "entry",
    "description": "$string:module_desc",
    "mainElement": "EntryAbility",
    "deviceTypes": [
      "phone"
    ],
    "deliveryWithInstall": true,
    "installationFree": false,
    "pages": "$profile:main_pages",
    "abilities": [
      {
        "name": "EntryAbility",
        "srcEntry": "./ets/entryability/EntryAbility.ts",
        "description": "$string:EntryAbility_desc",
        "icon": "$media:icon",
        "label": "$string:EntryAbility_label",
        "startWindowIcon": "$media:icon",
        "startWindowBackground": "$color:start_window_background",
        "exported": true,
        "skills": [
          {
            "entities": [
              "entity.system.home"
            ],
            "actions": [
              "action.system.home"
            ]
          }
        ]
      }
    ],
    "requestPermissions": [
      {
        "name": "ohos.permission.INTERNET",
        "reason": "$string:dependency_reason",
        "usedScene": {
          "abilities": [
            "EntryAbility"
          ],
          "when": "inuse"
        }
      }
    ]
```

      }
    }

13.5.2 通用模块

在本项目的"common"目录中,包含了项目中多个模块通用的代码和常量。其中"constant"目录存放了一些常量,如服务器地址、HTTP 请求路径等;"utils"目录则包含了一些通用的工具函数,比如处理下拉刷新和上拉加载的手势事件、发起 HTTP 请求等。这些通用的功能和常量可能在整个项目的多个模块或组件中都会用到,因此被放置在"common"目录以供其他模块引用和复用。

1. 定义常量

编写文件 src/main/ets/common/constant/CommonConstant.ets,定义一个包含各种常量的类,用于在 HarmonyOS 应用程序的前端中共享和使用。这些常量包括服务器地址、网络请求路径、页面布局参数、动画和刷新控件的参数,以及新闻列表中各个元素的样式和布局等。此外,文件还包含了一些枚举类型,用于表示刷新状态、页面加载状态和请求内容类型。这些常量和枚举类型的集合提供了统一的配置和规范,以便在整个应用程序中保持一致性和易于维护。

```
import { NewsTypeBean } from '../../viewmodel/NewsViewModel';

export class CommonConstant {

  // 服务器的主机地址。
  static readonly SERVER: string = 'http://**.**.**.**:9588';

  // 获取新闻类型的路径。
  static readonly GET_NEWS_TYPE: string = 'news/getNewsType';

  // 获取新闻列表的路径。
  static readonly GET_NEWS_LIST: string = 'news/getNewsList';

  // 请求成功的状态码。
  static readonly SERVER_CODE_SUCCESS: string = 'success';

  // 偏移系数。
  static readonly Y_OFF_SET_COEFFICIENT: number = 0.1;

  // 每页新闻数量。
  static readonly PAGE_SIZE: number = 4;

  // 刷新和加载的高度。
  static readonly CUSTOM_LAYOUT_HEIGHT: number = 70;

  //HTTP 请求成功的状态码。
  static readonly HTTP_CODE_200: number = 200;

  // 动画延迟时间。
  static readonly DELAY_ANIMATION_DURATION: number = 300;
```

```
// 延迟时间。
static readonly DELAY_TIME: number = 1000;

// 动画持续时间。
static readonly ANIMATION_DURATION: number = 2000;

//HTTP 超时时间。
static readonly HTTP_READ_TIMEOUT: number = 10000;

// 宽度占满全屏。
static readonly FULL_WIDTH: string = '100%';

// 高度占满全屏。
static readonly FULL_HEIGHT: string = '100%';

//TabBars 的常量。
static readonly TabBars_UN_SELECT_TEXT_FONT_SIZE: number = 18;
static readonly TabBars_SELECT_TEXT_FONT_SIZE: number = 24;
static readonly TabBars_UN_SELECT_TEXT_FONT_WEIGHT: number = 400;
static readonly TabBars_SELECT_TEXT_FONT_WEIGHT: number = 700;
static readonly TabBars_BAR_HEIGHT: string = '7.2%';
static readonly TabBars_HORIZONTAL_PADDING: string = '2.2%';
static readonly TabBars_BAR_WIDTH: string = '100%';
static readonly TabBars_DEFAULT_NEWS_TYPES: Array<NewsTypeBean> = [
  { id: 0, name: '全部' },
  { id: 1, name: '国内' },
  { id: 2, name: '国际' },
  { id: 3, name: '娱乐' },
  { id: 4, name: '军事' },
  { id: 5, name: '体育' },
  { id: 6, name: '科技' },
  { id: 7, name: '财经' }
];

//NewsListConstant 的常量。
static readonly NewsListConstant_LIST_DIVIDER_STROKE_WIDTH: number = 0.5;
static readonly NewsListConstant_GET_TAB_DATA_TYPE_ONE: number = 1;
static readonly NewsListConstant_ITEM_BORDER_RADIUS: number = 16;
static readonly NewsListConstant_NONE_IMAGE_SIZE: number = 120;
static readonly NewsListConstant_NONE_TEXT_opacity: number = 0.6;
static readonly NewsListConstant_NONE_TEXT_size: number = 16;
static readonly NewsListConstant_NONE_TEXT_margin: number = 12;
// static readonly NewsListConstant_ITEM_MARGIN_TOP: string = '1.5%';
static readonly NewsListConstant_LIST_MARGIN_LEFT: string = '3.3%';
static readonly NewsListConstant_LIST_MARGIN_RIGHT: string = '3.3%';
// static readonly NewsListConstant_ITEM_HEIGHT: string = '32%';
static readonly NewsListConstant_LIST_WIDTH: string = '93.3%';

//NewsTitle 的常量。
static readonly NewsTitle_TEXT_FONT_SIZE: number = 20;
static readonly NewsTitle_TEXT_FONT_WEIGHT: number = 500;
```

```
static readonly NewsTitle_TEXT_MARGIN_LEFT: string = '2.4%';
static readonly NewsTitle_TEXT_WIDTH: string = '78.6%';
static readonly NewsTitle_IMAGE_MARGIN_LEFT: string = '3.5%';
static readonly NewsTitle_IMAGE_WIDTH: string = '11.9%';

//NewsContent 的常量。
static readonly NewsContent_WIDTH: string = '93%';
static readonly NewsContent_HEIGHT: string = '16.8%';
static readonly NewsContent_MARGIN_LEFT: string = '3.5%';
static readonly NewsContent_MARGIN_TOP: string = '3.4%';
static readonly NewsContent_MAX_LINES: number = 2;
static readonly NewsContent_FONT_SIZE: number = 15;

//NewsSource 的常量。
static readonly NewsSource_MAX_LINES: number = 1;
static readonly NewsSource_FONT_SIZE: number = 12;
static readonly NewsSource_MARGIN_LEFT: string = '3.5%';
static readonly NewsSource_MARGIN_TOP: string = '3.4%';
static readonly NewsSource_HEIGHT: string = '7.2%';
static readonly NewsSource_WIDTH: string = '93%';

//NewsGrid 的常量。
static readonly NewsGrid_MARGIN_LEFT: string = '3.5%';
static readonly NewsGrid_MARGIN_RIGHT: string = '3.5%';
static readonly NewsGrid_MARGIN_TOP: string = '5.1%';
static readonly NewsGrid_WIDTH: string = '93%';
static readonly NewsGrid_HEIGHT: string = '31.5%';
static readonly NewsGrid_ASPECT_RATIO: number = 4;
static readonly NewsGrid_COLUMNS_GAP: number = 5;
static readonly NewsGrid_ROWS_TEMPLATE: string = '1fr';
static readonly NewsGrid_IMAGE_BORDER_RADIUS: number = 8;

//RefreshLayout 的常量。
static readonly RefreshLayout_MARGIN_LEFT: string = '40%';
static readonly RefreshLayout_TEXT_MARGIN_BOTTOM: number = 1;
static readonly RefreshLayout_TEXT_MARGIN_LEFT: number = 7;
static readonly RefreshLayout_TEXT_FONT_SIZE: number = 17;
static readonly RefreshLayout_IMAGE_WIDTH: number = 18;
static readonly RefreshLayout_IMAGE_HEIGHT: number = 18;

//NoMoreLayout 的常量。
static readonly NoMoreLayoutConstant_NORMAL_PADDING: number = 8;
static readonly NoMoreLayoutConstant_TITLE_FONT: string = '16fp';

//RefreshConstant 的常量。
static readonly RefreshConstant_DELAY_PULL_DOWN_REFRESH: number = 50;
static readonly RefreshConstant_CLOSE_PULL_DOWN_REFRESH_TIME: number = 150;
static readonly RefreshConstant_DELAY_SHRINK_ANIMATION_TIME: number = 500;

// 网格列模板。
static readonly GRID_COLUMN_TEMPLATES: string = '1fr ';
```

```
    // 列表偏移单位。
    static readonly LIST_OFFSET_UNIT: string = 'px';
}

// 刷新状态的枚举。
export const enum RefreshState {
    DropDown = 0,
    Release = 1,
    Refreshing = 2,
    Success = 3,
    Fail = 4
}

// 新闻列表状态的枚举。
export const enum PageState {
    Loading = 0,
    Success = 1,
    Fail = 2
}

// 请求内容类型的枚举。
export const enum ContentType {
    JSON = 'application/json'
}
```

2. 通用工具函数

在本项目前端的 "common/utils" 目录中，包含了通用的工具函数，这些工具函数可能被项目中的不同模块或组件共用。在软件开发中，通常将一些通用的功能或操作抽取成工具函数，以提高代码的复用性和可维护性。这个目录下的工具函数可能包括网络请求、日志记录、常量定义等，以便在整个项目中方便地引用和调用。

（1）编写文件 src/main/ets/common/utils/HttpUtil.ets，该 httpRequestGet 函数位于 "utils" 目录下，使用库 @ohos.net.http 实现了通过 HTTP 发起 GET 请求的功能。它创建 HTTP 请求对象，设置请求参数（如超时时间 TIMEOUT 和请求头 header），并处理异步 HTTP 响应，将服务器返回的数据封装在 ResponseResult 对象中，同时处理可能的错误情况，最终返回一个 Promise。这个函数的主要目的是简化和封装 HTTP GET 请求的处理过程，提供更便捷的接口用在 HarmonyOS 应用程序中进行网络请求。

```
import http from '@ohos.net.http';
import { ResponseResult } from '../../viewmodel/NewsViewModel';
import { CommonConstant as Const, ContentType } from '../constant/CommonConstant';

export function httpRequestGet(url: string): Promise<ResponseResult> {
    let httpRequest = http.createHttp();
    let responseResult = httpRequest.request(url, {
        method: http.RequestMethod.GET,
        readTimeout: Const.HTTP_READ_TIMEOUT,
        header: {
            'Content-Type': ContentType.JSON
```

```
      },
      connectTimeout: Const.HTTP_READ_TIMEOUT,
      extraData: {}
    });
    let serverData: ResponseResult = new ResponseResult();

    return responseResult.then((value: http.HttpResponse) => {
      if (value.responseCode === Const.HTTP_CODE_200) {
        // Obtains the returned data.
        let result = `${value.result}`;
        let resultJson: ResponseResult = JSON.parse(result);
        if (resultJson.code === Const.SERVER_CODE_SUCCESS) {
          serverData.data = resultJson.data;
        }
        serverData.code = resultJson.code;
        serverData.msg = resultJson.msg;
      } else {
        serverData.msg = `${$r('app.string.http_error_message')}&${value.responseCode}`;
      }
      return serverData;
    }).catch(() => {
      serverData.msg = $r('app.string.http_error_message');
      return serverData;
    })
  }
```

（2）编写文件 src/main/ets/common/utils/PullDownRefresh.ets，定义了一些用于处理下拉刷新操作的函数，主要包括 listTouchEvent 函数、touchMovePullRefresh 函数、touchUpPullRefresh 函数、pullRefreshState 函数和 closeRefresh 函数。

```
import promptAction from '@ohos.promptAction';
import { touchMoveLoadMore, touchUpLoadMore } from './PullUpLoadMore';
import {
  CommonConstant as Const,
  RefreshState
} from '../constant/CommonConstant';
import NewsViewModel, { NewsData } from '../../viewmodel/NewsViewModel';
import NewsModel from '../../viewmodel/NewsModel';

/**
 * 处理列表触摸事件，包括下拉刷新和加载更多的逻辑。
 *
 * @param newsModel NewsModel 实例，包含列表相关的状态和数据。
 * @param event 触摸事件对象。
 */
export function listTouchEvent(newsModel: NewsModel, event: TouchEvent) {
  switch (event.type) {
    case TouchType.Down:
      newsModel.downY = event.touches[0].y;
      newsModel.lastMoveY = event.touches[0].y;
      break;
```

```
        case TouchType.Move:
          if ((newsModel.isRefreshing === true) || (newsModel.isLoading === true)) {
            return;
          }
          let isDownPull = event.touches[0].y - newsModel.lastMoveY > 0;
          if ((((isDownPull === true) || (newsModel.isPullRefreshOperation === true))
&&  (newsModel.isCanLoadMore === false))
          {
            // 手指移动，处理下拉刷新。
            touchMovePullRefresh(newsModel, event);
          } else {
            // 手指移动，处理加载更多。
            touchMoveLoadMore(newsModel, event);
          }
          newsModel.lastMoveY = event.touches[0].y;
          break;
        case TouchType.Cancel:
          break;
        case TouchType.Up:
          if ((newsModel.isRefreshing === true) || (newsModel.isLoading === true)) {
            return;
          }
          if ((newsModel.isPullRefreshOperation === true)) {
            // 手指抬起，进行下拉刷新。
            touchUpPullRefresh(newsModel);
          } else {
            // 手指抬起，处理加载更多。
            touchUpLoadMore(newsModel);
          }
          break;
        default:
          break;
      }
    }

/**
 * 处理下拉刷新过程中的手指移动事件，根据滑动偏移量更新下拉刷新状态。
 *
 * @param newsModel NewsModel 实例，包含列表相关的状态和数据。
 * @param event 触摸事件对象。
 */
export function touchMovePullRefresh(newsModel: NewsModel, event: TouchEvent) {
    if (newsModel.startIndex === 0) {
      newsModel.isPullRefreshOperation = true;
      let height = vp2px(newsModel.pullDownRefreshHeight);
      newsModel.offsetY = event.touches[0].y - newsModel.downY;
      // 滑动偏移大于下拉刷新布局高度，满足刷新条件。
      if (newsModel.offsetY >= height) {
        pullRefreshState(newsModel, RefreshState.Release);
        newsModel.offsetY = height + newsModel.offsetY * Const.Y_OFF_SET_COEFFICIENT;
      } else {
        pullRefreshState(newsModel, RefreshState.DropDown);
```

```
      }
      if (newsModel.offsetY < 0) {
        newsModel.offsetY = 0;
        newsModel.isPullRefreshOperation = false;
      }
    }
  }

  /**
   * 处理手指离开屏幕时的下拉刷新逻辑，根据是否可以刷新发起刷新请求，并更新相关状态。
   *
   * @param newsModel NewsModel 实例，包含列表相关的状态和数据。
   */
  export function touchUpPullRefresh(newsModel: NewsModel) {
    if (newsModel.isCanRefresh === true) {
      newsModel.offsetY = vp2px(newsModel.pullDownRefreshHeight);
      pullRefreshState(newsModel, RefreshState.Refreshing);
      newsModel.currentPage = 1;
      setTimeout(() => {
        let self: NewsModel = newsModel;
        NewsViewModel.getNewsList(newsModel.currentPage, newsModel.pageSize,
  Const.GET_NEWS_LIST).then((data:
            NewsData[]) => {
            if (data.length === newsModel.pageSize) {
              self.hasMore = true;
              self.currentPage++;
            } else {
              self.hasMore = false;
            }
            self.newsData = data;
            closeRefresh(self, true);
        }).catch((err: string | Resource) => {
            promptAction.showToast({ message: err });
            closeRefresh(self, false);
        });
      }, Const.DELAY_TIME);
    } else {
      closeRefresh(newsModel, false);
    }
  }

  /**
   * 根据传入的状态值更新下拉刷新的文字、图片和相关状态。
   *
   * @param newsModel NewsModel 实例，包含列表相关的状态和数据。
   * @param state 刷新状态的枚举值。
   */
  export function pullRefreshState(newsModel: NewsModel, state: number) {
    switch (state) {
      case RefreshState.DropDown:
        newsModel.pullDownRefreshText = $r('app.string.pull_down_refresh_text');
        newsModel.pullDownRefreshImage = $r('app.media.ic_pull_down_refresh');
```

```
          newsModel.isCanRefresh = false;
          newsModel.isRefreshing = false;
          newsModel.isVisiblePullDown = true;
          break;
        case RefreshState.Release:
          newsModel.pullDownRefreshText = $r('app.string.release_refresh_text');
          newsModel.pullDownRefreshImage = $r('app.media.ic_pull_up_refresh');
          newsModel.isCanRefresh = true;
          newsModel.isRefreshing = false;
          break;
        case RefreshState.Refreshing:
          newsModel.offsetY = vp2px(newsModel.pullDownRefreshHeight);
          newsModel.pullDownRefreshText = $r('app.string.refreshing_text');
          newsModel.pullDownRefreshImage = $r('app.media.ic_pull_up_load');
          newsModel.isCanRefresh = true;
          newsModel.isRefreshing = true;
          break;
        case RefreshState.Success:
          newsModel.pullDownRefreshText = $r('app.string.refresh_success_text');
          newsModel.pullDownRefreshImage = $r('app.media.ic_succeed_refresh');
          newsModel.isCanRefresh = true;
          newsModel.isRefreshing = true;
          break;
        case RefreshState.Fail:
          newsModel.pullDownRefreshText = $r('app.string.refresh_fail_text');
          newsModel.pullDownRefreshImage = $r('app.media.ic_fail_refresh');
          newsModel.isCanRefresh = true;
          newsModel.isRefreshing = true;
          break;
        default:
          break;
      }
    }

    /**
     * 在一定延时后关闭下拉刷新的状态，包括动画效果和恢复初始状态。
     *
     * @param newsModel NewsModel 实例，包含列表相关的状态和数据。
     * @param isRefreshSuccess 是否刷新成功。
     */
    export function closeRefresh(newsModel: NewsModel, isRefreshSuccess: boolean) {
      let self = newsModel;
      setTimeout(() => {
        let delay = Const.RefreshConstant_DELAY_PULL_DOWN_REFRESH;
        if (self.isCanRefresh === true) {
          pullRefreshState(newsModel, isRefreshSuccess ? RefreshState.Success : RefreshState.Fail);
          delay = Const.RefreshConstant_DELAY_SHRINK_ANIMATION_TIME;
        }
        animateTo({
          duration: Const.RefreshConstant_CLOSE_PULL_DOWN_REFRESH_TIME,
          delay: delay,
```

```
        onFinish: () => {
          pullRefreshState(newsModel, RefreshState.DropDown);
          self.isVisiblePullDown = false;
          self.isPullRefreshOperation = false;
        }
      }, () => {
        self.offsetY = 0;
      })
    }, self.isCanRefresh ? Const.DELAY_ANIMATION_DURATION : 0);
  }
```

在上述代码中，这些函数通过处理触摸事件，实现了下拉刷新的交互效果。具体内容如下。

- listTouchEvent 函数根据触摸事件的类型执行相应的操作，包括处理下拉刷新和加载更多的逻辑。
- touchMovePullRefresh 函数处理下拉刷新过程中的手指移动事件，根据手指滑动的偏移量更新下拉刷新的状态。
- touchUpPullRefresh 函数处理手指离开屏幕时的下拉刷新逻辑，根据是否可以刷新发起刷新请求，并更新相关状态。
- pullRefreshState 函数根据传入的状态值更新下拉刷新的文字、图片和相关状态。
- closeRefresh 函数用于在一定延时后关闭下拉刷新的状态，包括动画效果和恢复初始状态。

（3）编写文件 src/main/ets/common/utils/PullUpLoadMore.ets，实现了列表上拉加载更多的逻辑，通过处理手指在列表上的滑动事件，根据滑动偏移量更新加载更多和下拉刷新的状态，同时包括发起加载更多请求和更新列表数据的逻辑。

```
import promptAction from '@ohos.promptAction';
import { CommonConstant as Const } from '../constant/CommonConstant';
import NewsViewModel, { NewsData } from '../../viewmodel/NewsViewModel';
import NewsModel from '../../viewmodel/NewsModel';

/**
 * 处理手指在列表上滑动时的加载更多逻辑，根据滑动偏移量更新加载更多状态。
 *
 * @param newsModel NewsModel 实例，包含列表相关的状态和数据。
 * @param event 触摸事件对象。
 */
export function touchMoveLoadMore(newsModel: NewsModel, event: TouchEvent) {
  if (newsModel.endIndex === newsModel.newsData.length - 1 || newsModel.endIndex === newsModel.newsData.length) {
    newsModel.offsetY = event.touches[0].y - newsModel.downY;
    if (Math.abs(newsModel.offsetY) > vp2px(newsModel.pullUpLoadHeight) / 2) {
      newsModel.isCanLoadMore = true;
      newsModel.isVisiblePullUpLoad = true;
      newsModel.offsetY = -vp2px(newsModel.pullUpLoadHeight) + newsModel.offsetY * Const.Y_OFF_SET_COEFFICIENT;
    }
  }
}
```

```
/**
 * 处理手指离开屏幕时的加载更多逻辑，根据是否可以加载更多，发起加载更多请求，并更新相关
状态。
 *
 * @param newsModel NewsModel 实例，包含列表相关的状态和数据。
 */
export function touchUpLoadMore(newsModel: NewsModel) {
  let self: NewsModel = newsModel;
  animateTo({
    duration: Const.ANIMATION_DURATION,
  }, () => {
    self.offsetY = 0;
  })
  if ((self.isCanLoadMore === true) && (self.hasMore === true)) {
    self.isLoading = true;
    setTimeout(() => {
      closeLoadMore(newsModel);
      NewsViewModel.getNewsList(self.currentPage, self.pageSize, Const.
GET_NEWS_LIST).then((data: NewsData[]) => {
        if (data.length === self.pageSize) {
          self.currentPage++;
          self.hasMore = true;
        } else {
          self.hasMore = false;
        }
        self.newsData = self.newsData.concat(data);
      }).catch((err: string | Resource) => {
        promptAction.showToast({ message: err });
      })
    }, Const.DELAY_TIME);
  } else {
    closeLoadMore(self);
  }
}

/**
 * 关闭加载更多的状态，包括重置相关状态和动画效果。
 *
 * @param newsModel NewsModel 实例，包含列表相关的状态和数据。
 */
export function closeLoadMore(newsModel: NewsModel) {
  newsModel.isCanLoadMore = false;
  newsModel.isLoading = false;
  newsModel.isVisiblePullUpLoad = false;
}
```

13.5.3 数据交互

在本项目中，"viewmodel"目录负责处理应用程序的业务逻辑和数据模型的状态，使其能够与用户界面进行交互。这包括从服务器获取新闻数据、管理新闻列表的展示和交互效果，以及定义了一些用于界面展示的数据类和自定义布局类。

（1）编写文件 src/main/ets/viewmodel/NewsModel.ets，定义类 NewsModel，这是一个用于管理新闻数据和页面状态的模型类。包含新闻数据数组、分页信息、下拉刷新和上拉加载状态等属性，用于在 HarmonyOS 应用中实现新闻列表的交互和展示。类 NewsModel 提供了对新闻数据的管理和页面状态的控制功能，支持下拉刷新和上拉加载功能。

```
import { CommonConstant as Const, PageState } from '../common/constant/CommonConstant';
import { NewsData } from './NewsViewModel';

// NewsModel 类用于管理新闻数据、分页、下拉刷新和上拉加载等状态
export default class NewsModel {
  // 新闻数据数组
  newsData: Array<NewsData> = [];
  // 当前页码
  currentPage: number = 1;
  // 分页大小
  pageSize: number = Const.PAGE_SIZE;
  // 下拉刷新文本
  pullDownRefreshText: Resource = $r('app.string.pull_down_refresh_text');
  // 下拉刷新图标
  pullDownRefreshImage: Resource = $r('app.media.ic_pull_down_refresh');
  // 下拉刷新布局高度
  pullDownRefreshHeight: number = Const.CUSTOM_LAYOUT_HEIGHT;
  // 是否显示下拉刷新布局
  isVisiblePullDown: boolean = false;
  // 上拉加载文本
  pullUpLoadText: Resource = $r('app.string.pull_up_load_text');
  // 上拉加载图标
  pullUpLoadImage: Resource = $r('app.media.ic_pull_up_load');
  // 上拉加载布局高度
  pullUpLoadHeight: number = Const.CUSTOM_LAYOUT_HEIGHT;
  // 是否显示上拉加载布局
  isVisiblePullUpLoad: boolean = false;
  // Y 轴偏移量
  offsetY: number = 0;
  // 页面状态
  pageState: number = PageState.Loading;
  // 是否还有更多数据
  hasMore: boolean = true;
  // 起始索引
  startIndex = 0;
  // 结束索引
  endIndex = 0;
  // 手指按下位置
  downY = 0;
  // 手指最后移动位置
  lastMoveY = 0;
  // 是否正在刷新中
  isRefreshing: boolean = false;
  // 是否可以刷新
  isCanRefresh = false;
```

```
    // 是否处于下拉刷新操作中
    isPullRefreshOperation = false;
    // 是否正在加载中
    isLoading: boolean = false;
    // 是否可以加载更多
    isCanLoadMore: boolean = false;
}
```

（2）编写文件 src/main/ets/viewmodel/NewsViewModel.ets，定义类 NewsViewModel，这是一个处理新闻数据请求和管理的模块，其中包括获取新闻类型列表、获取新闻列表等功能。通过调用 getNewsTypeList 方法和 getNewsList 方法，该类与服务器进行通信，获取新闻类型和新闻列表数据。同时，该模块定义了新闻数据的结构，包括标题、内容、图片 URL 等信息。另外还定义了 CustomRefreshLoadLayoutClass、NewsTypeBean、ResponseResult 等辅助类，用于表示自定义刷新加载布局、新闻类型信息和网络请求结果。整体而言，该模块提供了一个封装良好的接口，方便在 HarmonyOS 应用中获取和处理新闻相关数据。

```
class NewsViewModel {
  /**
   * 从服务器获取新闻类型列表。
   *
   * @return NewsTypeBean[] 新闻类型列表
   */
  getNewsTypeList(): Promise<NewsTypeBean[]> {
    // Implementation of fetching news type list from the server.
  }

  /**
   * 获取默认新闻类型列表。
   *
   * @return NewsTypeBean[] 新闻类型列表
   */
  getDefaultTypeList(): NewsTypeBean[] {
    // Returns the default news type list.
  }

  /**
   * 从服务器获取新闻数据列表。
   *
   * @return NewsData[] 新闻数据列表
   */
  getNewsList(currentPage: number, pageSize: number, path: string):
      Promise<NewsData[]> {
    // Implementation of fetching news list from the server.
  }
}

// 新闻列表项信息。
export class NewsData {

  // 新闻列表项标题。
```

```
  title: string = '';

  // 新闻列表项内容。
  content: string = '';

  // 新闻列表项图片 URL。
  imagesUrl: Array<NewsFile> = [new NewsFile()];

  // 新闻列表项来源。
  source: string = '';
}

// 新闻图片列表项信息。
export class NewsFile {

  // 新闻图片列表项 ID。
  id: number = 0;

  // 新闻图片列表项 URL。
  url: string = '';

  // 新闻图片列表项类型。
  type: number = 0;

  // 新闻图片列表项对应的新闻 ID。
  newsId: number = 0;
}

// 自定义刷新加载布局数据。
@Observed
export class CustomRefreshLoadLayoutClass {
  // 自定义刷新加载布局是否可见。
  isVisible: boolean;

  // 自定义刷新加载布局图片资源。
  imageSrc: Resource;

  // 自定义刷新加载布局文本资源。
  textValue: Resource;

  // 自定义刷新加载布局高度。
  heightValue: number;

  constructor(isVisible: boolean, imageSrc: Resource, textValue: Resource, heightValue: number) {
    // Constructor for CustomRefreshLoadLayoutClass.
  }
}

// 新闻类型信息。
export class NewsTypeBean {
  id: number = 0;
```

```
    name: ResourceStr = '';
}

// 网络请求结果信息。
export class ResponseResult {
    // 网络请求返回的代码：成功、失败。
    code: string;

    // 网络请求返回的消息。
    msg: string | Resource;

    // 网络请求返回的数据。
    data: string | Object | ArrayBuffer;

    constructor() {
        this.code = '';
        this.msg = '';
        this.data = '';
    }
}
```

13.5.4 视图界面

在本项目中，"view"目录实现了前端界面的核心功能，包括定义新闻列表项的布局（NewsItem）、新闻列表的展示与交互（NewsList）、自定义刷新与加载布局（CustomRefreshLoadLayout）、无更多数据提示布局（NoMoreLayout）、下拉刷新布局（RefreshLayout）、上拉加载更多布局（LoadMoreLayout）、标签栏（TabBar）以及通用自定义布局（CustomLayout）。这些组件协同工作，提供了完整的新闻浏览体验，支持多种交互操作，如下拉刷新、上拉加载更多和标签切换。

（1）编写文件 src/main/ets/view/LoadMoreLayout.ets，定义了加载更多布局的组件，引用了 CustomRefreshLoadLayoutClass 和 CustomRefreshLoadLayout，通过设置不同的属性，控制加载更多布局的显示状态、图标、文本和高度。

```
import { CustomRefreshLoadLayoutClass } from '../viewmodel/NewsViewModel';
import CustomRefreshLoadLayout from './CustomRefreshLoadLayout';

@Component
export default struct LoadMoreLayout {
    @ObjectLink loadMoreLayoutClass: CustomRefreshLoadLayoutClass;

    build() {
        Column() {
            if (this.loadMoreLayoutClass.isVisible) {
                CustomRefreshLoadLayout({
                    customRefreshLoadClass: new CustomRefreshLoadLayoutClass(this.
loadMoreLayoutClass. isVisible,
                        this.loadMoreLayoutClass.imageSrc, this.loadMoreLayoutClass.
textValue, this.loadMoreLayoutClass.heightValue)
                })
```

```
      } else {
        CustomRefreshLoadLayout({
          customRefreshLoadClass: new CustomRefreshLoadLayoutClass(this.
loadMoreLayoutClass.isVisible,
            this.loadMoreLayoutClass.imageSrc, this.loadMoreLayoutClass.textValue, 0)
        })
      }
    }
  }
}
```

（2）编写文件 src/main/ets/view/CustomRefreshLoadLayout.ets，定义了一个名为 CustomLayout 的组件，用于展示自定义的刷新或加载状态布局。该组件引用了 CustomRefreshLoadLayoutClass 和 CommonConstant，通过在布局中设置不同的属性，如图标、文本、边距等，以实现自定义的刷新或加载状态展示。布局采用了横向排列的行布局，包括一个图像和一个文本，并通过调整样式和属性来实现灵活的显示效果。

```
import { CommonConstant as Const } from '../common/constant/CommonConstant';
import { CustomRefreshLoadLayoutClass } from '../viewmodel/NewsViewModel';

@Component
export default struct CustomLayout {
  @ObjectLink customRefreshLoadClass: CustomRefreshLoadLayoutClass;

  build() {
    Row() {
      Image(this.customRefreshLoadClass.imageSrc)
        .width(Const.RefreshLayout_IMAGE_WIDTH)
        .height(Const.RefreshLayout_IMAGE_HEIGHT)

      Text(this.customRefreshLoadClass.textValue)
        .margin({
          left: Const.RefreshLayout_TEXT_MARGIN_LEFT,
          bottom: Const.RefreshLayout_TEXT_MARGIN_BOTTOM
        })
        .fontSize(Const.RefreshLayout_TEXT_FONT_SIZE)
        .textAlign(TextAlign.Center)
    }
    .clip(true)
    .width(Const.FULL_WIDTH)
    .justifyContent(FlexAlign.Center)
    .height(this.customRefreshLoadClass.heightValue)
  }
}
```

（3）编写文件 src/main/ets/view/NewsList.ets，定义了一个名为 NewsList 的组件，用于展示新闻列表。组件依赖于多个其他组件和工具函数，包括 NewsItem 函数、LoadMoreLayout 函数、RefreshLayout 函数、CustomRefreshLoadLayout 函数、listTouchEvent 函数、NoMoreLayout 函数等。组件通过 NewsModel 管理新闻列表的状态，包括页面状态、当前页数、每页数

量等。在构建函数中，根据页面状态的不同，分别展示加载中、成功加载和加载失败的布局。加载中布局包括一个自定义的刷新加载布局，成功加载布局包括一个下拉刷新布局、多个新闻项和一个上拉加载更多布局，加载失败布局包括一个图片和一段文本。通过 listTouchEvent 函数实现了下拉刷新的手势操作。组件通过监听滚动事件获取当前列表的起始和结束索引。

```
@Component
export default struct NewsList {
  @State newsModel: NewsModel = new NewsModel(); // 新闻列表的状态管理
  @Watch('changeCategory') @Link currentIndex: number; // 监听分类变化，关联当前索引

  // 当分类变化时，重新请求新闻数据
  changeCategory() {
    this.newsModel.currentPage = 1;
    NewsViewModel.getNewsList(this.newsModel.currentPage, this.newsModel.pageSize, Const.GET_NEWS_LIST)
      .then((data: NewsData[]) => {
        this.newsModel.pageState = PageState.Success;
        if (data.length === this.newsModel.pageSize) {
          this.newsModel.currentPage++;
          this.newsModel.hasMore = true;
        } else {
          this.newsModel.hasMore = false;
        }
        this.newsModel.newsData = data;
      })
      .catch((err: string | Resource) => {
        promptAction.showToast({
          message: err,
          duration: Const.ANIMATION_DURATION
        });
        this.newsModel.pageState = PageState.Fail;
      });
  }

  // 组件即将出现时，请求新闻数据
  aboutToAppear() {
    this.changeCategory();
  }

  // 构建组件
  build() {
    Column() {
      if (this.newsModel.pageState === PageState.Success) {
        this.ListLayout(); // 成功加载数据的布局
      } else if (this.newsModel.pageState === PageState.Loading) {
        this.LoadingLayout(); // 加载中的布局
      } else {
        this.FailLayout(); // 加载失败的布局
      }
```

```
    }
    .width(Const.FULL_WIDTH)
    .height(Const.FULL_HEIGHT)
    .justifyContent(FlexAlign.Center)
    .onTouch((event: TouchEvent | undefined) => {
      if (event) {
        if (this.newsModel.pageState === PageState.Success) {
          listTouchEvent(this.newsModel, event); // 处理下拉刷新手势事件
        }
      }
    });
}

// 加载中的布局
@Builder LoadingLayout() {
  CustomRefreshLoadLayout({
    customRefreshLoadClass: new CustomRefreshLoadLayoutClass(true,
      $r('app.media.ic_pull_up_load'), $r('app.string.pull_up_load_text'),
this.newsModel.pullDownRefreshHeight)
  });
}

// 成功加载数据的布局
@Builder ListLayout() {
  List() {
    ListItem() {
      RefreshLayout({
        refreshLayoutClass: new CustomRefreshLoadLayoutClass(this.newsModel.
isVisiblePullDown, this.newsModel.pullDownRefreshImage,
          this.newsModel.pullDownRefreshText, this.newsModel.pullDownRefreshHeight)
      });
    }

    ForEach(this.newsModel.newsData, (item: NewsData) => {
      ListItem() {
        NewsItem({ newsData: item });
      }
      .height($r('app.float.news_list_height'))
      .backgroundColor($r('app.color.white'))
      .margin({ top: $r('app.float.news_list_margin_top') })
      .borderRadius(Const.NewsListConstant_ITEM_BORDER_RADIUS);
    }, (item: NewsData, index?: number) => JSON.stringify(item) + index);

    ListItem() {
      if (this.newsModel.hasMore) {
        LoadMoreLayout({
          loadMoreLayoutClass: new CustomRefreshLoadLayoutClass(this.newsModel.
isVisiblePullUpLoad, this.newsModel.pullUpLoadImage,
            this.newsModel.pullUpLoadText, this.newsModel.pullUpLoadHeight)
        });
      } else {
        NoMoreLayout(); // 没有更多数据的布局
```

```
        }
      }
    }
    .width(Const.NewsListConstant_LIST_WIDTH)
    .height(Const.FULL_HEIGHT)
    .margin({ left: Const.NewsListConstant_LIST_MARGIN_LEFT, right: Const.
NewsListConstant_LIST_MARGIN_RIGHT })
    .backgroundColor($r('app.color.listColor'))
    .divider({
      color: $r('app.color.dividerColor'),
      strokeWidth: Const.NewsListConstant_LIST_DIVIDER_STROKE_WIDTH,
      endMargin: Const.NewsListConstant_LIST_MARGIN_RIGHT
    })
    // 移除弹性效果
    .edgeEffect(EdgeEffect.None)
    .offset({ x: 0, y: `${this.newsModel.offsetY}${CommonConstant.LIST_OFFSET_
UNIT}` })
    // 监听滚动事件,获取当前列表的起始和结束索引
    .onScrollIndex((start: number, end: number) => {
      this.newsModel.startIndex = start;
      this.newsModel.endIndex = end;
    });
  }

  // 加载失败的布局
  @Builder FailLayout() {
    Image($r('app.media.none'))
      .height(Const.NewsListConstant_NONE_IMAGE_SIZE)
      .width(Const.NewsListConstant_NONE_IMAGE_SIZE);
    Text($r('app.string.page_none_msg'))
      .opacity(Const.NewsListConstant_NONE_TEXT_opacity)
      .fontSize(Const.NewsListConstant_NONE_TEXT_size)
      .fontColor($r('app.color.fontColor_text3'))
      .margin({ top: Const.NewsListConstant_NONE_TEXT_margin });
  }
}
```

（4）编写文件 src/main/ets/view/NoMoreLayout.ets，定义了一个名为 NoMoreLayout 的组件，用于显示"没有更多数据"的布局。它通过引入常量和在布局中设置文本样式，创建一个居中显示的、高度为预定义常量值的文本提示，用于表示当前列表没有更多可加载的数据。

```
import { CommonConstant as Const } from '../common/constant/CommonConstant'

@Component
export default struct NoMoreLayout {
  build() {
    Row() {
      Text($r('app.string.prompt_message'))
        .margin({ left: Const.NoMoreLayoutConstant_NORMAL_PADDING })
        .fontSize(Const.NoMoreLayoutConstant_TITLE_FONT)
        .textAlign(TextAlign.Center)
```

```
      }
      .width(Const.FULL_WIDTH)
      .justifyContent(FlexAlign.Center)
      .height(Const.CUSTOM_LAYOUT_HEIGHT)
    }
  }
```

（5）编写文件 src/main/ets/view/RefreshLayout.ets，定义了一个名为 RefreshLayout 的组件，用于显示刷新布局。通过引入视图模型中的自定义刷新加载布局类（CustomRefreshLoadLayoutClass）和对应的 CustomRefreshLoadLayout 组件，创建一个用于表示刷新状态的布局。根据 isVisible 属性，当刷新状态可见时，加载并显示自定义刷新加载布局。

```
import { CustomRefreshLoadLayoutClass } from '../viewmodel/NewsViewModel';
import CustomRefreshLoadLayout from './CustomRefreshLoadLayout';

@Component
export default struct RefreshLayout {
  @ObjectLink refreshLayoutClass: CustomRefreshLoadLayoutClass;

  build() {
    Column() {
      if (this.refreshLayoutClass.isVisible) {
        CustomRefreshLoadLayout({ customRefreshLoadClass: new CustomRefreshLoadLayoutClass
          (this.refreshLayoutClass.isVisible, this.refreshLayoutClass.imageSrc, this.refreshLayoutClass.textValue,
            this.refreshLayoutClass.heightValue) })
      }
    }
  }
}
```

（6）编写文件 src/main/ets/view/TabBar.ets，定义了一个名为 TabBar 的组件，用作选项卡栏。它引入了 NewsList 组件、常量定义以及 NewsViewModel 中的新闻类别相关信息。通过 Tabs 组件创建了一个选项卡栏，其中 ForEach 循环遍历新闻类别数组（tabBarArray），为每个选项卡设置 TabContent，并使用 TabBuilder 构建选项卡内容。在 aboutToAppear 生命周期方法中，请求新闻类别数据，并在 build 方法中通过 Tabs 组件和相关配置渲染了选项卡栏，实现了新闻类别的展示和切换功能。

```
import NewsList from '../view/newslist';
import { CommonConstant as Const } from '../common/constant/CommonConstant';
import NewsViewModel, { NewsTypeBean } from '../viewmodel/NewsViewModel';

@Component
export default struct TabBar {
  @State tabBarArray: NewsTypeBean[] = NewsViewModel.getDefaultTypeList();
  @State currentIndex: number = 0;
  @State currentPage: number = 1;

  @Builder TabBuilder(index: number) {
    Column() {
```

```
            Text(this.tabBarArray[index].name)
              .height(Const.FULL_HEIGHT)
              .padding({ left: Const.TabBars_HORIZONTAL_PADDING, right: Const.
TabBars_HORIZONTAL_PADDING })
              .fontSize(this.currentIndex === index ? Const.TabBars_SELECT_TEXT_
FONT_SIZE : Const.TabBars_UN_SELECT_TEXT_FONT_SIZE)
              .fontWeight(this.currentIndex === index ? Const.TabBars_SELECT_
TEXT_FONT_WEIGHT : Const.TabBars_UN_SELECT_TEXT_FONT_WEIGHT)
              .fontColor($r('app.color.fontColor_text3'))
          }
      }

      aboutToAppear() {
        // Request news category.
        NewsViewModel.getNewsTypeList().then((typeList: NewsTypeBean[]) => {
          this.tabBarArray = typeList;
        }).catch((typeList: NewsTypeBean[]) => {
          this.tabBarArray = typeList;
        });
      }

      build() {
        Tabs() {
          ForEach(this.tabBarArray, (tabsItem: NewsTypeBean) => {
            TabContent() {
              Column() {
                NewsList({ currentIndex: $currentIndex })
              }
            }
            .tabBar(this.TabBuilder(tabsItem.id))
          }, (item: NewsTypeBean) => JSON.stringify(item));
        }
        .barHeight(Const.TabBars_BAR_HEIGHT)
        .barMode(BarMode.Scrollable)
        .barWidth(Const.TabBars_BAR_WIDTH)
        .onChange((index: number) => {
          this.currentIndex = index;
          this.currentPage = 1;
        })
        .vertical(false)
      }
    }
```

（7）编写文件 src/main/ets/view/NewsItem.ets，定义了一个名为 NewsItem 的组件，用作新闻列表中的每一项。通过引入常量定义和 NewsViewModel 中的新闻数据相关信息，构建了一个包含新闻标题、内容、图片、来源等元素的组件。在 build 方法中，使用 Column 组件和 Row 组件设置了新闻项的整体布局，包括标题、内容、图片和来源等部分。通过 Text 组件和 Image 组件设置了各个元素的样式和布局，其中使用了 Grid 组件展示了新闻图片的网格布局。整个 NewsItem 组件通过 alignItems 和 margin 等属性实现了灵活的布局和样式控制，用于在新闻列表中展示每一条新闻的详细信息。

```
import { CommonConstant, CommonConstant as Const } from '../common/
constant/CommonConstant';
import { NewsData, NewsFile } from '../viewmodel/NewsViewModel';

@Component
export default struct NewsItem {
  private newsData: NewsData = new NewsData();

  build() {
    Column() {
      Row() {
        Image($r('app.media.news'))
          .width(Const.NewsTitle_IMAGE_WIDTH)
          .height($r('app.float.news_title_image_height'))
          .objectFit(ImageFit.Fill)
        Text(this.newsData.title)
          .fontSize(Const.NewsTitle_TEXT_FONT_SIZE)
          .fontColor($r('app.color.fontColor_text'))
          .width(Const.NewsTitle_TEXT_WIDTH)
          .maxLines(1)
          .margin({ left: Const.NewsTitle_TEXT_MARGIN_LEFT })
          .textOverflow({ overflow: TextOverflow.Ellipsis })
          .fontWeight(Const.NewsTitle_TEXT_FONT_WEIGHT)
      }
      .alignItems(VerticalAlign.Center)
      .height($r('app.float.news_title_row_height'))
      .margin({
        top: $r('app.float.news_title_row_margin_top'),
        left: Const.NewsTitle_IMAGE_MARGIN_LEFT
      })

      Text(this.newsData.content)
        .fontSize(Const.NewsContent_FONT_SIZE)
        .fontColor($r('app.color.fontColor_text'))
        .height(Const.NewsContent_HEIGHT)
        .width(Const.NewsContent_WIDTH)
        .maxLines(Const.NewsContent_MAX_LINES)
        .margin({ left: Const.NewsContent_MARGIN_LEFT, top: Const.NewsContent_MARGIN_TOP })
        .textOverflow({ overflow: TextOverflow.Ellipsis })

      Grid() {
        ForEach(this.newsData.imagesUrl, (itemImg: NewsFile) => {
          GridItem() {
            Image(Const.SERVER + itemImg.url)
              .objectFit(ImageFit.Cover)
              .borderRadius(Const.NewsGrid_IMAGE_BORDER_RADIUS)
          }
        }, (itemImg: NewsFile, index?: number) => JSON.stringify(itemImg) + index)
      }
      .columnsTemplate(CommonConstant.GRID_COLUMN_TEMPLATES.repeat(this.
newsData.imagesUrl.length))
```

```
            .columnsGap(Const.NewsGrid_COLUMNS_GAP)
            .rowsTemplate(Const.NewsGrid_ROWS_TEMPLATE)
            .width(Const.NewsGrid_WIDTH)
            .height(Const.NewsGrid_HEIGHT)
            .margin({ left: Const.NewsGrid_MARGIN_LEFT, top: Const.NewsGrid_MARGIN_TOP,
              right: Const.NewsGrid_MARGIN_RIGHT })

          Text(this.newsData.source)
            .fontSize(Const.NewsSource_FONT_SIZE)
            .fontColor($r('app.color.fontColor_text2'))
            .height(Const.NewsSource_HEIGHT)
            .width(Const.NewsSource_WIDTH)
            .maxLines(Const.NewsSource_MAX_LINES)
            .margin({ left: Const.NewsSource_MARGIN_LEFT, top: Const.NewsSource_
MARGIN_TOP })
            .textOverflow({ overflow: TextOverflow.None })
        }
        .alignItems(HorizontalAlign.Start)
      }
    }
```

13.5.5 入口界面

编写文件 src/main/ets/pages/Index.ets，这是本项目前端鸿蒙程序的入口页面（Entry Page）的定义，使用了 @Entry 和 @Component 注解。在页面的构建函数 build 中，一个垂直排列的 Column 包含了 TabBar 组件，占据整个页面宽度，背景颜色为列表颜色，并在垂直方向上居中对齐。

```
@Entry
@Component
struct Index {

  build() {
    Column() {
      TabBar()
    }
    .width(Const.FULL_WIDTH)
    .backgroundColor($r('app.color.listColor'))
    .justifyContent(FlexAlign.Center)
  }
}
```

13.6 调试运行

（1）运行后端

进入后端目录"HttpServerOf News"，运行以下命令安装依赖：

```
npm install
```

然后运行以下命令启动后端服务，运行成功后的界面，如图 13-2 所示。

```
npm start
```

图 13-2　后端运行成功

（2）运行前端

使用 HarmonyOS 开发工具 DevEco Studio 打开本项目，编译并运行。执行效果如图 13-3 所示。

图 13-3　运行前端执行效果